The Human Body in the Age of Catastrophe

**The Human Body in the Age of Catastrophe:
Brittleness, Integration, Science, and the Great War**

Stefanos Geroulanos and Todd Meyers

The University of Chicago Press :: Chicago and London

The University of Chicago Press, Chicago 60637
The University of Chicago Press, Ltd., London
© 2018 by The University of Chicago
All rights reserved. No part of this book may be used or reproduced in any manner whatsoever without written permission, except in the case of brief quotations in critical articles and reviews. For more information, contact the University of Chicago Press, 1427 E. 60th St., Chicago, IL 60637.
Published 2018
Printed in the United States of America

27 26 25 24 23 22 21 20 19 18   1 2 3 4 5

ISBN-13: 978-0-226-55645-1 (cloth)
ISBN-13: 978-0-226-55659-8 (paper)
ISBN-13: 978-0-226-55662-8 (e-book)
DOI: https://doi.org/10.7208/chicago/9780226556628.001.0001

Names: Geroulanos, Stefanos, 1979– author. | Meyers, Todd, author.
Title: The human body in the age of catastrophe : brittleness, integration, science, and the Great War / Stefanos Geroulanos and Todd Meyers.
Description: Chicago ; London : The University of Chicago Press, 2018. | Includes bibliographical references and index.
Identifiers: LCCN 2017055602 | ISBN 9780226556451 (cloth : alk. paper) | ISBN 9780226556598 (pbk. : alk. paper) | ISBN 9780226556628 (e-book)
Subjects: LCSH: Medicine—History—20th century. | Physiology—History—20th century. | World War, 1914–1918—Influence. | Europe—Intellectual life—20th century. | Human body—Symbolic aspects.
Classification: LCC R149 .G47 2018 | DDC 610.9/04—dc23
LC record available at https://lccn.loc.gov/2017055602

♾ This paper meets the requirements of ANSI/NISO Z39.48-1992 (Permanence of Paper).

Contents

Prologue: "Why Don't We Die Daily?"   vii

Part One

1. The Whole on the Verge of Collapse: Physiology's Test   3
2. The Puzzle of Wounds: Shock and the Body at War   34
3. The Visible and the Invisible: The Rise and Operationalization of Case Studies, 1915–1919   78

Part Two

4. Brain Injury, Patienthood, and Nervous Integration in Sherrington, Goldstein, and Head, 1905–1934   111
5. Physiology Incorporates the Psyche: Digestion, Emotions, and Homeostasis in Walter Cannon, 1898–1932   138
6. The Organism and Its Environment: Integration, Interiority, and Individuality around 1930   161
7. Psychoanalysis and Disintegration: W. H. R. Rivers's Endangered Self and Sigmund Freud's Death Drive   207

Part Three

8	The Political Economy in Bodily Metaphor and the Anthropologies of Integrated Communication	247
9	*Vis medicatrix*, or the Fragmentation of Medical Humanism	292
10	Closure: The Individual	316

Acknowledgments	323
Abbreviations and Archives	327
Notes	331
Index	409

Prologue: "Why Don't We Die Daily?"

In a draft of his 1926 lecture "Some Tentative Postulates Regarding the Physiological Regulation of Normal States," the American physiologist Walter Bradford Cannon endeavored to outline how an organism holds itself. It functions, he contended, not organ by organ or part by part but as a stable, self-regulating whole. And yet within his thesis he could not help but contemplate the fragile complexity of self-regulation. In the midst of all the bodily labor to maintain a delicate equilibrium, Cannon wondered, "Why don't we die daily?"[1]

Cannon probably wasn't trying to be clever. If anything he was being momentarily unselfconscious: having only recently returned from the study of war-related injury, he could not easily ward off a lingering astonishment that organisms do not constantly fall apart. Eager though he was to turn attention away from topics of wartime physiopathology and traumatic injury, perhaps, like many of his peers he simply could not move out from under the pall that the war had spread. Physiological normality had become unimaginable without a clear accounting for catastrophe. Homeostasis, the concept Cannon invented to relay his solution to this puzzle, presented the body as first and foremost a dynamic process of integration meant to avert the constant danger of collapse in an indifferent or hostile world.

The path to homeostasis began in earnest for Cannon with the questions he raised in his 1915 book on the emotions, *Bodily Changes in Pain, Hunger, Fear, and Rage*. The project was as much a discrete scientific endeavor concerned with bodies pushed to their limits as it was a form of social witnessing. Its central claim was that the physiological production of great emotional states is brought about by confrontation with particular aggressions and causes the discharge of chemical and nervous messengers throughout the body, forcing vital actions of survival. A stimulus from the outside prompts the arousal of overwhelming emotions that change, in the short and long term, the inner composition—a change of posture necessary to restore balance even in extreme circumstances. World War I provided many such circumstances. Despite having a common physiology, whether a person fought or fled in the face of environmental aggressions was unpredictable. Some men froze in the face of the unspeakable carnage of battle. Others ran toward the fight. Others ran away. When some soldiers were injured and wounds ravaged their bodies and minds, life clung to them, while others succumbed to injuries that seemed minor.

What did physicians and medical researchers make of all this physiological chaos, this inconsistency? In their conceptual and experimental efforts, medical thinkers repeatedly attempted to measure up to the confusion produced by the war and to overcome their frustrating inability to develop a lasting semiotics of injuries and behaviors. They produced a new web of concepts that splintered existing paradigms in the social sciences and became essential for thinking about integration and collapse in economics, social organization, psychoanalysis, symbolic representation, and international politics. To show the fine detail of this labor in a conceptual-historical and anthropological fashion is to stage thought and experimentation as they became attuned to, and tried to anticipate—even outdo—specific conditions of the lived body in states of extreme exertion. The ability of researchers to recalibrate their thinking about this body deployed particular logics, notably of integration, which led to further engagements, concepts, and theories, but also to a plethora of unstable or problematic assumptions, assertions, and systematizations.

Integration, collapse, self-preservation, crisis, catastrophe, witnessing, evidence, unknowing—in this book we pull at a tangled thread of ideas and problems that spread across the registers of the individual and the social. In early twentieth-century medical science a model emerged of a body that was both brittle and tightly integrated—integrated because brittle and vice versa. We write about how Cannon's question and his answer became possible, not just for him but for his colleagues across a series of related fields. We explain how particular kinds of injury emerged during World War I and in its aftermath, and how such injuries found their place in the bodies and lifeworlds of wounded soldiers and doctors, and in the thinking and actions of theorists and policy makers. We describe how medical care came to be understood, however tenuously, as care of the whole individual, and how the body and its medical management stirred various rethinkings of the social role of clinical medicine. We show how these conceptual threads form around and through individual, predominantly male, bodies, how they weave together different medical thinkers and medical problems, and how they sometimes fray and are replaced by others.

As Paul Veyne once suggested apropos history writing, "In this world, we do not play chess with eternal figures like the king and the fool; the figures are what the successive configurations of the chessboard make of them."[2] Indeed, we are drawn to the arrangements on a chessboard made up of physiological theory, wartime and postwar therapeutics, psychiatry, and conceptions of the body; these changed, and the figures made up new groupings over and over, and along with them created new tangles in the conceptual threads that bind them together. The ensuing knots became visible to contemporaries across several fields, just as they now are to us.

"Case histories" presented one such knot. Across disciplines cases proliferated, commanded new directions for doctors, and became both instructive and fraught. Sometimes the case was an anecdote or narrative to illustrate the effects of disorder on an individual level. For the American neuropsychiatrist E. E. Southard, cases came from the clinical accounts of others, and had to be compiled, aggregated, condensed, and sorted to refine the taxonomy of disorder toward clinical and administrative ends. At other times, as the English neurologist Henry Head and the German neuropsychiatrist Kurt Goldstein recognized, each case, male or female, young or old, represented an individual, a damaged world in need of meticulous care, and even the category "case" needed to be reinvented almost with each patient they studied.

Another knot to untie was "shock," a major contributor to mortality during World War I and perhaps the exemplary whole-body injury. Shock seemed at first to bind together everything from open wounds and cerebellar lesions to psychic trauma as it ravaged the injured body that sought to stabilize itself. But this was not so. Cannon, the English physiologist and pharmacologist Henry Dale, and their counterparts on the Royal Army Medical Corps Committee on Shock found in it an object of contestation that forced multiple, often quite divergent, reappraisals of the entire system of organic functioning.

Central to all of this was the individual: the story of a body on the verge of collapse is a story about the horizon of the individual soldier, patient, human being. Meanwhile, Head and Goldstein looked to the conditions of nervous integration and the pressures of particularized environments to develop new ideas of individuality for patients whose devastating brain injuries had made them paradoxically more individual than uninjured people seemed.

By the war's aftermath, these three objects—shock, cases, and brain injury—created an altogether new arrangement of medical thought around the body. Physiologists disagreed on the details, but their perspective on the body had undergone a fundamental change, and they pursued various conceptual gambits to understand integration and collapse together. Cannon resolved a series of issues raised by alternate theories by investing in homeostasis, which he declared to be "the wisdom of the body." Goldstein developed new theories of norms, normality, disintegration, and catastrophe. Dale turned to what he called autopharmacology, a complex set of organic operations that tracked the effects of external stimuli on the organism's inner chemical ecology but easily turned the organism against itself.

Individuality as a concept was up for grabs, and physiologists and neurologists claimed that their logics put some very real flesh on its abstract bones. Perhaps individuality as constructed by medicine elided other understandings of human individuality—but against what indices could physiological individuals be distinguished? We are fascinated by the radicalism and precision with which physiologists and physicians reorganized ideas about the individual human being; by their humanist devotion to patients as individuals; and by their will to theorize individuality not as a liberal category to be shared among all but an effect of the difference inherent to the very bodies that could so casually disintegrate. Derived at once from the military and medico-military situation, the soldier-patient was the absolute object of a doctor's care: for perhaps the first time, it was less the soldier's will or speech that attested to the soldier's individuality than the physiological condition, the totality of the interwoven data recovered from the study of systemic and functional aspects of the broken organism. Physiological and medical agency nestled within a bodily nonconscious.

It should come as no surprise that the body was a matter not merely of biological but also of social study. It contributed to the establishment of a nonliberal individualism fundamentally attached to social welfare, and it facilitated analogies of the body to society, which came itself to be imagined as an integrated and fragile whole in need of care. Psychiatrists were the first among several groups that confronted neurologists' and physiologists' schema of integration, collapse, and individuality. By 1925 social theories developed by political economists, anthropologists, and international lawyers echoed these concerns. The interwar period finds in this image of the body its most compelling avatar.

We do not aim to establish precedents for present-day situations, but it is difficult not to hear echoes of this past today. Integrative medicine continues to push for broader conceptualizations of the whole body in medical practice. Personalized medicine takes a different approach to the individual, using medical genetics, gene therapy, and predictive pharmacotherapy to shape treatment and prevention around a highly particularized picture of the patient, one that is neoliberal in character and always future tense in orientation. Personalized medicine is a project to anticipate and guide the patient's destiny. The welfare state that tends toward the widest possible inclusion of its citizens in medical schemes under its umbrella clashes even more intensely than it did after World War I with the realities of the marketplace and the needs of each individual who demands a cure. But perhaps most notable is the thin line that threads together the aftereffects of war-related injuries from World War I to the present. Limb loss, permanent disability, traumatic brain injuries, the social and behavioral changes associated with them, and the challenges of reintegration into civilian life all remain as critical now as ever. Today, fields such as physical and occupational therapy expend hefty amounts of conceptual and clinical labor to help the individual reconcile her cognitive and physical realities with the particular demands of her world. Beyond the battlefield (be it Fallujah or Chicago), epigeneticists and public health scientists have turned their attention to a multitude of stressors and embodied traumas, especially environmental and in utero exposures, which have in turn blurred the lines between the psychological, the neurological, the endocrinological, and the sociological in the quest to understand the locus of injury. In many ways, these new frontiers of medicine resettle old ground.

At the other end of the spectrum are dead-end New Age medical humanisms, which also found and continue to find inspiration in the thinkers we discuss here. Without a proper sense of the integration-crisis duet as it established whole-body ideals at the end of World War I, it is impossible to understand the rapid focalization of medical humanism around the patient that took place at the time and has continued down its own staggered path. All the more so because, since the 1920s, this medical epistemology has cornered us between the utopias of life and the social demands of an afterlife of injury, rehabilitation, and bodily modification. Without understanding integration and crisis in medicine, it is just as hard to wrap our heads around the way that later-dominant cybernetic and structuralist theories located the individual in the world. It is, finally, impossible to tell why intellectuals and politicians, lawyers and economists, insisted so strongly on using its exact bodily metaphors when advocating broader and denser integrations of domestic societies and the international scene in the face of the great crises that began with the 1929 crash and continued through World War II.

Technologies and theories developed by medical experimenters, clinicians, and other thinkers were not responses to the scale of human destruction visited on soldiers and civilians; nor can researchers' efforts be reduced to simple theories of triage. Rather, these researchers' aim was to reconcile novel findings on human physiology with a flood of patient-soldiers who would challenge their presuppositions and claims. So much was uncertain. To tame this uncertainty they would need new tools for assessment and treatment, new ways to talk with and about patients, and new theories to account for patient-soldiers' myriad and highly individualized reactions to injuries. The organism's fragility, precarity, and struggle for self-regulation did not begin because of World War I, and it certainly did not end when the fighting stopped. The organism has always teetered on a razor's edge; the war gave it—and its concept—a very particular push. Indeed, the integrated, disintegrating human body became a new site of meaning and care, the subject for new analogies of body biological and body politic, as well as the object of direct experimentation: its disintegration was the order of the day. In the pages that follow we find ourselves, as these medical scientists did, telling a story of the human body as a doubled form, at once social and biological, describing its shape along the contours of discovery, through the pressures of refinement, and, at times, within an atmosphere of utter confusion.

Part One

1 The Whole on the Verge of Collapse: Physiology's Test

On or around June 1, 1905, Sigmund Freud published *Three Essays on the Theory of Sexuality*, which announced his concept of the drive: "The drive is . . . one of the concepts on the frontier delimiting the psychic from the physical. What differentiates drives from one another and endows them with specific qualities is their relation to their somatic sources and their aims. The source of a drive is an excitatory process in an organ, and the immediate aim of the drive lies in the annulment of this organic stimulus."[1] Around the same date, perhaps that very evening, the thirty-six-year-old English physiologist Ernest Starling was dining at Caius College, Cambridge. Ever since he and his brother-in-law William Bayliss had carried out experiments on the chemical control of the body a year earlier, Starling had become a star scientist. The experiments resulted in the invention of "secretin," which forced the revision of Ivan Pavlov's famous theory on the nervous control of the gut, favoring chemical control instead.[2] Yet Starling remained unconvinced by existing names for the internal secretions that enabled the body's chemical governance, and, the anecdote goes, his fellow diner W. T. Vesey proposed a Greek-language alternative.[3]

Three weeks later, in the first of four Croonian Lectures to the Royal College of Physicians, Starling adopted Vesey's idea. "The Chemical Correlation of the Functions of the Body" was due, he said, to hormones: "These chemical messengers, . . . or 'hormones' (from ορμάω, I excite or arouse), as we might call them, have to be carried from the organ where they are produced to the organ which they affect by means of the blood stream and the continually recurring physiological needs of the organism must determine their repeated production and circulation through the body."[4] The coincidence is astonishing. There was Freud, publishing his first words on the drives that underpin mental activity, noting the drive's independence from consciousness, its "representation" of somatic processes, its organic and integrative origins, its residence on the border of soma and psyche, and its purpose in rebalancing energy. At that same moment, here was Starling, seeking a term to represent what Freud called a "continuously flowing, endosomatic" force, capable of "correlating" chemical functions, and opting for an active verb denoting "to excite or arouse," as he says, but also to rush someone, to overwhelm, to attack.[5] Harkening to a tradition dating to Claude Bernard, who in the 1860s had first argued that what conditioned physiological activity was the body's organized internal environment,[6] Freud defined drive in terms of a continuous internal response to an internal excitation or disequilibrium, a regulatory operation seeking "the removal of this organic stimulus." Starling, even better aware of his discipline's tradition, and committed in heart and mind to the development of a systematic, scientific physiology, practiced much the same nominalism, dreaming up the body as a proto-unity calibrated and held in check by a fluid system responsive to bodily needs and regulation.

Over the past century, Freud and Starling have been read in very different intellectual estates, where drives and hormones do very different work. But while any serious reader of Freud would chafe at too close a comparison between drives and hormones, there ought to be little doubt that Freud and Starling imagined fundamentally isomorphic processes, or that they produced mirroring concepts to represent what to them seemed like bodywide, unconscious activities that were responsible for internal regulation as well as for the subject's behavior. Starling aimed at biological laws of messaging and organization, relying on animal experimentation to craft a model of the human body. Freud proposed a metapsychology that could bring his theory of the normal maturation of human behavior into conformity with his pathological cases and his hermeneutic and therapeutic practices. Each of them attempted a theory of the individual, insisting on the subject's individuality while granting agency not to each subject but to internal forces almost entirely outside the individual's control.

The tradition that followed Starling's discovery remains to this day a pillar of medical and biological thought, but much remains to be learned about integrationist physiology, about its epistemological transformation during World War I into a full-fledged theory of the body, behavior, pathology, and society, and about its conceptual reconstruction of humanity for a new century. In the chapters that follow, we find purposes that were common to Freud and Starling intermingling in the work, the promises, and the failures of neurologists, endocrinologists, and surgeons. We explore their stories, which gave new meaning to integration and disintegration, and developed a new ontology of the soldier, the patient, and the human subject in general.

Drawing Together the Whole

The identification of hormones in 1905 marked a definitive advance in human physiology and testified to a moment when the broader interpretation of the human body began to shift. Over the subsequent three decades, medical thinkers in the English-, French-, and German-speaking worlds rethought the body in terms of the integration of its different parts, organs, and systems. "Instead of taking an organ, such as the liver, and talking about all the different things that it does," Bayliss wrote, "we intend to discuss the processes in which it plays its part with the other organs."[7] It was a curious kind of integration, motivated and articulated above all by the danger that an integrated, self-regulated body often faced: that certain events leading to even minor disequilibrium could cause systemic collapse. What in the restricted domain of physiology began to be called regulatory physiology, what gradually came to involve the bodily and conceptual duets of integration and radical crisis, we describe as integrationism.

At University College London, Starling, Bayliss, and their students dissected dogs, cats, and guinea pigs to study stomachs, hearts, and adrenal glands and to explain how secretions controlled bodywide systems. Meanwhile, at Harvard Medical School, Walter Bradford Cannon and his collaborators had begun using newly invented X-rays to determine the mechanisms of swallowing and of the stomach. Their work gradually led to a novel perspective on the influence of emotions, hunger, and pain on these functions.[8] By taking advantage of new technologies and experimental protocols, British and American physiologists tested and reworked their epistemic universe, and in so doing they developed a theoretical and experimental corpus to match the French and German physiological work that until then had been canonical—the schools begun by Claude Bernard in Paris and Johannes Müller in Berlin. This influential body of work contributed to the creation of entire laboratory objects and new theoretical fields, from the physiology of emotion to the toxicity of chemical substances normally coursing through the body. As it disentangled the chemical, nervous, and emotional threads of interdependence that held the body together, the new physiology also generated an ever-mutating collection of demands for theoretical revisions and practical and laboratory problems.

With the explosion of World War I, these newly emerging promises and problems were abruptly seized and given new direction by the urgent need for a therapeutic conceptual arsenal to deal with the physical and social consequences of wartime violence. As the war dragged on, physiologists, neurologists, psychologists, and clinicians came face-to-face with diseases, injuries, and whole-body responses to violence for which traditional models of the organism could not account. High rates of mortality and permanent disability resulting from bodily reactions to apparently minor, nonlethal wounds were disruptive to both regimes of care and medical theories of bodily functioning.

Wartime researchers' objective was no longer to describe pathological phenomena on paper; it was to intercede in the war's disordering momentum on the body, reverse that momentum, alter its sequelae, and devise acts of repair in the face of seemingly impossible crisis. The processes of collapse, exhaustion, disequilibrium, and shock exceeded physiologists' knowledge just as they strained the human body to its limits. There was no consistent, systematic precedent for conceptualizing them because they tore at both the whole body and its discrete systems. Only because of recent advances did they consider this whole body available to them. Put another way, the earlier lurch toward general biological laws had careened into an immense multitude of "cases" that could not be covered by those laws and that demanded a staging of their differentials. The mechanistic understanding of the human body's composition that had underpinned research and experimentation encountered functional, or whole-body, conditions that could not be attributed to lesions or localizable derangements of the young, male, human machine, and that at times seemed to pit organs or organ systems against each other.[9]

By 1914, physiologists had only partly formed a new paradigm. They could identify the failure of traditional approaches, and they were obliged to work in new directions with a very different sense of urgency. By shattering and disfiguring the body, the war placed demands on medical thought that the laboratory had not. New physiological advances offered ciphers of a different understanding of this body, which had been revealed to be easily torn asunder. Violence—extensively studied in more recent years in cultural-historical perspectives focused on the injured soldiers' suffering[10]—is for us a central object because of the interdisciplinary elaboration of new scientific theories that sought and failed to understand the conduits of a body reacting to its tutelage. We are concerned with the engagements that caused scientists and clinicians to animate their existing systems of knowledge in order to arrest and contravene the brutal emerging norm of bodily breakdown and to move toward a new terrain of biological understanding.

This sought-after terrain mapped quite well onto the work of neurological, psychiatric, and ethnographic researchers—especially in the ways that this work changed direction during and after World War I. None of these fields was subsumed by physiology, although they were not distant from physiological concerns either. Neurology and physiology had been closely tied, especially since Charles Sherrington's 1906 *The Integrative Action of the Nervous System*, but wartime concerns radicalized integrationist priorities with an eye to both nervous and physiological concerns.[11]

In the specific instance of aphasia, patients with brain injury brought neurologists to the conclusion that most past research on the subject was all but useless: scientists had sought the precise location of functions in the brain, but now it appeared that the brain and mind were fundamentally dynamic, indivisible, and nonlocalizable constructions that responded elaborately to intrusions. The brain and mind would have to be rethought as neurologic and symbolic totalities. Among a number of innovative neurophysiologists from Germany to the United States, including Albrecht Bethe in Frankfurt, Alexander Luria in Moscow, and Karl Lashley at Chicago—particularly innovative were Henry Head in London and Kurt Goldstein in Frankfurt, the latter partly through his collaboration with the gestalt psychologist Adhémar Gelb. Head and Goldstein approached brain injury and aphasia by studying individual cases carefully and by considering not so much the injury as the disturbance of a symbolic universe or the destruction of well-ordered, integrated behavior; they demonstrated a capacity for intellectual abstraction that correlated with freedom from and within one's environment. The Soviet psychologist Alexander Luria similarly dedicated a book in 1932 to the psychological understanding of neurophysiological "disorganization and organization," focusing on aphasia and other disorders as pathologies demonstrating strong neuropsychological integration, and emphasizing that his purpose was to offer "an objective and materialistic description of the mechanisms lying at the basis of the disorganization of human behavior and an experimental approach to the laws of its regulation."[12]

: : :

Physiology—a field rarely hailed as inspired, and later eclipsed by the progressive subdivision of the organism in genetics and molecular biology—needs to be understood as a star player in these developments. As the threat of bodily collapse, breakdown, or disintegration became more widely shared and prevalent across disciplines, it obliged a reconceptualization of the body, most of which involved studying the ways in which integration occurred through the intertwining of the body's agencies, organs, and systems. Through its newfound disposition toward corporeal integration, physiology emerged around World War I as an umbrella for this research, stretching the connective tissue between sciences like neurology, endocrinology, and surgery. Together, scientists from these disciplines could look at the wounded differently and could derive from the profusion of cases and the differences among them explanations hinging on the interlacing of the body.[13]

Severely tested and profoundly transformed by the war, the hormonal self that was brought forth in physiology created an array of research objects, from emotions to bodily shock and from histamine poisoning to brain injury and the symbolic self of integrative neurophysiology. This hormonal self of 1905–1914, radicalized by its research promise and viewed as a major contributor to the experience of harm due to injury, interacted with other, similar selves, from the aphasic to the traumatic to the anaphylactic, in what had become by then new sciences of the individual. Together these sciences delineated the bodily and psychic systems that at once guaranteed completeness and health, and also staged disease, violence, and suffering, while becoming far more attentive to the patient-specific qualities of dissolution.

: : :

In the development of a new epistemology—in the rethinking of how the body works, lives, breaks down, recuperates, fails to recuperate, harms itself, and perishes—we find the play of experiment, therapeutics, and philosophy. We find the development of a "style of thought," one that gradually coalesced into this consistent epistemology.[14] We find experimental innovations and novel technological possibilities responding to demands placed on bodies in the theater of conflict. The continually revised regimes of experimentation, care for the injured, technology, and therapeutics offered repeated openings for concept building and theoretical revision.

Here we focus on the process of how, across different disciplines, new concepts were worked out—concepts that affected human beings at that most fundamental level where the sciences of embodiment interact viscerally with physical care and its failure. Our task is to trace the refinement of such concepts in formation and reformulation, as lenses were retrained and meanings were tested, mobilized, advanced, dispensed with, overwritten, or conflated with seemingly similar problems, only to present dead ends or new domains of inquiry. Our scientist-actors recognized the fullness of pathology's effect on human functioning, whether in the laboratory, in the field hospital, or in the society that would reabsorb the affected men.

Among the concepts that endured repeated revision were ones concerning disease and pathology, disarrangement and disintegration, health and recovery, and norms. Perhaps the main concept that this generation of thinkers restyled from the 1910s to the 1930s was that of the individual, and before we look at the way physiology and other disciplines conceived wholeness and integration, it serves to anticipate some of the questions concerning individuality that will arise throughout this project. The individual was suddenly freed up for clinicians and researchers as a problem, and not only in simple terms of the subjective clinical attention given to the single patient. As each individual seemed to suffer somewhat differently, the need for categorizations could no longer obviate differences in suffering and in the complex corporeal impressions of injury. Such a rethinking of the pathological organism and of the value of therapeutics imposed itself as the question of how individual organisms reacting to particular aggressions obviated, or at least displaced, the question of how laws and physiological meaning could be generated. Highlighting this is essential: this was a particular kind of individuality, one premised on a need to treat each human being regardless of gender, race, age, or social status as *singular* while, perhaps surprisingly, actively depriving that human being of agency, voice, and subjectivity. As we shall see, agency came to apply to the individual as a totality of internal systems and not to the individual as subject. One might give several reasons why, in a war famous for the literature of the soldier, physicians did not listen to the soldier-patient's voice. For more than a century the thrust of the patient narrative had been at odds with the interpretation of symptoms,[15] and because of administrative conventions there was no place for the soldier's story within the clinical ledger. It was a question of value: the patient-narrator speaks askance to the priorities of physiology, whereas the body simply speaks. The diagnostician, rightly or not, looks for signs and symptoms along the lines of physiology's priorities above and beneath what the patient reports. In other words, there was no convention for recording because there was little apparent value in recording: only for a body that stuttered in confusion did the voice help clarify. Because bodily testaments to injury and emotional disequilibrium far superseded spoken narratives, soldier-patients were not the adjudicators of their own will and conditions in the way that memoirists and novelists have come to be regarded. Instead they were cases.

This led to a particular and unstable ontology of the soldier, which in the 1920s would expand to a general medico-physiological ontogeny of the human body. On the one hand, the erasure of subjective agency benefited the conceptual design of an integrated organism composed of systems of physical markers that were joined together through mechanisms of regulation and stabilization. The subject became impossible to imagine as an aggressive, violent force of war. He became a stitched-together group of systems, fragile because material forces of war such as bullets and shrapnel intruded into him, and because his own constitution was such as to facilitate his collapse.

On the other hand, the subject had to be conceived as an individual because both in health and in suffering he reacted differently from others, and this difference was due to his wholeness. In contrast to the fear of disindividuation, of persons as "human matériel," to which innumerable memoirs and memorials to the fallen responded, the medical self was profoundly individuated. Ruptured from the outside, the organism fell apart as a result of internal movements and activities. The appearance of pathology or disorder caused the organism to face a potentially "catastrophic situation," to use Goldstein's pregnant expression.[16] In this imagined crisis, as in Cannon's theorization of the fight-or-flight response triggered when an organism was faced with the demand that it respond immediately to a threatening external stimulus, the patient responded along lines that were profoundly individualized; his reaction was impossible to anticipate except in the most general terms. To discover individuality became a matter of studying how the body internalizes and attempts to compensate for—to control—this generic "environmental" situation.

The paradox, perhaps even paroxysm, was thus that the individual was at once indivisible and eminently divisible. This was as far from a language of race and social degeneration that denied individuality as it was from any liberal language—per John Stuart Mill, for example, according to whom "over himself, over his own body and mind, the individual is sovereign."[17] The understanding of the body as integrated yet brittle could be co-opted by these languages, but more importantly, it contributed to the establishment of a nonliberal individualism fundamentally attached to social welfare. It also empowered an often unspoken humanist presupposition in therapeutic practice that has long presented the patient as a being whose pain amounts to profound disarray—a disarray that undermines one's subjectivity and control of one's speech while marking a new, ever more particular course for one's life.

The Emergence of Physiology

What did physiology look like around 1900? What was new about this old yet ostensibly fertile science? Why study physiologists at all?

At the turn of the century human physiology claimed a peculiar but hardly clear-cut place among the biological sciences. Physiology could declare superior hermeneutic power over complex biological phenomena in the living body, yet it was also obliged to concede that it was unable to work directly on the very organisms and functions it studied. The first textbook Starling published, *Elements of Human Physiology* (1892), begins as follows:

> Physiology is the science of the phenomena of living organisms, and of the laws regulating those phenomena. In its wider sense it will thus include the phenomena of all vegetable and animal life. In this work, however, our immediate object is the physiology of man: but in physiology, as in all other sciences, the only sure foundation of knowledge is that gained by experiment; and since ethical considerations prevent our experimenting on our fellow-creatures, we find ourselves again and again forced to judge of the functions of men by analogy with those of lower animals on whom we can experiment. We can, however, learn many things from experiments which we may make on ourselves. . . . We find means, moreover, of checking the results of our experience in lower animals by studying the disorders of function caused in man by lesions of the various parts of the body which we may observe in the wards and post-mortem room. Nature, however, rarely limits her experiments on our vile bodies to one function or organ, so that in most diseases we have such a complexity of disturbances that this method of investigation used by itself is apt to lead to many erroneous deductions.[18]

The labor of disciplinary competition is staged front and center in this account. Starling—then only twenty-six—begins with an expansive definition of physiology that asserts its disciplinary and scientific standing. He then immediately moves outward to claim for physiology the entire kingdom of biology, before triumphantly placing the specific subject of human physiology on the throne. Next, having declared physiology capable of promulgating laws, he runs into an obstacle: direct experimentation on human beings, "the only sure foundation" for physiology's self-sufficiency, is impossible in light of ethical concerns. The organismic complexity of "our vile bodies" —and the rhetorical revulsion toward them—further hampers deductions from pathology.

The entire scientific paradox is housed in this dilemma: physiology wishes to conquer the sciences, yet it can barely occupy its own lands. Starling resolves this problem by proposing a different practice that does not appear to endanger the pretensions he has just asserted: he advocates for animal experimentation and speaks of complementing it with marginal human experimentation—namely self-experimentation and pathological derivation.[19] Human physiology may thus operate and consider itself well founded without being deterred by a fundamental dependence on the analogical study of lower organisms or on the dubious practice of abstracting rules for the normal out of the pathological.[20]

Starling's confidence that physiology could appropriate the entire field of biology and tower over it was thus belied by physiology's dependence on other biological domains—evolutionary biology, anatomy, and pathology. British physiologists, despite tracing their work back to William Harvey and the early seventeenth century, believed their discipline was very young. Institutionally it had received its "first outward mark of recognition by the official and intellectual world" in 1872 by being included in the Prince Albert Memorial at Kensington Gardens— "betokened by a female figure with a microscope."[21] Emancipated from anatomy in the course of the nineteenth century, it was still in relative infancy at the publication of Starling's *Elements of Human Physiology* (1892), when John Burdon-Sanderson and Edward Sharpey-Schafer were the leading figures in the development of experimental protocols, institutional structures, and technical innovations in the discipline. These early protocols relied to an outsized degree on the German school of physiology emanating in part from Johannes Müller's students and collaborators—notably Hermann von Helmholtz, Carl Ludwig, and Emil du Bois-Reymond, whose neurophysiological work Sharpey-Schafer piloted in Britain. The German school dominated the second half of the nineteenth century and trained physiologists and neurologists across the continent and in the United Kingdom.

By 1914, the picture had changed dramatically. "For some years," Bayliss noted in a 1915 letter to Cannon, "it had seemed to many of us that the German 'Zentralblätter' have not been altogether satisfactory; and from letters I have received from various physiologists in other countries, they have felt the same. But it cannot be denied that it has been due to the terrible outbreak of war and its inevitable results, for which we do not regard ourselves as responsible, that we have been forced to realize that we have been too content to rely on Germany for this kind of work."[22] German physiology was no longer the standard-bearer; the discipline had relocated. University College London, the Harvard Medical School, and Cambridge University had usurped, in originality, productivity, and fame, the place that had been long held by the Physiological Institute in Leipzig, Helmholtz's labs in Heidelberg and Berlin, and also Claude Bernard's lab at the Collège de France in Paris.[23] A renovated department and laboratory at University College London was completed in 1909, as Cannon's at Harvard had been around 1900. Lushly illustrated textbooks were being published in England and the United States at an astonishing rate and were being reissued in frequent revised editions, establishing the young science as dynamic, self-sufficient, and influential, particularly on medicine and medical teaching.[24] Cannon's *A Laboratory Course in Physiology*, for example, first appeared in 1910 and was in its fifth edition in 1926; Starling's *Principles of Human Physiology*, first published in 1912, was in its fourth edition by 1926 and would continue to be updated by his successors, notably Charles Lovatt Evans, into the mid-1960s; by 1920, its fifth year in print, Bayliss's *Principles of General Physiology* was already in its third edition.

New protocols for experimentation hardened around 1900, with crucial results, such as Starling's configuration of the fluids of the body and Cannon's X-ray studies of the stomach and intestines. Meanwhile, self-experimentation became a more common practice, just as Starling had proposed. It allowed for researchers to directly feel the effect of particular pathologies in ways that untrained patients would not be able to convey. Head had his radial nerve sliced in 1904 to experience its regeneration and theorize the functioning of the nervous system, J. S. Haldane carried out respiratory poisoning experiments on himself, and Cannon tested gastric effects by having an assistant swallow tubed balloons full of fluid and observing the change in liquid levels in the tube coming out of his mouth.

No surprise, then, that once results of consequence poured in, histories of physiology became invariably triumphalist. By 1913 at the latest, the efforts of the London group of physiologists headed by Bayliss and Starling and including Sherrington at Oxford were credited with contributing to a remarkable transformation of physiology.[25] By 1927, Sharpey-Schafer, at that point a respected elder in the discipline, could write article- and book-length histories of the Physiological Society, in which he noted Britain's erstwhile backwardness vis-à-vis Germany and France but added—as if physiological laws and principles had not existed before—that "it is generally recognized at the present time that the British school of physiology occupies a foremost position in the biological sciences. . . . This great progress has been due not only to individual discoveries, . . . but has depended still more on the discoveries being of such a character as to lead to the enunciation of general principles."[26] Starling, in his Harveian Oration of 1923, "The Wisdom of the Body," went even further, summoning the sun's rise, the rebirth of humanity, and a new, historically unprecedented destiny to exult the meaning of physiology's achievement: "When I compare our present knowledge of the workings of the body, and our powers of interfering with and of controlling those workings for the benefit of humanity, with the ignorance and despairing impotence of my student days, I feel that I have had the good fortune to see the sun rise on a darkened world, and that the life of my contemporaries has coincided not with a renaissance but with a new birth of man's powers over his environment and his destinies unparalleled in the whole history of mankind."[27]

Behind the successes, practitioners and historians saw three main developments: Bayliss and Starling's hormones; the neurological work of Ivan Pavlov on conditional reflexes and of Charles Sherrington on the integrative hierarchy of the nervous system; and Cannon's research with the X-ray, which allowed him look at inner workings of the body in a manner that did not violate it.[28] For the historian Charles Singer, hormones and nervous integration participated, alongside relativity and quantum physics, in a radical critique of "the nature of objective reality itself."[29]

Celebrations aside, the new generation had begun by 1914 to present its efforts in terms of a domestic, generally self-sufficient, forward-looking body of work, no longer derivative of evolutionary thought and German biology. A handful of figures could be bestowed the status of major precursors to the emergence of physiology as a discipline, but even they were treated mostly as precursors, gentlemen-scholars, amateur geniuses, forebears often to be invoked but rarely to be seen as current, given the experimental protocols and epistemological priorities of the new century.[30] The physiologist who was now awarded parenthood over the entire discipline was Claude Bernard of France, who had articulated physiology as an experimental, mechanistic discipline at the fount of medicine itself. Much work had taken place since Bernard, in many countries and toward different ends; suddenly, in the first decade of the century it seemed possible to present the discipline as having reached a new shore, its different trajectories standing tall next to one another and working together in new directions using new procedures and priorities. Some—Cannon, for example—cited almost entirely from works published in the new century.[31]

For all the paeans sung to it, the self-sufficiency of physiology was precarious. The field's breadth remained vast—so much so that Starling, in the same year as he pronounced the law of the heart (which is still named after him), declared in a letter to a colleague that he was "absolutely out of touch with physiology" and could not revise his *Principles of Human Physiology* except for the chapter on dietetics.[32] Experimentally and conceptually, the fragility resided in the analogical and evolutionary logic involved in basing evidence on animals. Experiments generally involved a violent induction of pathology in animals—including spinal injury, decortication, and the utterly routine removal of organs; results relied on the assumption that complexity rises along with an ascent of the evolutionary ladder and that it rises only marginally, not constitutively. Only thus could physiologists indicate the isomorphism and identity between the systems they studied experimentally and the humans who formed their purview. This ferocity and instability was laid bare during the vivisection controversies at the turn of the century. Most famous was the Brown Dog controversy in Britain, which erupted in 1903 when observers of a vivisection carried out by Bayliss charged him publicly with inadequately anaesthetizing a dog and with carrying out a cruel procedure on the animal.

Precisely because they couldn't experiment on humans, in the decade that followed physiologists spent an enormous amount of time publishing newspaper articles, brochures, medical articles, and manifestos on the benefits of animal experimentation.[33] (Cannon's texts alone number twenty-nine, most of them between 1911 and 1916.) Even as the particulars of the debate can correctly be posed in the troubling terms of a macho Edwardian medical culture that supposedly pitted the grand arbiters of science and truth against emotional, weak women, the vivisection controversy should also be understood as the discipline's founding myth in Britain and the United States. Bayliss's win in court at the close of 1903 and subsequent riots by medical students who clashed with, among others, suffragettes gave researchers and students a common purpose and a bonding experience: they would establish animal experimentation, despite its insufficiencies, as the necessary basis of research for humans.[34] The myth institutionally legitimized a set of procedures that still evaded a direct confrontation with the body for understandable, largely moral reasons that even Starling's new dawn could not justify. The war would offer just the right institutional and moral opening.

No less significant for its precarious position, physiology was, as we have already seen, intertwined with other disciplines. "Physiology retained a peculiarly lonely position among the biological sciences during the nineteenth century," wrote Singer, whereas "since the dawn of the twentieth century there has certainly been a breakdown of this isolation."[35] Only recently emancipated from anatomy, physiology had remained subservient to other fields; for example, it had been used as a technique for the development of the experimental psychologies of Wilhelm Wundt and Hugo Münsterberg. Only with Cannon's 1915 *Bodily Changes in Pain, Hunger, Fear, and Rage* did it become possible to demonstrate the *physiological* basis for emotional reactions and to show those reactions to be fundamentally bodily ones, thereby declaring hermeneutic independence from psychology. The discipline was, thanks to Pavlov and Sherrington, intertwined with neurology, even as neurologists beyond Britain remained profoundly ambivalent about physiology. Like psychology, anthropology was deeply involved in physiological concerns. Ever since the mid-nineteenth century anthropology had settled at the frontiers of physiology, establishing routes between physical (including craniometric) and sociocultural characteristics. In the words of the historian Andrew Evans, anthropology had been valuable to physiology as "a substratum of medicine."[36] Ethnology, sociology, and compound fields like ethnopsychology had sought to ground psychology and physiology as both natural and social sciences,[37] while morphology, social hygiene, and racial science had been instrumental in grounding (possibly exposing) physiology's social-political commitments.

To give one example, only fifteen years before the war, the British psychologists Charles S. Myers, William McDougall, and W. H. R. Rivers had accompanied the anthropologist Alfred Cort Haddon on a Cambridge expedition to the Torres Strait, New Guinea, and Borneo. After a few years of writing at home in neurology, psychology, and anthropology, Rivers returned for more fieldwork in the South Pacific, this time the Solomon Islands. He, Myers, and McDougall had grounded their ethnography in German psychophysics, physiology, and psychotechnics; but with the coming of the war, they found their experimental work too textbook for the field and moved to shed the rigidity of their approaches—proposing, for example, that shell shock required new training in psychoanalysis, neurology, physiology, and symbolism—rather than continue the commitment to psychological laws. In so doing, they attended to how their patients individually inhabited their conditions instead of regarding these same patients as psychophysical entities.

In becoming an object of anthropology, or at the very least aligned with its concerns, physiology contributed to its transformation, and in the process also profoundly affected social psychology. Whereas anthropology had long been obsessed with social cohesion—as in Marcel Mauss's theory of sacrifice, Freud's *Totem and Taboo*, and Franz Boas's refutation of racial determinism in favor of social variation and environmental pressures—World War I established the occurrence and threat of social dissolution, modeled on bodily dissolution, as the standard of concern.[38] This was a worry already at the turn of the century: Haddon had made clear that the object of anthropological endeavor was in danger of vanishing. In reference to the decay of social institutions that was resulting from rapid Europeanization, he described the Torres Strait expedition as an effort to capture and save the "memory of a vanished past."[39] The rhetoric became far more dramatic after the war. The British-trained Polish anthropologist Bronisław Malinowski, who spent the war years in the Trobriand Islands, opened his magnum opus *Argonauts of the Western Pacific* (1922) with a note that indigenous societies were disappearing: "Ethnology is in the sadly ludicrous, not to say tragic, position, that at the very moment when it begins to put its workshop in order . . . the material of its study melts away with hopeless rapidity."[40] Rivers found social disarray such a pivotal issue that he returned to it time and again during the last months of his life in 1922. Having sought in his study of shell shock to render his patients autonomous once again, he wrote that colonial practices had caused the depopulation of Melanesia in a way analogous to shock but operating at the social level of "racial suicide": "If people who are interested in life and do not wish to die can be killed in a few days or even hours by a mere belief, how much more easy it is to understand that a people who have lost all interest in life should become the prey of any morbid agency acting through the body as well as through the mind."[41] Anthropologically influenced theorists of social psychology also became obsessed with the dissolution of societies and groups and the effects of that on individual psychology.[42] From Gustave Le Bon's theory of the crowd in panic to McDougall's fading group mind and especially Freud's *Group Psychology and the Analysis of the Ego*, the concept of the dissolution of an integrated set was on the rise.

World War I bridged all these disciplines anew, redrawing their domains, threading them together anew. Last but not least, physiology was hardly separate from the broader medico-therapeutic, metaphysical universe of biology, as the war would make evident by turning new discoveries on their head. The new physiology invoked a much older split, between Hippocratic physicians for whom "physiology was essentially the knowledge of man's place in the general cosmological order" and Galen, who was concerned with the "usefulness of the bodily parts."[43] Here again, the novelty of the field coincided with a potentially profound shift in the meaning and capacity of medicine.

This alloy of aspiration, promise, and limitations was molded and remolded in the transformation of two central conceptual problems of early twentieth-century physiology: the debate between mechanism and vitalism, and the structure of bodily integration.

Beyond Mechanism and Vitalism

The first problem concerned the competition between mechanism and vitalism. Regardless of their commitment to mechanism (i.e., to the idea that organisms are no more than complex machines that can be studied part by part), physiologists since Bernard had emphasized the differences between inanimate matter and living matter as structural ones. Histories of the discipline written after the delegitimation of vitalism, following World War II and the advent of DNA and molecular biology, often fail to appreciate the leeway—differently put, the degree of ambiguity—between vitalism and mechanism that was available in the first half of the century, especially in physiology.

Most participants in and historians of biology at the time described traditional vitalism as sterile.[44] But in 1931 mechanism seemed almost as "unsatisfying," as Charles Singer put it in prefacing his history of biology, because of "the conditioning of any one form of vital activity by innumerable concurrent forms, and this not only in the organism as a whole but in each part [that is] susceptible of independent investigation."[45] The integration of the body and the "relativity of [its] functions" were central to this reasoning: mechanism was too strict to make sense of a body whose different parts influenced one another in so complex a fashion. Along a different line, Erwin Ackerknecht, a major figure in midcentury history of medicine, declared that Bayliss and Starling's hormones should be understood to unexpectedly rejuvenate humoralism against solidism—traditional concepts in physiology that were conventionally allied against and with mechanism, respectively.[46] Georges Canguilhem, who relied greatly on Singer for his own history of biology, and who identified the thrust of physiology since Bernard with the main conceptual problems of biology, similarly came down in favor of a limited vitalism.[47]

These authors agreed on the need for mechanist proof, but they did not accept mechanism as a broader theory of organismic functioning and behavior. Physiology had to preserve complexity, life's irreducibility to physicochemical or merely mechanical processes, while also relying on mechanistic "laws."[48] Part of the problem was structural: although attached to animal experimentation and evolutionary extrapolation, to be self-sufficient physiology needed to argue that it was not altogether derivative of evolutionary thought, that its object was itself adequate. Yet to claim that it was not derivative was equivalent to arguing for a top-down approach, going from more complex to less complex organisms. This is precisely what was unacceptable at the turn of the century. Physiology could not argue for a top-down approach either experimentally or theoretically without opening itself to the accusation of crude, unscientific vitalism.

The fragility of mechanism dated to the mid-nineteenth century. Promoting a mechanistic explanation of life, Bernard had famously argued for the existence of an "internal environment" in the individual.[49] But, faced with "hard" forms of mechanism, Bernard hedged, on the one hand equating the organism to a machine and describing living or "vital" properties as merely unexplained, and on the other defining life as creation, retaining the language of the vital and the living, and speaking of an essence of life that belongs to the "creation of this machine . . . under conditions proper to it."[50] Sherrington, in the same year that he published his influential *The Integrative Action of the Nervous System* (1906), delivered a lecture on physiology as a science in which he traipsed around the mechanistic problem: "Physiology may be described as the study of the working of living things. . . . The living machine, like those of the engineer, produces work moving itself and things—and heat—hence our body's warmth—and electricity, &c. The living machine wears too as it works, but, more complex than the machines of the engineer, it restores its parts as it works and much of its own labour goes in renewing its own living fabric from suitable dead material which we call food."[51] One could read this passage as suggesting that the living machine was merely more intricate and complicated than the engineer's, but the difference seems to be in the order, not merely the degree, of complexity.[52] Note the sense of the body reconstructing itself: without accepting any vitalist principles, Sherrington was using a language far less attached to mechanism than one might expect. This awkwardness was of paramount importance, and it was directly linked to methods of experimentation and the limitations of the discipline. From the discovery of hormones and nervous integration onward, neither mechanism nor vitalism satisfied the expectations of physiology.

The compromise language of complexity spread. Bayliss insisted, like Sherrington, on bodies as "extraordinarily complicated machines."[53] Asking in 1915, "What is the range and position of Physiological Science as a branch of knowledge, and what is its value as a means of mental discipline?" Bayliss quoted Thomas Henry Huxley focusing on physiology's complexity "mid-way between the physico-chemical and the social sciences" and its "training of common sense," and then added, in his own voice, an appeal to "the great experimental skill demanded, owing to the complexity of the phenomena studied."[54] It was this complexity that could be wrought together. Alexander Luria, in *The Nature of Human Conflicts* (1932), stressed that physiological complexity meant that the nervous system could not be studied by analogy to machines or from an accumulative perspective that cast it as a mosaic of elements.[55] Similarly, Lashley, prefacing Goldstein's neurophysiological work, railed against the "oversimplification of facts" in mechanism and the "postulation of forces which cannot be investigated" in vitalism, praising Goldstein for threading the needle of complexity, function, and integration.[56]

This language of complexity directly undercut the mechanist position. As writers began to consider bodily devices, reactions, forms of influence, and functions that, however mechanistic in explanation, they came face-to-face with mechanisms that could no longer be explained in terms of traditional reflexes alone. These involved messaging, direct and indirect influence of one system on another, bodywide fluid stability, coordination, feedback loops, and sudden emotional activation. The question remains: were these holistic or not?

The issue troubled medical thought deeply: mechanism was identified with "good" science, but with the entry of physiology into the war, and the new possibilities this offered for studying pathologies and understanding integration, the old mechanism could be jettisoned. In 1916 the Oxford physiologist J. S. Haldane delivered four Silliman Lectures at Yale titled "Organism and Environment as Illustrated by the Physiology of Breathing," and titled his conclusion "Organic Regulation as the Essence of Life; Inadequacy of Mechanistic and Vitalistic Conceptions."⁵⁷ Haldane was unrelenting against physicochemical reductionism.⁵⁸ The living organism "has, in truth, very little resemblance to a machine," but vitalism too meets "insuperable objections":⁵⁹ "It is in vain that the mechanistic theorists endeavour to exorcise what du Bois-Reymond called the 'spectre of vitalism.' This spectre is nothing but the shadow cast by the mechanistic theory itself—a shadow which has only become, and could only become, deeper the longer the mechanistic theory has lasted. . . . So long as vitalism seemed the only alternative to mechanistic interpretations, they were driven towards the latter. In the din of controversy between vitalists and mechanists there was, however, a complete failure to go to the root of the matter, and enquire into the validity of the assumptions as to physical reality which were accepted by both sides." Haldane declared the "new physiology" to be attempting a far more complex experimental operation than physicochemical reductionism ever allowed, and he denounced the latter for its failure to comprehend integration and endurance in the respiratory and circulation systems.⁶⁰ He expanded that denunciation into a wholesale rejection of mechanism:

> The normals of a living organism are no mere accidents of physical structure. They persist and endure, and they are just the expression of what the organism is. . . . The ground axiom of biology is that they hang together and actively persist as a whole, whether they are normals of structure, activity, environment or life history. In other words organisms are just organisms and life is just life, as it has always seemed to the ordinary man to be. Life as such is a reality. Physiology is therefore a biological science, and the only possible physiology is biological physiology. The new physiology is biological physiology—not bio-physics or bio-chemistry. The attempt to analyze living organisms into physical and chemical mechanism is probably the most colossal failure in the whole history of modern science. It is a failure, not, as its present defenders suggest, because the facts we know are so few, but because the facts we already know are inconsistent with the mechanistic theory.⁶¹

Neither vitalism nor mechanism, then, even if Haldane was still attached to a concept of life that was traditional enough to look like sleight-of-hand vitalism. Cannon, among others, agreed with Haldane's complaint that physiologists had been reducing the body to its "scraps and fragments"—which meant at best its organs and more often the interaction of its chemical elements.⁶² Haldane insisted, moreover, that even hormones did not tell the whole story, not least as they were not the only "messengers" that crossed different organs and systems.⁶³ The organic identity of each body, its stability, and its maintenance of stability required a properly biological method entirely reliant on integration and individuality and not on physicochemical calculations.⁶⁴

At one degree of remove from Haldane, Starling, the godfather of hormones, remained convinced into his last years that mechanisms were at stake, yet the mechanisms he studied in hormones were forms of chemical integration. His therapeutically oriented convictions went well beyond classical mechanism and explicitly advocated integrationism as a way to lessen human suffering. In "The Wisdom of the Body" (1923), he spoke of the therapeutic successes offered by integration:

> The fundamental means for the integration of the functions of the body ... are not merely interesting facts which form a pretty story, but they are pregnant of possibilities for our control of the processes of the body and therewith for our mastery of disease. Already medical science can boast of notable achievements in this direction. The conversion of a stunted, pot-bellied, slavering cretin into a pretty attractive child by the administration of thyroid, and the restoration of normal health and personality to a sufferer from Graves's disease by the removal of the excess of thyroid gland, must always impress us as almost miraculous.⁶⁵

Knowledge and experimentation battled suffering; it fell to physicians and experimental physiologists to stand against "the sorrow of life," which was not death but "the pain, mental and physical, associated with sickness and disability, or the cutting off of a man by disease in the prime of life." This moral claim on the prevention of disease and suffering could blame "ignorance or disregard of the immutable working of the forces of Nature which is being continually revealed to us by scientific investigation."⁶⁶ Starling's "medical utopia," to use Christiane Sinding's term, rested precisely on language that blended ever more detailed experimentation with therapeutic care for a well-integrated, regulated, emotionally and physically suffering body.⁶⁷

The neuropsychiatrist Goldstein agreed as well. Too often interpreted in the context of German Romantic holism—that is, as a technophobic vitalist opposed to mechanization because of its forceful breakdown of the human into function and parts—Goldstein proposed in *The Or-*

ganism (1934) a highly particular integrative holism grounded in the individual, in which he presumed a need for mechanist explanations that did not look at parts alone.[68] Goldstein offered strong critiques of atomism and localizationism (effectively synonyms for *mechanism*), and he instead took stock of what he saw as a new kind of integral individuality that in situations of disease finds itself broken down, divided, and yearning for any reestablishment that promises to stabilize and preserve. To study the organism was to study different elements and string them together on the basis of the organism's total behavior, not to think that through atomism one had solved the bigger problems. This was particularly important with brain injuries, which threatened the organism with "systemic disintegration."[69]

Goldstein was not the first neurophysiologist to manage a top-down conception of the body whose integrationism remained largely mechanistic in terms of its explanations and causality. Luria, in 1932, offered a very similar critique that was all the more significant for the materialism to which Luria was bound in the Soviet Union, despite which he managed to argue against mechanistic reduction and in favor of "structural organization" and order: "The conception of 'organization' is to a certain degree opposed to a mechanical conception of the organism [understood] as an equilibrium of its component parts, in that it is adequate for an analysis of some of the more complicated processes of human behaviour."[70] This claim reverberated across Europe.[71] Goldstein's own explicit, systematic, and detailed critique of conventional physiology, to the benefit of a neurophysiology capable of comprehending higher-ordered behavior, is highly instructive:

> The approach which is attempted by present-day physiology, and which is regarded as the ideal, is to examine the organism by physical and chemical methods, and to form a concept of the functioning of the organism on the basis of results thereby obtained. . . . Of course, there are authors to whom it is a foregone conclusion that life processes can ultimately be resolved into physico-chemical processes, and to whom the failure of this explanation of the biological process is only due to the incompleteness of our present state of research. . . . Those who do not believe that life can be comprehended by physico-chemical methods regard the physico-chemical facts solely as the necessary fundament from which an understanding of the functioning of the organism must ultimately arise, even if additional fac-

tors have to be included. This entire view seems problematic to us, to say the least, because it could be questioned whether anything at all that will clarify the *performances* of the organism can be discovered by this method.[72]

The Structure of Bodily Integration

We have already seen hints of the second major conceptual problem of the period, which concerns the genesis, structure, and meaning of organismic integration. Until the beginning of World War I, organismic integration had been a conceptual problem that the new generation of physiologists had explained in terms of evolution and phylogenesis: organisms were integrated because their long historical evolution had been an evolution into greater complexity, such that more recently evolved animals had developed systems of self-regulation unavailable to ones further down the evolutionary ladder. Any integration could in this sense be explained as the result of adaptive history—the result, that is, of genesis and not structure or meaning. Research on human pathology during the war created a very different ground from that which animal experimentation, with its evolutionist extrapolation, had allowed.

Three other changes facilitated the advance of an integrationist paradigm. First, physiology warmed to the idea of an integration premised on the interior of the organism functioning as a regulated unit; second, it was affected by competition and engagement with other disciplines, especially physiology's siblings neurology, psychology, and pathology; and third, it changed parallel to—and perhaps under the influence of—political and socioeconomic metaphors of integration and collapse.

By playing in a new key Bernard's idea of the internal environment as a prerequisite for the body's tendency toward self-regulation, physiologists began to sideline their reliance on evolution as the definitive heuristic and argumentative logic. Already in Sherrington's paradigm-setting *The Integrative Action of the Nervous System*, the evolutionary rationale had been reduced to an explanation of why integration had come about. But evolution could not address the physiologist's question: thanks to what stabilization processes did the organism actually hold together? Sherrington described the integrative hierarchy of reflexes, reflex arcs, combinations, and coordination in terms that did not use the Darwinian explanation as a rationale, although he recalled that evolution had been "incorporated among the methods of physiology."[73] Indeed, he turned to Darwin only a full two-thirds of the way into the book.[74] It was instead Bernard's internal environment that provided the necessary citation for nervous integration.[75] Thus, as Canguilhem would later affirm, Sherrington's work on integrative action "allowed him to confirm and rectify [John] Hughlings Jackson's guiding idea on the sole terrain of physiology."[76] Jackson, writing some thirty years earlier, had famously established the nervous system on evolutionary grounds.

Even contemporary histories sharply distinguished physiological writings from, say, those of Thomas Henry Huxley or Charles Davenport (even if some, like Bayliss, did occasionally quote Huxley).[77] Starling, in his 1907 *The Fluids of the Body*, retained evolutionary thought in writing of "higher animals," but he otherwise pirouetted to an internal explanation:

> In all the higher animals the cells, of which their bodies are composed, are bathed by an internal medium, from which the cells derive their nourishment, and into which they discharge their waste products. By the provision of such an internal medium, the cells obtain an average constancy of environment and are withdrawn from the buffeting and constant changes to which the animal as a whole is exposed. They are thus enabled to devote their entire energies to the discharge of their particular functions in the commonwealth of the body. By this means moreover provision is made for maintaining the solidarity of the component parts of the commonwealth. The products of any one set of cells are able to influence the activity of the cells in remote parts of the body, and in some cases it has been shown that special chemical messengers are poured out into the tissue fluids for the express purpose of effecting this chemical integration and of bringing about, in widely diverse tissues, the united action to a common end which is characteristic of all the higher types of living organisms.[78]

The articulation of a "commonwealth" based on nervous and chemical integration marked a major conceptual shift, and the first occasion for an autocephalous physiology. Bodily integration—the very notion or idea that seemed to resolve and go a step beyond the mechanism-vitalism dialectic—had changed rationales.

: : :

A contributing shift occurred in physiological integration thanks to a newfound sense of its relation to medicine. Ethnology and ethnography provided some direction in this: they were the first to indicate that social and psychological factors had to be taken into account in the establishment of biological stability and that the conventional order of the disciplines had been disrupted. Until the later nineteenth century, anatomical physiology had offered a strict positivist logic for its power to explain the body, dismissing anthropology, for example, as a derivative science whose physical and medical claims presupposed a physiological rationale. This was exemplified by brain anatomist Paul Broca in his founding and direction of the French Société d'anthropologie and by Adolf Bastian's disregard of "primitive medicine." British physiology no longer needed this schema of the order of the sciences to claim primacy; nor could it rigorously maintain its positivism any longer.[79]

Rivers, who had first arrived in Melanesia bearing psychophysical ideas and technologies, changed his thinking over the course of the following decade. By his 1908 return from the Solomon Islands, he argued instead that "medicine is a social institution. It comprises a set of beliefs and practices which only become possible when held and carried out by members of an organized society." This suggested that medicine could not rely on mechanistic principles alone, and Rivers argued strongly against contempt for "primitive medicine," paving the way for an interlacing of psychological, neurological, ethnological, and medical concerns.[80] By 1915, Cannon, who had relied almost exclusively on laboratory work in the years 1900–1910, was citing ethnographic anecdotes and travelers' reports, even grounding a full chapter on emotional excitement on them.[81] Showing the social life of the emotions suddenly became as essential as demonstrating the physiological processes that explained the causes of divergences from the norm and how these amounted to pathological disarrangements. Particularly curious—as Rivers made clear and as Cannon would, as well, in his 1944 essay "Voodoo Death"—was the effect of magic on the organism, an effect that the psychopathology of war neuroses made strangely familiar within medicine. How these highly integrated bodily systems could be externally as well as internally affected became a priority for research and clinical work.

Meanwhile, physiological innovations in the first quarter of the century established new connections between medicine, pathology, and physiology, especially with the entry of medical thinkers and practitioners into the war. The development here was ambiguous: while the clinic and the laboratory were institutionally walled apart, in medical education as elsewhere, physiological writings and case reporting bound them ever closer.[82] This point was well summarized by Haldane in his 1916 lecture "The New Physiology":

> What is the practical object of medicine? It is to promote the maintenance and assist in the reestablishment of health. But what is health? Surely it is what is normal for an organism. . . . . Medicine needs a new physiology which will teach what health really means, and how it is maintained under the ordinarily varying conditions of environment. We also need a pathology which will teach how health tends to reassert itself under totally abnormal conditions, and a pharmacology which will teach us, not merely the "actions" of drugs, but how drugs can be used rationally to aid the body in the maintenance and reestablishment of health. The new physiology, new pathology, and new pharmacology are growing up around us just now.[83]

What matters most in this argument is Haldane's turn to an organism-specific conception of health and normality; his insistence that only the new physiology could provide this; and conversely, that only a properly conceived pathology could influence physiology by imposing proper conceptions of health and normativity. At the time Haldane delivered these lines, his colleagues working for the Royal British and US Army medical corps, both in research and in the field, were thinking through the same questions.

Conditions and diseases that were psychoneurological in origin similarly came to demand a new kind of attention, particularly when they appeared to be too complicated to explain in terms of traditional localized or limited traumas. In a neuropsychiatric compendium, the American psychologist E. E. Southard collected hundreds of case studies on shock, drawing material from records in Germany, France, England, and the United States. His work (like Goldstein's neurology and Rivers's psychiatry) built bridges between physicians and experimenters who otherwise often came to startlingly different conclusions, even about the essential characteristics of a disease (shell shock being a case in point). In particular, neurologists who studied patients with brain injury attended to the need for clinical care to ground neurological theory. At the end of the war, while compiling his massive clinical research into a theory of brain disorders, Head charged aphasiologists with being "compelled to lop and twist their cases to fit the procrustean bed of their hypothetical conceptions."[84] Goldstein leveled these same charges even at Sherrington; their argument involved the absorption of the experience of suffering and patienthood into medical and physiological theory. Pathology confirmed integration and was essential to explaining its functioning precisely because environmental inadaptation and organismic disarrangement showcased the difference between individuals and therefore affected and restated the overall theory.

In other words, over the period 1905–1935 integration shifted to a perspective of brittleness, disorganization, fallibility, failure, and collapse. The body, so well ordered when healthy, was constantly on the verge of implosion. Cannon, who in 1915 opened his *Bodily Changes in Pain, Hunger, Fear, and Rage* by appealing to an evolutionary rationale for the origin of the emotions, reduced that rationale to tertiary status when he finally organized his theory of homeostasis in 1932. His *The Wisdom of the Body*, named in homage to Starling's Harveian Oration, begins: "Our bodies are made of extraordinarily unstable material."[85] Goldstein's *The Organism*, theorizing the individual as a constituted whole—not merely a sum of mechanistic parts—also postulated the individual as capable of autonomy vis-à-vis his or her milieu, except that, in disease, a patient was resituated on the verge of what he called catastrophe. New concepts of shock, emotions, and anxiety became characteristic of the new direction and expansion of physiological integration. Not only was organismic integration explained by disarrangement and disintegration; the dyad of integration and disintegration had become the basis for a broader theory of the body's possible reactions while under extreme duress.

In integration and disintegration fused also the different pressures we have seen already regarding the individual patient as a category for medical thought. Precisely because of the complex forms that integrated or integrative disorders were taking, and the need to attend to these at the same time as injuries and emotional reactions, physiologists were offering physicians a more complex confrontation of physician and patient. As the English physician Walter Langdon-Brown argued, for the physician-patient relationship, the twentieth-century conglomerate of pathology, neuropsychology, and physiology amounted to a move away from a structured Galenic medicine and toward a "return of Aesculapius." Langdon-Brown celebrated this kind of care as "the recognition of the importance not only of the disease which the patient has, but of the patient which has the disease—his reactions as an individual, his environment, and his hereditary trends."[86] To care for a patient was to change the patient, and by the same token, to theorize a patient was to link the patient's pathology to the whole of the patient's being, including emotional, psychological, and behavioral response, and to think of these together in a dynamic sense.

The patient was no sculpture, no automaton, no embalmed thing to which (once it was damaged) care could be arrogated as if by chemical fixes or mechanical repairs; the patient was a dynamic form that perfused the violence inflicted on it throughout its existence, and that would experience medical care as a blend of violence, anguish, care, and hope. Medicine lived on, and off of, this blend, and it had supposedly found a conceptual web that allowed it to work as much on the whole body as on the lens through which the patient viewed the world. This cut both ways: it supported medical humanism while creating the utopian expectation of sustained health *for every individual*, which was quite impossible for the welfare state. Well beyond the original pronunciations of an "internal environment," this marked a pressure in Britain to ground in integration a medical ontology, a set of therapeutic principles, even a social theory.[87]

: : :

Integration, crisis, and collapse were, finally, figures that went far beyond the medico-physiological field, and as major interpretive concerns they spanned other fields. We have already seen this to be the case with anthropologists and psychologists, from Rivers to McDougall, who put social disintegration in close proximity with physiological concerns. But the language of integration also had close ties with political, legal, and economic thought that ordinarily would not seem relevant to biology. It would be stretching the matter to argue that these fields caused change in physiology or vice versa, or even that these fields could be woven seamlessly into one another thanks to integration. But certainly political and economic uses were closely aligned with physiology's metaphorical and semiotic universe precisely because they discussed the collapse and reconstitution of bodies politic and economic.[88] With this in mind we can note that the language of integration rose in frequency and utility after World War I, and that theories binding social integration to violent disruption became de rigueur by the financial "crash" of 1929.[89]

These problems spanned fields as broad as international law, sociology, economics, international cooperation, and welfare-state governance; they would render the contrast between integration and violent anarchy into a core principle of international relations as well as anthropological diagnosis. The internationalism of the 1920s provided a major and understudied parallel for bodies politic, and fields that incorporated it routinely made reference to the whole, single body. In this, they relied in part on existing biological and political similes and in part on the spread of the whole as a dominant category, not a mere topographic or descriptive reference, but an organized reality that fundamentally affected relations between its parts. Normalcy meant that the experience of wholeness passed without comment. Social and political languages of integration, crisis, homogeneity, and individuality suggested that the problems faced and the language employed by physiology remained central to liberal, socialist, juridical, and political-economic theories in Britain, Germany, and elsewhere.

Having begun in 1914 and reemerged in 1919, these concerns were brought back to the fore with the economic crisis of 1929, leading some physiologists—notably Starling, Cannon, and Goldstein—to emphasize the metonymy of the body physical and the body politic and to write at length on the importance of social and economic integration because of the perils of crisis. Throughout, these different social and political languages closely paralleled physiology's new direction on individuality, collapse, and wholeness. In the 1930s and 1940s, communism, cybernetics, and new approaches to stress would bring the new physiology explicitly into political conversations. Thus, the sense of disintegration that defined our protagonists' approaches extended also to matters of philosophy and social thought, and they too often proposed entire social theories and conceptual as well as anthropological remedies. Intellectuals from Mauss, Norbert Wiener, and Claude Lévi-Strauss through Canguilhem, Ernst Cassirer, and Hans Jonas produced a whole world of mostly unstudied refigurations of the neurophysiological universe discussed here.

From ideas of fight or flight that continue to define psychology, to cybernetic conceptions of integrated wholes and feedback, to neurodynamic approaches and the opposition to deterministic trends in neuroscience and psychiatry, and even to new age and hipster fantasies of self-actualization and stress, the conception of the organism devised in the 1920s and 1930s became as potent beyond medical thought as it was essential within it. Other ideas, too, benefited and gained purchase from holistic integration: the Swiss linguist Ferdinand de Saussure opened his famous *Course in General Linguistics* (1907–1911) by arguing that the "integral object of linguistics" was nowhere to be found in the traditional approaches, whether localized or general; historical and evolutionary linguistic efforts had also failed: "Everywhere we are confronted with a dilemma: if we fix our attention on only one side of each problem, we run the risk of failing to perceive the dualities; . . . on the other hand, if we study speech from several viewpoints simultaneously, the object of linguistics appears to us as a confused mass of heterogeneous and unrelated things."[90] Language, Saussure insisted, "is a self-contained whole and a principle of classification," a homogeneous system, a system of signs from which a set of shifting relations could be retrieved.[91] For his follower Roman Jakobson, Saussure had to be read alongside Kurt Goldstein. Structuration and pathological frailty would remain intertwined concepts in linguistics as well.[92]

Generations at War

In what follows, we look first to the entry of physiology into the war, and of the war into physiology, with a chapter on whole-body conditions focused on wound shock, and another chapter on individuality and case studies. In part 2, we turn to broader theoretical and experimental developments in integration in the period 1905–1930. Chapter 4 follows the integrative neuropathologies of Sherrington, Head, and Goldstein, and chapter 5 the development of the concept of homeostasis in Cannon's physiology. Chapter 6 elucidates the model of the body adjudicated in the major debates of post–World War I physiology in the lead-up to a full-fledged homeostatic theory. Chapter 7 is a study of the model's influence on psychoanalysis, and particularly on Freud's revision of the theory of the drives in the early postwar period. Social theory and the influence of our protagonists in and beyond the medico-biological domain are the subject of part 3 of the book. Chapter 8 looks at their force and influence in international political-economic thought, in cybernetic theory, and in anthropology, and chapter 9 traces the implications in medical individualism and the doctor-patient relationship.

We have titled this book *The Human Body in the Age of Catastrophe: Brittleness, Integration, Science, and the Great War* to foreground the crises brought about by World War I in terms of its effects on the bodies of soldiers and civilians: bodily integrity came under threat by forces seen and unseen, and the human organism struggled to regulate itself and to survive under the threat. No other site in modern European history had precipitated as consequential a transformation of the popular and scientific understandings of the human body and its selfhood.

To demonstrate the stakes and the conceptual foundations of the new physiology and its disciplinary kin in the early twentieth century, let us conclude by foregrounding a case that both illustrates the medical rush into World War I and serves as a counterpoint to the physiological engagement at the center of our attention: the case of George Washington Crile. A surgeon and, from the early moments of the war, a major in the Officers' Training Corps, Crile was committed to appraising the broad problems of bodily integration. In 1914, more than two years before the United States entered the war in Europe, Crile told his surgical staff at Lakeside Hospital in Cleveland to pack their bags and "ready themselves" for a journey to France.[93] His official mission was to update antiquated surgical equipment and operating suites of the Ambulance américaine, a military medical corps established during the Franco-Prussian War.[94] But Crile emphasized to the trustees of Lakeside Hospital—arguing with considerable zeal and a self-possessed sense of heroic purpose—that Europe was experiencing a humanitarian crisis and that the war was leaving a double wound on the body politic of Europe as well as the bodies of young men.[95] If the latter wound could be remedied by surgical intervention and innovation, so too, perhaps, could the former.

Crile's staff joined him at the Lycée Pasteur in Neuilly-sur-Seine, a suburb of Paris that had been transformed into the base of the ambulance corps.[96] Crile was soon followed by a close friend, Harvey Cushing from Harvard Medical School, along with Cushing's own staff. Together they introduced changes in surgical priorities, like Crile's "anoci-association" anesthesia by nitrous oxide–oxygen, in an effort to address the most dire medical conditions on the battlefield: sepsis, frostbite, gangrene, and fractures—all of which resulted in systemic effects on the injured soldier, often turning the human organism irreversibly toward death.[97]

Crile recognized the war as an opportunity to advance his own research priorities, namely on exhaustion and surgical shock. Previously, he had carried out surgical work and "simple animal experiments," but "then came the war! This . . . allied scientists into the crucible and put the whole problem [of shock] to a clinical test on a vast scale."[98] The bodies of the injured, a once limited resource, had become available in seemingly limitless numbers—and Crile's mission in France, as Ackerknecht asserted, fulfilled "a faustian urge toward scientific truth" set forth in "a spirit of brazen self-advertisement."[99] The medical historian's language is unambiguous. Crile belonged to an elder generation of researchers who made strong claims on the very domains of injury and disability with which a new generation of physiologists were preoccupied in their laboratories. Yet both the story that Crile told of his own self-invention and the attitude he took toward his younger and often admiring colleagues yield for us precisely the attitude of a surgeon and physician who thought little of laboratory and theoretical approaches. Only a decade earlier, Cannon had written to him in friendship and admiration, crediting him with "arousing" students of physiology at Harvard;[100] but once Cannon arrived first in Britain and then in France in 1916, Crile cooled considerably. For Crile, the field hospital was a site of experimental knowledge, but one set in opposition to the tedium of the physiologist's laboratory. Put another way, for him, physiology ended where surgery began. Crile wrote scathingly about the impotence of the laboratory-dwelling physiologist lost in his experimental preoccupations.[101] Lab scientists would return the compliment by brushing Crile aside as obsolete: they needed and engaged with human clinical material at length and with resolution.

This image of researchers working in parallel—sometimes in moments of contact, sometimes decidedly disconnected—will become central to our purpose.[102] While nomenclature—of exhaustion and collapse, for example—was often shared, a sharp generational turn moved researchers like Cannon away from the surgical conceptualization of traumatic and wound shock, advancing ideas of the individual human organism that took the injured and their suffering in a very different direction. This is also where Crile's generation, so central to the history of diagnosis, surgery, and therapeutics, begins to fade from the picture, to be replaced by the systemic, integrated effects of injury and infection that were central for a group of researchers who continued under the banner of physiology. Scraping the canvas and painting and repainting again and again an evolving picture, gouache after gouache, the new generation would dream up laws of the individual as much as it would invent laws of medicine.

2   The Puzzle of Wounds: Shock and the Body at War

The Great War ended on November 11, 1918. Four days later, the British Expeditionary Force held a conference in Boulogne for its leading surgeons and pathologists. The subject was "the treatment of shock and haemorrhage," and almost the entire meeting was spent deliberating the merits of a specific treatment for the condition known as wound shock.[1] Chaired by British surgeons Anthony Bowlby and Cuthbert Wallace, the conference was attended by some of the most influential physiologists of the day, including Harvard Medical School professor Walter B. Cannon, who represented the American Expeditionary Forces, and professors William M. Bayliss and Henry Dale on behalf of the London Shock Committee. Minutes from the meeting highlighted the collaborative nature and vastness of the endeavor to bring shock under medical control. One by one the participants reported experiences with a gum acacia solution (a mixture of acacia and sodium bicarbonate) that Bayliss had developed in London with the intent of boosting blood volume. Detailed debates followed: on the solution's amount and concentration, on the temperature at which it had to be administered, on adverse effects that had been observed, on circumstances in which its use was ill advised.

The disagreement over the merits and dangers of the gum acacia solution, compared to blood transfusions and simple saline solutions, was heated. Cannon's opinion, biased by his proximity to Bayliss, was decidedly positive, yet he also reported that George Washington Crile at the American Ambulance Hospital wanted "to condemn the gum injections as dangerous and wholly undesirable."[2] At the end of the meeting, the gum acacia solution was cautiously recommended for continued use. The war had come to an end, but participants at the conference acted out their disagreements as if wound-shocked soldiers were still coming in from the battlefront.

Of course, the signing of the Armistice mattered little to those who had been seriously injured in the war. The highly technical nature of the debate hides the stakes of the Boulogne conference: a new kind of suffering, a largely new concept of the body, a ferocious epistemic reality. Wound shock was a "new," little-understood disorder. Its symptoms were fairly obvious, "unmistakeable," but unlike most traditional frontline injuries, its onset worked very differently in different soldiers.[3] Its cause and the way it ravaged the wounded remained subjects for speculation and research. No sure consensus was forthcoming regarding its treatment and long-term consequences.[4] Shock was at once the most individualized disorder and the disorder that most definitively affected the whole body. Its effect on individuals could not be anticipated; it could not be ascribed a single or clear cause; it could not be treated a priori across a patient group; and in each patient it demanded immediate and inventive treatment, lest the organism go into irreversible collapse.

This chapter launches our study of the body at war—the injured soldier's body during and after World War I, and specifically the body that, once injured, appeared to be at war with itself. It focuses on the fierce debates around shock that took place in 1916–1919 and that brought together many of the protagonists of this book who have since receded into historical oblivion. It also locates shock in a group of daunting whole-body conditions: "soldier's heart," shell shock, sepsis, shallow breathing, and exhaustion.[5] Such disorders belonged to the category of "functional disease"—according to the 1913 edition of *Webster's* dictionary, "a disease of which the symptoms cannot be referred to any appreciable lesion or change of structure; the derangement of an organ arising from a cause, often unknown, external to itself; opposed to organic disease in which the organ is affected."[6] Functional diseases straddled the physiological-psychological boundary, with important consequences for the understanding of the body.[7] They provided sites for pursuing, adapting, and applying research, particularly on hormones and the interaction between different systems within the body, and consequently they supplied a framework for understanding how each organism behaves as a unit.

Thus, while tropes such as "limb from limb" and "deeper wounds" are often front and center in historical writing on war injuries, we direct attention to the body's responses to injury, responses that were totalizing and often devastating, paradoxically far more destructive than the injury itself, which was often minor, even near invisible. The discrepancy vis-à-vis the response posed crucial epistemological questions.

The conception of wound shock exemplified the new devotion to the whole body: through its response to an injury, the shocked organism entered a vicious circle that often led to death. Surgery and physiology were summoned to pull soldiers out of the circle and ultimately nurse and rehabilitate them.[8] Physicians and physiologists were called on to articulate this situation, isolate its effects on different systems, and explain the compounding effects. With multiple groups of scientists working in parallel—often with little contact with one another—we get a sense of different research priorities converging toward an understanding of the body as a single object fundamentally threatened in its wholeness. Amid wartime brutalization and the rationalization of slaughter, no condition better described the body's arrest in injury than its struggle for self-preservation, which contributed to, and sometimes even caused, its destruction.[9] To restore wound shock to its centrality in wartime medical discussions is to display the motivational and conceptual tapestry that wove together a model of the organism whose proof lay in the organism's experience of catastrophic decline.

This model of the organism brought together a constellation of medical, political, and anthropological concepts, practices, and even moralities. It speaks to the ontology of the body at war, the subject with which we begin. We then turn to the unfolding of wound shock as a condition; to its physiological representation, its histories, and the meaning for it of trauma, survival, healing, and rehabilitation; and more broadly to the effects of violence and its assumptions within the body. In closing, we argue that the physiology of wound shock also enables us to appreciate the place of shell shock in wartime discussions very differently than the psychopathological model of trauma that dominates its historiography has allowed.[10]

War and the Body

For physiologists, war immediately solved the major institutional and ethical dilemma that Ernest Starling had identified as the key problem of human physiology: the inability to experiment directly on human beings.[11] Battle provided a ready-made experimental situation in which research and physiological experimentation came under the heading of care, therapy, and rehabilitation. "In my laboratory we have studied the effects of fear and exhaustion in the bodies of animals, but in a peaceful community, there is almost no opportunity for the study of human material," Crile wrote in 1914, praising the promise of wartime research: "In Europe such opportunities are now abundant and they may never again be available on such a scale."[12]

The negligence of research and experimentation on the newly available human subjects was immoral. Alexis Carrel, in France on behalf of the Rockefeller Institute for Medical Research, presented casualties in the language of experimental effects and described poor conditions at hospitals as "the failure of Medicine,"[13] as if medicine were itself at war and losing. The obverse—interpreting the war in biomedical terms—was no less frequent. For Crile, Germany had subjected the Belgians to systematic exhaustion in the fall of 1914: disarrangement, neurasthenia, a population-wide "surgical shock."[14] This "vivisection of Belgium" was a giant experiment that inflicted physiological damage on the whole population, damage of phylogenetic proportions that upended long-established population-wide norms, even affecting "action patterns in the brains of the children."[15] Much as Crile claimed to judge the "vivisection" dispassionately, his moral accusation helped legitimize Entente information gathering, fitness and aptitude testing, and experimental trials whose aim was not simply humanitarian but also military: to restore the injured to battle-ready health.[16]

Not only were soldiers constantly being wounded on a scale that defied even the "barbaric" precedent of the Balkan wars; improved medical care meant they might survive injuries that previously would have proved fatal. Even when at rest, soldiers lived in acute physical exertion, which posed interesting questions: Was the soldier's body the same as the civilian's, or did it undergo a wholesale transformation? Did war precondition soldiers for certain diseases or conditions? How could normalcy, anxiety, and care now be theorized?

: : :

Almost every literary account of the trenches stages scenes in which the body itself testifies to a transformative experience. Humphrey Cobb opened *Paths of Glory* (1935) with a forceful narrative of the difference between the civilian's and the soldier's bodies:

> "Look at their faces. See that sort of greyish tint to their skin? That's not from sitting in a café on a Sunday afternoon. . . . See how the lower jaw looks sort of loose, how it seems to hang down a bit? . . . It shows they've been clenching them. . . . Their eyes are glazed. They're nearly all of them constipated." . . . "Now I know you're fooling me. Everybody always says the front line acts on you just the other way . . ." "Is that so?" "Yes it is. Why only the other day I went to the medical officer for a pill. He said . . . 'that's the last one you'll need till the war's over. The German artillerymen will keep your bowels open for you from now on.'" "That doctor was a fool . . . he's never been near the line or he wouldn't talk that way."[17]

In this account, constipation, glazing of the eyes, and deformation of the jaw are synonymous with a self that attempts to protect itself by closing its orifices, letting nothing go, hoarding its vulnerabilities, entrenching itself against a world in shards, shards that threaten to tear it apart. Two sites take priority for the experienced soldier-narrator: the fearful, contorted face of a soldier who tries to control himself and the endocrine body whose normal mechanisms have been disrupted. The soldier persists in a state of semipermanent modification. When the questioner challenges him, invoking medico-physiological authority to the effect that diarrhea should be the norm in the trenches, the first insists, with Freud and after Charcot, "La théorie c'est bon, mais elle ne nous empêche pas d'exister." He goes on to explain how the difficulty of actually going to the toilet brought about the soldier's bodily mutation.[18] Letting nothing in also meant letting nothing out.

The fact of a shift—and the difference between theory and practice—was hardly lost on physicians, who were concerned with identifying the ways the body was transformed by trench life and battle well before actual injury.[19] Captain E. M. Cowell of the Royal Army Medical Corps (RAMC) proposed in 1917 that conditions at the trenches diverged from peacetime bodily norms and "afford important pre-wound factors in the initiation of wound shock":

> The soldier is subjected to long spells of hard physical labor, often accompanied by profuse sweating. Sleep is short, and generally interrupted. . . . Water is very often only available as a limited ration. In the firing line and during battle all these factors become accentuated. For a large part of the year exposure to wet and cold must be taken into account. The man,

> who may be hit at any time, is likely to be in a state of fatigue, with a tendency to concentration of plasma, sluggish peripheral circulation, and accumulation of waste products of muscular metabolism. The urine of such men is dark, scanty, and loaded with phosphates.[20]

These factors, Cowell added, semipermanently raise pulse and blood pressure even at relatively calm times.

Crile, in 1915, pointed to a different bodily transformation. While the soldier awaits orders to charge,

> his brain is activated by the approach of the enemy. The activated brain in turn stimulates the adrenals, the thyroid, the liver. In consequence, thyreoiodin, adrenaline, and glycogen are thrown into the blood in more than normal quantities. These activating substances are for the purpose of facilitating attack or escape, heat and the muscular actions of shaking and trembling produced. The rapid transformation of energy causes a correspondingly rapid production of acid byproducts. These increased acid byproducts stimulate the rest to greater activity to eliminate the carbonic acid gas. The increased adrenaline output mobilizes the circulation of the limbs, withdraws blood from the abdominal area, causes increased heart action and dilatation of the pupils. In addition, the increased acidity causes increased sweating, increased thirst, increased urinary output, all of these water phenomena being adaptations for the neutralization of acidity.[21]

Crile's image of the fatigued, burdened, distended body differs greatly from Cowell's depiction of the emotionally and hormonally activated, tense body. The creation of a new trench normal amid such emotional and physical duress was a matter of grave interest, especially given all the recent work on hormonal influence.[22] Where such transformations pathological in themselves? Did they precipitate pathologies? How did emotions and hormones influence the body's original norms? Could this body ever return to peacetime norms?

Answers to these questions diverged considerably: in opposition to Crile, William Townsend Porter argued that the "unprecedented violence of the bombardments" at the front directly predisposed the wounded to surgical shock.[23] Other studies, for example by Cowell and his colleagues on the London Shock Committee, interpreted excitations and deformations as factors precipitating many, but not all, cases of shock.[24] Theorists of war neuroses also diverged on the issue of preparation. Freud would later argue that the absence of preparation was a key precondition for a frightful event to become traumatic.[25] By contrast, Charles S. Myers remained indifferent to the soldier's preparation for a traumatic experience.

"The great outstanding causes of death are shock and infection," Théodore-Marin Tuffier, then president of the French Société de chirurgie, told Crile in January 1915.[26] Studies of infection dominated the early stages of the war, prompted by the disproportionate danger posed by sepsis even in minor injuries and complicated by the infested battlefield environment:[27] "In Flanders in 1914, practically every wound was septic and surgery was completely helpless."[28] Already during the war, the best-celebrated successes of medical research concerned the staving off of infection; most famous among them was Carrel and Henry Dakin's work, under the auspices of the Rockefeller Institute, which led to development of the Carrel-Dakin antiseptic solution.[29]

Though often carried out by the same physicians, research on infection, hemorrhage, and intrusions (gunshot and shrapnel injuries) differed profoundly from research on organism-wide conditions, among them shallow breathing, "soldier's heart," and wound shock.[30] In his 1919 manual on the treatment of war wounds, the British surgeon H. M. W. Gray called attention to the inverse relation between the two types: "the greater the progress that is made in other directions, the more does 'shock' stand out as the great unsolved riddle of military surgery."[31] Indeed, sepsis—itself a whole-body condition—could be conceptualized as the internal effect of a direct intrusion from the outside into the body, an infection that could be averted by the prompt use of a disinfectant.[32] Though externally sparked, functional or organism-wide diseases were fundamentally internal; what caused them was the body's response to an intrusion, not the intrusion itself.

Before attending more closely to wound shock, we look here at the hermeneutic imperatives involved in two other conditions: exhaustion and soldier's heart.[33] Crile had sought funding and institutional support to travel to France to study "the effect on the human body of the emotions, exertion, and exhaustion," especially because he understood exhaustion as a predominant factor in surgical shock.[34] He described the Great Retreat of the British Expeditionary Force after the Battle of Mons as an "extraordinary human experiment."[35] Continuous marches across some 180 miles exhausted soldiers, who, Crile claimed, entered a somnambulistic state in which they staggered without sleep for days.

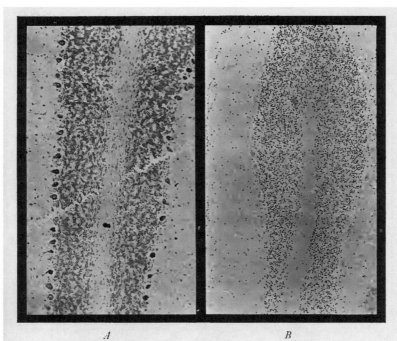

A   *B*
SECTION OF NORMAL CEREBELLUM   SECTION OF CEREBELLUM AFTER INSOMNIA
(× 100)   — 100 HOURS (× 100)
Compare the well-stained clearly defined Purkinje cells along the margin of section *A* with the faint traces of the Purkinje cells which are barely visible along the margins of section *B*.

FIG. 2.1. Images from Crile, *A Mechanistic View of War and Peace*, insets after 20, 22 and 24. Reproduced from George W. Crile, *A Mechanistic View of War and Peace* (New York: Macmillan, 1915) (public domain).

Crile's epistemological priorities are on full display in his way of reporting on physiological and phenomenological changes caused by protracted and overwhelming demands. To show the effect on organs and organ systems, Crile reproduced microphotographs of sections of the cerebellum and the adrenal glands that he pretended were taken after one hundred hours of marching without sleep.

To represent the effect on comportment, he added "faces of exhaustion," photographs whose value resided less in their indexical quality (most depict very little) than in their visual corroboration of the narrative.[36] To further accentuate the violence of exhaustion, Crile reported that soldiers who had finally achieved rest suffered from dreams: "The dream is always the same, always of the enemy: . . . a dream of the

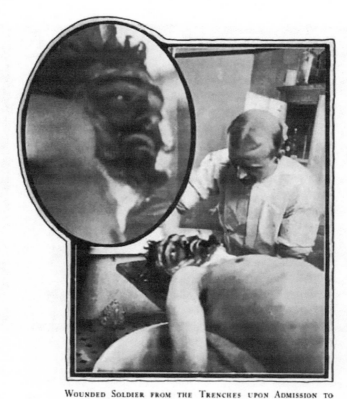

WOUNDED SOLDIER FROM THE TRENCHES UPON ADMISSION TO THE AMBULANCE. NOTE THE FACIES OF EXHAUSTION.

FIG. 2.2. Image from Crile, *A Mechanistic View of War and Peace*, 26. Reproduced from George W. Crile, *A Mechanistic View of War and Peace* (New York: Macmillan, 1915) (public domain).

charge, of the bursting shell, of the bayonet thrust! Again and again in camp and in hospital wards, in spite of the great desire to sleep, a desire so great that the dressing of a compound fracture would not be felt, men sprang up with a battle cry and reached for their rifles, the dream outcry startling their comrades, whose thresholds were excessively low to the stimuli of attack."[37] That Crile should bring up the scene of traumatic nightmares today famously associated with psychic trauma to describe physiological exhaustion speaks to the complexity of functional, multisymptom, multicause disorders at the beginning of the war, before diagnostic categories like shell shock and neurasthenia, itself linked to exhaustion, had been refined.[38] That he should further illustrate his account with microimages that he could not have obtained from the Great Retreat—they were actually from rabbits—gives a hint of the growing demand for clinical specificity and the urgency of casting complex conditions in diagnostic and illustrative terms.[39]

Analogical reasoning from animals to humans remained as necessary as the fast identification of symptoms. What counts in this case is Crile's hermeneutic imperative, his need for these tools to illuminate an organism-wide disorder. In sleepless exhaustion the patient was locked into a protracted, whole-body implosion. Affected organs (adrenal glands, liver, cerebellum) were coparticipants in this implosion. Extrapolating from accounts of the British army's exhaustion during the retreat and tying those to images derived from experiments on animal exhaustion helped him accentuate the urgency of direct study and intervention given the broad, perhaps irreversible effects on the organism. Crile carried into physicians' diagnostic and hermeneutic arsenal a far-reaching physiological language that was far more common in discussions of work, energy, and fatigue, here both for the survival and efficiency of the army and for the well-being of soldiers.[40]

"Soldier's heart," also known as disordered action of the heart, Da Costa's syndrome, irritable heart, or neurocirculatory asthenia, posed similar difficulties, although in this case the priorities were of an inverse variety.[41] Its expression in patients was made explicit by conditions at the front, and with its expression came a determination that it be accurately described, a determination not easily satisfied.

A report published by the Medical Research Committee identified soldier's heart as a group of cardiovascular conditions relating to infection and shock, and offered an extensive list of symptoms focused on the heartbeat's irregularity, the exaggerated visibility of the symptoms, the effects on blood composition, the irritability of respiration, and the look of the extremities.[42] It still treated the condition as largely indeterminate, its hierarchy of symptoms as ambiguous: despite all its symptoms, one basically knew soldier's heart when one saw it.[43] In the sixth edition of his *Dictionary of Treatment* (1920), William Whitla couldn't even decide whether soldier's heart was organic, systemic, or "psychic."[44] Adolphe Abrahams wrote in 1917 that the accepted place of soldier's heart within the general "nomenclature of diseases" demanded refinement, lest it become shorthand for all battlefield-related maladies. Without care to distinguish the condition, tightness in the chest and mild tachycardia could just as likely come from excessive smoking ("a common vice amongst soldiers") as from "genuine" soldier's heart, which was the result of prolonged exposure to conflict and bombardment.[45]

The specific mechanisms of cardiac strain and rapid, shallow breathing produced effects in soldiers that constituted the "physical" and "psychic" whole, in both the short and the long term.[46]

The capacity of war to bring forth large numbers of cases and thus make comparison and categorization possible foregrounded general pathologies, and some functional disorders emerged not simply as overarching, general, unknown conditions. Physiology's newfound abilities to study the blood, respiration, and nervous and endocrine systems both generated and frustrated the demand that these pathologies be precisely explained. They made traditional categories problematic, threatened determination, and made the functional disease and the whole-body condition the decisive epistemological problem of the era. By the early postwar period, the difficulty of using these kinds of nomenclature to describe conditions or forms of suffering was affecting the very possibility of diagnosis and even the stability of the doctor-patient relationship. The patient was becoming not just a problem of medical knowledge, ignorance, and categorization, but an often unresolvable player in the meaning of embodiment, transformation, and suffering. By 1916, wound shock had emerged as the exemplary condition of this sort, generating a veritable medico-physiological industry: a theoretical apparatus, a canon and its critics, a genealogy, and a set of therapeutic demands.

Wound Shock

Already in the early stages of the war, a disproportionate percentage of femur wounds resulted in death—even in cases of relatively minor wounds in which death could not be attributed to hemorrhage. The rate was sometimes pronounced to be 90 percent, and a major reason for this was deemed to be the shock endured by the organism either swiftly at the moment of injury or several hours later, even in "apparently simple flesh wounds of the thigh or buttock."[47] Alarmed by such data, the RAMC established the London Shock Committee. Chaired by Starling, the committee involved some of the brightest stars in British and American physiology and endocrinology: Bayliss, Dale, Cowell, and, in a consulting capacity, Cannon.[48] Their joint research resulted in the adaptation and widespread use of one of the earliest traction splints, the Thomas splint.

Traction splints had been common since the Balkan Wars, mostly as stabilization devices in hospitals.[49] Now, however, Hugh Owen Thomas's 1875 splint was repurposed so it could be applied at the site of injury following a complex procedure to secure the injured leg.[50] Besides immobilizing the wound, as was Major-General Grey Turner's original idea,[51] it deployed a second device: a system of three blankets folded over and under the wounded soldier so no fewer than four layers would cover him on each side.[52]

The application of the splint contributed to a dramatic decline in wound shock deaths from femur wounds. Cowell would later report a drop of 30 percent; another source indicated that femur deaths in cases where the splint was used had dropped from 80 percent to 15.6 percent by the end of the war.[53] The shift was to a considerable degree premised on new studies linking blood sugar levels to the temperature of the organism. The blankets addressed the entire organism's reaction: even minor wounds could generate a systemic response that involved a drop of blood volume and pressure, quickly sending the body into the state of collapse generally known as shock. This was the decisive benefit of the blankets. Besides stabilizing the leg, including the bone and shell fragments in the flesh, which researchers worried released toxins into the bloodstream and worsened the shock, medics needed to stabilize body heat and blood pressure. The simplest tool, the blanket, proved in 1918 to have been the most essential of all.

But let us return to the beginning of the war, when wound shock was noticed as a dangerous state in which the injured organism entered cardiovascular decline and often experienced a rapid "collapse."[54] Crile wrote to Harvey Cushing in early 1915 that "The great outstanding problem of this war is infection; next comes shock—shock from overwork, loss of sleep, and injury."[55] With the development of disinfectants, shock became a paramount concern for the allied armies of the western front, precisely because it was an elusive, unstable, organism-wide condition with an unusual clinical picture and a profoundly complex symptomatology in which no single or predominant cause could be identified.[56]

Although shock affected different patients very differently, it looked like this:

> (1) An ambulance driver ... was hit in the abdomen by a shell fragment. He fell down, was carried in, and put on the dressing-table at once. As he was being attended to he drew the attention of the medical officer to his profuse sweating. I saw him 30 minutes later as he passed the next relay post on his way to the operating centre. He then complained of severe pain, looked pale, and was still sweating

profusely. His hands were cold and clammy, pulse 96, pressure 100-70. On arrival at the clearing station an hour later, his pulse was only 100, but the pressure had further dropped to 82-70. Operation was immediately performed, haemorrhage stopped, and ten rents in the bowel repaired. The man's condition was serious for some hours, but he responded to treatment and eventually recovered.

(2) On a cold, wet, muddy night a man was seriously wounded by a shell while digging a new trench. He was brought to the advanced dressing-station 50 minutes later and found to have sustained severe multiple wounds, including compound fractures of femur and humerus. The exposed lacerated muscle looked like dead tissue, there were no vessels of any size bleeding and hardly any capillary oozing. The blood pressure was 40 mm. Mentally the patient was quite bright and responsive, so that the medical officer in charge of the case remarked how wonderfully fit he was. The man was dead, however, within the hour.[57]

These two cases, which Cowell reported as exemplary in his Arris and Gale Lecture of 1919, offer a sense of the primary concerns, notably a drop in blood pressure disproportionate to the bleeding and to the prima facie threat posed by the wounds. By this point Cowell had great authority because of his direct study of the wounded on the front and even in no-man's-land—activities for which he was decorated for bravery in the name of science.[58]

The typical wound shock patient was injured in the thigh or calf, either from a bullet wound involving substantial tissue damage or from mutilation.[59] In the second of Cowell's cases, the effect of shock appears all the more pernicious given that the principal wound was cauterized by the shell fragment and was not bleeding, and given the apparent strength and focus of the patient.

Injuries to abdominal organs could certainly cause shock, and in those cases shock was almost certain to accelerate death, but the extremities seemed a more surprising site for the initiation of systemic damage.[60] Abdominal and facial injuries invariably threatened the organism with massive hemorrhage and organ failure first, and only concomitantly with shock; when noticed in these cases, shock either passed quickly or hastened death. Instead, wounds that involved tissue injury to the arm, thigh, or calf—including especially the fracturing and even fragmentation of the femur, knee, tibia, and/or fibula—offered the principal image of the wound-shocked soldier. In Cowell's description, acutely low blood pressure, long known as a factor in surgical shock,[61] was the condition's main physiological marker. The other anticipated markers and their ordering were the subject of controversy, even among surgeons well experienced with shocked patients.

Let us trace the original picture of the shocked patient as it appeared to relatively naïve physicians, surgeons, and physiologists at the beginning of the war. Tuffier wrote: "I have seen men with slight wounds suffer considerably from shock; I have seen others severely wounded who did not experience shock."[62] Another French surgeon emphasized the apparent contradiction in 1916: "Disturbing shock in a wounded man with no visceral lesions; no shock in another who suffered a section and five perforations of the small intestine."[63] The injury might be to a vital organ, or it might lack visible signs altogether. The onset of shock might be immediate or might postdate the injury by several hours. The organism might be endangered by already-existing conditions in the battlefield or the trench. Patients suffering from shock looked pale and discolored, sometimes with a gray or "dirty brown color," and not necessarily because they had lost blood.[64] They were cold; their vital statistics were diminished in strength; their blood pressure was low; and they had a very slow or soft pulse, although it often sped up irregularly or slowed to pulselessness. Many screamed of their need for air. The decline of vitality was as obvious as the causality was confusing.

The patient might survive, but only after several hours of a struggle, or he might die by his doctor's hand; surgeons at the front called wound shock their "bête noire."[65] More often than not, surgery on shocked wounds precipitated patient collapse by directly causing what was named surgical shock (shock caused by and during surgical intervention), which doubled on the original shock and often caused death.[66]

The "learned"—that is, physiologists who came prepared having spent years on wound shock—were no less troubled by the condition than those who merely happened upon it. Crile relied on vague categories like exhaustion, and Bayliss still hesitated in May 1918, after years of direct experimental and clinical work: "Wound shock is probably, in most cases, due to a combination of several factors, each in itself not necessarily serious. The part played by each . . . naturally varies in individual cases."[67] Eighteen months later, Dale declared in his Harvey Lecture that, despite his conclusions, he made "no claim to give an explanation of shock in general."[68] Too often the condition was simply decried as obscure.[69]

Some of the mystery concerned its uneven symptomatology: as Crile again remarked in 1921, some thirty years after his first publications on surgical shock, "extremely diverse factors are synergistic," and particular causes could not be depended upon.[70] Combinations of symptoms had to be investigated without most symptoms appearing dominant, or specific to the condition, and worse, with some symptoms being in theory impossible. For example, blood was "disappearing" from the arteries to somewhere in the body. Porter surmised that "the man is bleeding to death in his own veins!"—but Cannon and Dale gradually corrected him: blood was pooling in the patient's capillaries.[71] Yet capillaries were not supposed to contain much blood. Also, existing theories, notably of surgical shock, reflected only in broad strokes the clinical picture that wounded soldiers exhibited upon their arrival in the casualty clearing station. Wound shock derived historically from surgical shock, and thus it seemed important to draw on theories of the latter. Still, Crile's dominant anoci-association theory of surgical shock was quickly criticized for confusing causes, coefficients, symptoms, and effects.[72]

This produced a second form of instability, this time theoretical, pertaining to the very order and claims of physiology. The difficulty in coming up with a proper name for what they were encountering was characteristic of the problem: wound shock was ostensibly named as such only in 1917, "following a suggestion by E. M. Cowell;" before that the condition was mostly known as shock, in some cases as nervous shock. These terms persisted into the 1920s.[73] Cannon sought to name it *exaemia* (the condition of being drained of blood), creatively stretching the genealogy for the condition back to Hippocrates.[74] This term didn't stick any better than the London Shock Committee's earlier *oligaemia* (a term deriving from Yandell Henderson's work, which the committee also cast aside).[75] Not one to abandon his Grecophilia, Cannon turned to *traumatic shock*, another problematic term that was quickly forgotten.[76] Surgical shock also appeared plausible, although the analogy was difficult, and surgical shock was now reconceived as a surgery-induced subcategory of shock in general.[77] Moreover, although British and American researchers could agree that wound shock differed from shell shock,[78] the French did not always do so;[79] even so, how they differed remained unclear.[80]

What exactly the term *shock* referred to and why was no clearer. Bayliss punted: "Although the condition is generally recognized and unmistakable, the primary cause is still obscure, and I do not feel called upon to define or explain what is the actual nature of the conditions underlying it, except to mention . . . the most obvious symptoms."[81] Definitions were heuristic, multicausal, and often speculative. In 1923,

Cannon offered a general definition of shock that was based on its "easily recognizable features": "We may say that shock is a general bodily state which occurs after severe injury and which is characterized by a persistent reduced arterial pressure, by a rapid thready pulse, by a pallid or grayish or slightly cyanotic appearance of the skin which is cold and moist with sweat, by thirst, by superficial rapid respiration, and commonly by vomiting and restlessness, by a lessened sensibility and often by a somewhat dulled mental state."[82]

So why did shock precipitate a near-complete collapse of the organism, which researchers identified as a vicious circle? Was this merely a matter of the blood pressure drop? Of a release into the bloodstream of toxic substances (as per Dale) or subcutaneous fat (as per Porter)? What was the role in causality of weakened, aggravated respiration, or "air hunger," as it became known? What kind of temporal fold did shock create for the patient, out of the injury, the organism's reaction, and the attempt at treatment? As a privileged condition that exposed problems with the state of medical and physiological knowledge, wound shock facilitated the understanding of the interconnection of the body's different systems and directed researchers' attention again to the organism's regulation of the blood and to substances released within it. Wound shock became a site of knowledge that defined physiology, science, and war.

The Space and Time of Wound Shock

To British, American, and French researchers, wound shock—whatever it was—had a geography and a distinct temporality.

The geography of wound shock was established parallel to the emergence of the disorder itself, and it involved as much aberration as planning. First, as far as the shocked were concerned, their basic path was largely the same as for other injured soldiers: from the site of injury back through dressing stations at the trench, then casualty clearing stations, and then, when necessary, on to hospitals further behind the front.[83] Transportation posed an important problem, as did treatment at the casualty clearing stations.

A second geography, no less meaningful, emerged with the medical engagement of shock. It included experimental and clinical centers, including in Béthune, just west of the front line; an experimental lab in Hampstead on Bayliss's own estate; and as of early 1918, an American-run hospital and experimental center in Dijon. Military physicians traveled back and forth to the front to establish research and treatment protocols.[84] Experimenters in Hampstead reported to the London Shock Committee.

American researchers were involved in several, often competing groups. American physiologists, including Crile, Porter, and Cannon, came to the Western Front under different auspices to study shock even before the arrival of the US Army and its medical corps. Crile—the foremost authority on surgical shock—wrote to the American Ambulance Hospital in Paris in late 1914 and arrived in France soon thereafter.[85] He stayed for the most part in Paris and Neuilly-sur-Seine, although he did travel to Compiègne and the front for his studies of shock and exhaustion, thanks in part to Carrel's help.[86] The Rockefeller Institute selected Porter to work under Carrel's direction in 1915 after Cannon, his archenemy at the Harvard Medical School, demurred on the invitation.[87] Porter crossed the Atlantic twice and spent months at Carrel's clinical and experimental sites in La Penne and Compiègne. Cannon gradually changed his mind and followed a quite complex trajectory. On February 14, 1916, he wrote to Charles Sherrington, horrified by reports by the Dardanelles Campaign and by a letter from Sherrington detailing the deaths and injuries of physiologists:

> I dreamed that I was dwelling in a great house not far from a battle line. We heard the booming of the guns begin, and then the ambulances began to bring back the wounded who were temporarily laid on the ground outside. It was pitch dark. Instead of going out to succor them, I went upstairs to the roof and began looking about among the chairs and other things that were there, for a blanket which I could not find. My search was occasionally interrupted by a glance down at the flaring torches carried by the orderlies and by the hoarse cries of anguish that arose from the dark field. There was a Freudian symbolism about this vision, for it seems to me that as I stay here in the laboratory I am quite truly "looking for a blanket in the dark," while there is great need for more direct and practical work elsewhere.[88]

"One can only study what one has first dreamed about," Gaston Bachelard wrote, and indeed, later that year Cannon got to work.[89] First he joined the Committee on Physiology at the US National Research Council, then he traveled in May 1917 to the British lines in France with the Harvard University Hospital Unit and spent several months in Casualty Clearing Station 33 of the British Expeditionary Force in

Béthune. He also spent time in London with the Shock Committee, and in February 1918 in Dijon with the American Expeditionary Force laboratory, which he helped set up.[90]

That much of this engagement took place in France was not lost on Anglo-American researchers, even as many disparaged their French colleagues as poorly organized.[91] Such assessments are excessive. The French followed a separate conceptual and medical geography and French descriptions of shock were at least as detailed. For them, too, it was a totalizing condition that was difficult to dissociate from other conditions that were far more intense.[92] Édouard Quénu, credited by Cannon and Dale with being the first to identify toxemia as a factor in shock, worked at the Medical Faculty of the University of Paris and relied on assistants and other physicians for secondhand reports.[93] The surgeon René Leriche, moving between Paris and the front, addressed shock in the context of fractures, discussing it in *The Treatment of Fractures* as a grave threat that complicated every injury dramatically.[94]

The geography of shock—on and in the body, on the battlefield or in the clinic, and in the distance between—is troubled further by its temporality. By "temporality of shock" we mean two things. First, shock was not a major concern early in the war, and the long history of surgical shock and railway spine had been largely left aside as ancient history. Then for the Entente powers on the Western Front, shock became a pressing matter in 1916, and even more so in 1917 and 1918. The failed attempt to resolve shock by using saline solutions at the Battle of the Somme contributed to the growing urgency.[95] The US National Research Council established a subcommittee on shock in the fall of 1916. Britain's Medical Research Committee similarly prioritized shock in late 1916. After publishing a first appeal for information in *The Lancet* and the *British Medical Journal* in March 1917, it brought out its principal publication on shock, including policy and treatment recommendations, on Christmas Day in 1917.[96] Major publications followed in 1918 and 1919 by the principal researchers associated with the committee. Shock receded from view around 1920, with the development of different pathologies among ex-soldiers. By then it also had grown to an impossible, unmanageable category, and as the clinical urgency of the condition faded, so did its epistemic clarity.

A second temporality was specific to the injured people suffering from shock: the tricky matter of its onset and course, which was the core issue involved in treatment. Time was recognized as a predominant problem in wartime surgery to the point that delay was occasionally directly identified with "the enemy," as if from the surgeon's perspective the two were the same.[97] On temporal matters, shock complicated an already-difficult situation. Wound shock could follow immediately after the injury, although it was usually preceded by a lag of unclear duration, and once it set in, it could proceed rapidly and devastatingly.[98] Even if shock did not kill them quickly, many soldiers had a few hours at most for medical treatment to begin.

In 1917, the physiologists of the London Shock Committee concurred on a terminology of primary and secondary shock, making the distinction largely on the grounds of the speed with which it developed, and also in relation to the injury inflicted on vital organs.[99] By early November of that year, this terminology was standard in the British Expeditionary Force, and Cowell's instructions for the prompt treatment of shock were forwarded even to field ambulances.[100] Primary shock was identified as occurring quickly, generally as a result of damage to a vital organ; it either resulted in death quite quickly or "yielded to simple restorative measures at the dressing stations and created no serious problem,"[101] as if it were "analogous to faint."[102] Secondary shock, which until Crile's 1899 book on surgical shock had been known as collapse, had a trickier, more "insidious" temporality; it arose slowly and gradually, and it often involved a recurrence of primary shock.[103]

Secondary shock was the major concern, "the condition we were urged to investigate," and it became the archetypal form studied.[104] Secondary shock was deemed especially dangerous, "causing heavy loss of life among men whose wounds, from the purely surgical aspect, presented no hopeless problem."[105] In the case of secondary shock the symptoms and causes of shock generally could be studied in greatest detail: it allowed for both observation and treatment. Protracted secondary shock involved definitive and irreversible effects on the organism that one could not access in primary shock, so the temporality of secondary shock became the focus of research.[106] The distinction between primary and secondary shock was quickly adopted also among the French, who retained a third category, of immediate shock.[107]

As was repeatedly stated at the time, conditions from the battlefield to the clearing station were an essential element in survival. The treatment of shock didn't cooperate with the route of delivering the patient to the hospital quickly and performing an operation. Transportation tended to aggravate the patient's condition. It subjected him to further exposure and facilitated further hemorrhage, tissue damage, and toxemia, not least because of broken or splintered bones in the flesh. Pain and anguish worsened and burdened the organism emotionally.

Early on in the war Carrel recommended that shock sufferers be treated at the site of the injury rather than moved to a hospital;[108] subsequent clinical studies linked cold weather and the nature and long duration of transportation across trenches and danger zones to the very incidence of shock. All the while, sepsis and gangrene threatened to set in.[109] Thus, surgeons faced a concurrence of morbid conditions with divergent temporalities that lay largely beyond their control. Intervention was urgent, but intervention to what end—relief from the injury or relief from shock? Surgery, usually essential for countering the injury, was known to *cause* shock and thus compound the problem. The French surgeon Pierre Delbet argued that "instead of waiting for the patient to exit the state of shock to operate on him, one must operate to get him out of the state of shock."[110] To the British and the Americans, this was utter folly: shocked patients would simply die in their doctor's hands from the resurgence of shock.

The concurrence of morbid conditions thus posed the central temporal problem within treatment. Shock required that the organism be warmed up, which also clashed with the urgency of surgery and amputation; it also masked and complicated prophylactic efforts against infection and septicemia.[111] Some French surgeons and physiologists hedged that the presence of shock cautioned against the use of anesthetics. But without them, how much could be done for the injury?[112] Crile maintained that general anesthesia was essential.[113] As surgery became a theater of experiment, calculations as to waiting, anesthetizing, warming, operating, disinfecting, and so on, became ever more complex. If the patient recovered from the shock, he would be largely safe from its threat, and treatment could focus on the rest of his injuries. Now a new kind of long-term suffering could set in, as many survivors would suffer from chronic functional disease, myocardial conditions, tachycardia, fatigue, soldier's heart, and vascular complications.

The definition and even the very being of shock were profoundly affected by the geography of its study and the temporality of suffering and treatment. Not only was shock rendered a "new" disease, but the confusing wealth of clinical information collected with urgency and often at risk to the investigator was shared mostly *within* each individual group of researchers. Partial and competing theories and approaches emerged, and the fact that study and treatment were decentered encouraged competition for a solution. At the same time, this decentering contributed to the establishment of the London Shock Committee's particular interpretation, which, thanks to the committee's reach within the RAMC, became by and large canonical.

Theory and Treatment

The questions explicitly raised by late 1916 relayed a developing hermeneutics of bodily integration and brittleness. To address them is to begin to understand the newfound priority of the clinical domain and experimenters' efforts to address the individual body as a whole that is mechanistically fashioned.

Across very different approaches, this remained a central theme. Both Crile and Porter, especially Crile, had clinical and experimental experience with surgical shock. This led them to downplay the need for new experimental work or elaborate clinical data. In late 1914 and early 1915 at the American Ambulance Hospital, Crile lectured on shock both publicly and in private, and his notes indicate that his attention to shock was mostly surgery related. His "kinetic" theory, developed before the war,[114] sought to appreciate the linked systems of kinetic function (digestive, respiratory, circulatory, urinary, genital) and to pinpoint the effects of their obstruction on the production, conservation, and loss of energy.[115] His ensuing concern with the nervous exhaustion of the vasomotor center in the brain, which he deemed a primary factor in shock, followed closely from the effort to understand the "close resemblance between infection-exhaustion and traumatic-exhaustion."[116]

Porter, who worked in Carrel's group, agreed with Carrel's early suggestion that the cause of wound shock lay in the fracture itself.[117] Like Carrel, he was contemptuous of French researchers, and he was hostile to and competitive with Cannon. He evidently felt authorized by his earlier work on blood pressure and the vasomotor center to discard interpretations premised on them. In his reports, he also celebrated his access to the front. Compiègne and La Penne, where he was stationed, were at times within shelling distance of the German positions, and Porter repeatedly requested and received visitation rights in order to gauge the condition of soldiers soon after they were hit. Porter mixed his laboratory work with the knowledge he developed from these visits. He staged his memoir *Shock at the Front* as a genre-bending medical epic and police thriller in which he heroically overcame the idiocy of French authorities as well as the dangers of enemy action to alone uncover the truth of shock in the trenches. Together with his mostly unnamed lab partners, he settled on two causes of shock: femur fractures and fat embolism: "The surgeons at Compiègne had noticed that shock came on chiefly after wounds of the great bones, such as the thigh-bone, and after multiple wounds through the skin and subcutaneous fat, as from a shower of shell-fragments."[118]

Other authors would also study toxemia, but Porter focused exclusively on the transport of fat through the veins. "The femur, or thighbone, is the largest in the body. When it is broken by a shell fragment, the rich bone-marrow is exposed. Perhaps some potent chemical substance is thereby set free, to be absorbed into the blood-vessels, through which it might reach the brain and spinal cord, and by poisoning the nerve-cells produce the phenomena of shock."[119] For this potent chemical substance, Porter homed in experimentally on fat and bone marrow coursing through the blood, which displayed, according to him, precisely the effects found in shock. Unlike Crile, for whom the holistic character of wound shock could be mitigated only if shock were treated as a matter of surgical care, Porter resolved this holistic character of shock by denying it and opting for a largely monocausal solution.

The London Shock Committee also had to jockey extensive theoretical and laboratory prowess in a field that had come to belong to the clinic and the experience of war, and by and large it wrote the history of shock anew. Bayliss, Starling, Dale, and Cannon were less interested in surgery than Crile or Porter were, and they proposed a balance, outsourcing clinical work and squaring it with physiological results. In its second and third meetings, the committee offered a preliminary definition: "Shock is a failure of the circulation owing to the deficiency of entry of blood into the heart."[120] In its first report, "Memorandum upon Surgical Shock and Some Allied Conditions," published in late March 1917, the committee appealed to anyone able to offer clinical information, while presenting at length the experimental results of Dale, A. N. Richards, and Patrick P. Laidlaw's research in histamine.[121] Cannon was well aware of his own lack of direct clinical experience and sought out field research during his trips to France to match the laboratory work he carried out with the London Shock Committee and at the American Expeditionary Forces laboratories.[122]

By the fall of 1917, the Shock Committee was ready to flaunt its extensive clinical experience, notably N. M. Keith's six months and Cowell's three years "at various points in the firing line, and often actually in the front trench."[123] Although it aimed for a physiological model of the body undergoing this collapse, it insisted on the importance of raising clinical individuality to the level of experimentally confirmed rules: "Criticism has been leveled against the preponderance in this report of clinical data over practical deductions for treatment. But Mr. Cowell very rightly points out that once the knowledge of wound shock is brought on to a sound clinical and pathological basis the application of therapeutic principles will follow as a matter of course."[124]

This turn toward the clinical standardized a number of technologies that had become available or portable only since the beginning of the war. These included an inconsistently used array of devices including the Riva-Rocci instrument, the Tycos instrument, and other spring and mercury sphygmomanometers.[125] Such apparatuses were much easier to transfer to hospitals near the front. They could measure blood pressure and volume as well as sugar and carbon dioxide levels, making it possible to study effects on the organism as a whole. They allowed for physicians and surgeons to match the patient's appearance and comportment to interpretable numbers, and often to record measurements precisely while taking more general notes and hence to produce a complex body of information. By 1918 it was possible to measure blood and plasma volumes to understand where the "vanished" blood would have gone.

No less significant was the medical and literary technology of the case study, which was the clinical unit most frequently brought forth, and which quickly became a device for both understanding and narration. One article in the Shock Committee's report of December 25, 1917, referenced ninety-eight cases of shock and hemorrhage and compared them to fourteen control cases; Cannon based another article on a study of forty-seven cases.[126] Although case studies had been used in reports on surgical shock before the war, such uses were incidental and short.[127] Now, just as organism-wide data was becoming possible to record on the spot, it was also becoming easier to amass case studies that would give a sense of the immediate and theoretical needs of the condition in a manner that engaged closely the significant differences among patients. For Cannon, Bayliss, and Dale, the shift was considerable, from a focus on experimental biology, in which for biological laws one had to extrapolate from research on animals or use X-rays and self-experimentation, to the clinical study of an individual, who presented a set of experimental factors that could be studied as a case and could in turn be replicated on animals in the lab. The correct purposing of animals became a matter to accentuate in reports.[128] It is not just that researchers turned to the clinical picture; they used clinical knowledge to define and redirect their experimental work, simulating effects known from case data in their laboratory work on animals, often seeking to bring about and then delay or avert the animals' organismic collapse.

The Shock Committee's priorities were three: first, to discover a treatment, or set of treatments, that could prevent or alleviate shock, notably secondary shock; second, to make possible a theory that would explain the considerable divergence of shock's course and effects across different individuals; and third, to trace the effect of shock on the specific organism and understand how the condition worked its damage across different systems (cardiovascular, nervous, endocrine) for its effects to be so extensive and mutually reinforcing—that is, how they generated a vicious circle that brought about the organism's collapse.[129]

So what was the Shock Committee's new and soon-to-be canonical understanding of shock? The central symptomatic reference was the patient's very low blood pressure. This was accompanied by an unstable, often rapid pulse that felt negligible to the touch; a frequent demand for air; an overall look of pallor; and agitation. The committee's reports on the blood complicated this well-known picture considerably. Cannon, John Fraser, and A. N. Hooper emphasized that under shock the red blood content in the capillaries rose considerably while the blood count in the rest of the body dropped.[130] This meant that capillaries dilated to fill with arterial blood and did not "push" blood into the veins, and overall blood pressure declined.[131] Cold especially aggravated this pooling into the capillaries,[132] and so did any drop in body temperature resulting from the drop in blood pressure.

The hemoglobin count also dropped, and this was greatly accentuated in cases of hemorrhage; hemoglobin carries oxygen to the different parts of the body.[133] The alkali reserve in the blood (i.e., the presence of salts, notably sodium bicarbonate) also decreased considerably;[134] this was not harmful by itself, but in the context of other disturbances, notably the hemoglobin drop, it amounted to a potentially dangerous acidosis of the blood and caused a rise in the blood's hydrogen ion content (perhaps at further expense to its oxygen load).[135] Acidosis endangered the patient by depressing the flow of oxygen and inhibiting its uptake, thereby troubling tissue respiration and causing "air hunger," which, Cannon pointed out, was at times symptomatic of impending death.[136] Acidosis also greatly endangered a patient who was undergoing surgery. Cannon argued that acidosis, and not Crile's "exhaustion," was responsible for shock during surgery.[137]

These pathologies compounded one other, like forces generating torque.[138] Cold, capillary stagnation of the blood, gradual asphyxiation of tissue, and spread of acidotic blood across the body (and potentially of toxic substances as well, whether histamine or gas bacillus) resulted together in lower blood volume reaching the heart,[139] which gradually became incapable of maintaining even the already-depressed blood pressure (the "head").[140] The organism, already cold, became colder, and in turn, disturbances in circulation worsened. Cannon called this a "state equivalent to hemorrhage" and emphasized that actual hemorrhage worsened the condition dramatically.[141] Starting with Bayliss in 1918, it became routine to cite experiments by Robert Gesell in California that showed that even a small loss of blood led to a profound loss of pressure.[142] To accentuate the analogy of invisible hemorrhage, Cannon and Bayliss cited an experiment in which 10 percent blood loss correlated with no less than 60 percent hypotension.[143]

Normally, the body would attempt to compensate for the injury, but the equilibration of even minor injuries was inhibited by the disordered cardiovascular system. The very attempt to bring forth equilibrium became deeply pathological.[144] As a matter of fact, many cardiovascular actions noted previously were precisely compensatory actions, aimed at diminishing the blood loss occurring at the wound site.[145] In counteracting the injury, the body was harming itself.

It is worth emphasizing that the entire description offered thus far concerns the cardiovascular system *alone*. Although this system was the one accentuated in the Shock Committee's report of Christmas Day 1917, it was by no means the only one affected. The committee's argument on other systems and their involvements was spread out across different reports and displayed divergent views among committee members. For example, Cannon's prewar research had rendered emotional reaction to injury central to its imbrication in the general endocrine situation. This was confirmed by Cowell on the significance of his physiology of excitement in understanding endocrine factors affecting shock.[146] Cannon then prioritized "nervous action," temperament, and the organization of the individual in "the development of shock or exaemia." He wrote: "Primary wound shock—dusky pallor; rapid, thready, low-tension pulse; hypotension; sweating; thirst, and restlessness—may come on so soon after injury as to be accounted for only as the result of nervous action. The organization of the individual (for example, a 'high-strung' temperament), fear and fatigue probably provide favorable conditions for the nervous response. . . . Sweating and exposure lead to rapid loss of heat from the body; previous sweating, wetness of the clothing, and low external temperature favor the process. Inactivity of the wounded man and absence of shivering lessen heat protection."[147]

The shared idea that temperament and excitement were significant—including for Cowell, who postulated that "the excitement of battle probably comes first, converting the normal human machine into a fighting mechanism"[148]—reminded Cannon that a release of adrenaline facilitated a higher heart rate and greater strength in the short run that, once prolonged, could be poisonous. Because adrenaline coursed through the wounded man's restless body, it "may prove to be one of the factors responsible for wound shock."[149] Bayliss agreed.[150] Dale, Laidlaw, and Richards, meanwhile, insisted that a substance like histamine must be involved in shock, and they identified histamine shock as at the very least analogous to wound shock.[151]

The respiratory system was as crucially involved. In Bayliss's cats, the vasomotor and respiratory centers suffered from the drop in blood pressure.[152] Bayliss surmised that the rise in hydrogen ions, a coefficient of the drop in alkali reserves, badly troubled the respiratory center.[153] While maintaining that the drop in oxygen was more consequential than acidosis,[154] Bayliss thought he had shown that acidosis and oxygen uptake were tied together in a causal connection. The two then built on each other: the kidneys, for example, depended on normal blood pressure and without it could not cleanse the blood of toxic substances. In turn, the decline in the oxygen uptake at different tissue points forced a rapid, shallow respiration, thereby setting off another vicious circle as panicked breathing brought even less oxygen into the already-starved body. Cannon insisted experimentally that the loss of oxygen facilitated by acidosis led to the formation of lactic acid, which, in the absence of oxygen, could not be burned down to carbon dioxide and water; it thus accumulated and worsened circulation further.[155] Lack of oxygen also contributed to cell death in different parts of the body and caused often irreparable harm to the central nervous system.[156] Injuries to the vasomotor center, to peripheral nerves, and to the central nervous system had at once functional and morphological effects depending on blood drainage at specific points in the body. Thus, the organism met with both organism-wide and local effects, and neither the location nor the buildup of local effects nor the systemic or functional effects could be predicted.[157]

Finally, shock was rarely the sole danger to the organism; threats of infection, sepsis, gangrene, and so on, generated intersecting morbidity threats. Hemorrhage was so frequent that separating shock from it was impossible.[158] As it had done in its first memorandum of March 31, 1917, the Shock Committee reported again in March 1919 that toxemia, the release of a certain toxic histamine-like substance, was not merely a side effect but a major cause of the body's response.[159] In 1922, Cannon extended his earlier analogy of shock to hemorrhage to suggest that the body in shock was also similar to that in gas gangrene, emphasizing a comparison Dale had pursued in his experimental work.[160] Proposals for treatment—notably by Leriche and Gray—focused on the overlapping dangers.

From blood pressure drop to respiratory, vasomotor, nervous, and cardiac problems, the body collapsed across them all in a manner that was frequently irreversible. Besides the sheer multiplicity of possible symptoms, several factors contributed to the uniqueness of each patient's clinical picture: the injury itself, the prewound factors, the cold, the mixed effect of sweating and shivering, the adrenaline, and the nervous response to fear.

Thus, at the same time that the pathology of shock was being described as multisystemic, the patient was reconceived as profoundly individualized. Cowell: "The observations I have been able to make on the effects of excitement on soldiers show that a stimulus such as exposure to the danger of enemy fire produces a reaction which, as might be expected, varies greatly according to the individual."[161] Bayliss: "The part played by each of [a combination of factors] naturally varies in individual cases."[162] Gray: "Different individuals vary widely in their capacity to withstand shock."[163] The body was becoming a singularized object—each body a single being, a singular object of study—at the very moment of its collapse.

"No sovereign remedy exists for the treatment of shock," concluded Gray, in his 1919 *The Early Treatment of War Wounds*.[164] Cowell and the London Shock Committee agreed: as befitted a condition that affects individuals differently, for all the clinical and experimental detail and all the conceptual work that went into understanding the shocked body, there was relatively little that could be done for individual patients once they were shocked. Dale concurred: "Efforts to prevent shock, by measures designed to keep the patient warm, to supply abundant water, to immobilize injured parts during transport, were much more successful than the treatment of the developed condition."[165] Doctors were severely limited in their treatment options; with shock as with sepsis, it was preventative measures that mattered most: maintaining body warmth paralleled preventing infection. The approach to treatment favored by the London Shock Committee and accepted by the British Expeditionary Forces was largely conditioned by this understanding that shock needed to be treated early, before the patient arrived at the clearing station, if possible even on the battlefield.

Actually, this was known well before the war: patients needed to be blanketed, warmed with tea, and, according to some surgeons, given morphine.[166] For Leriche, who despaired of the possibility of saving someone who was shocked, the patient "must be treated by means of the usual stimulants (oil and camphor, artificial serum)—morphine is very useful in severe cases of shock—and then early operation must be undertaken without waiting for the clinical signs of established septic poisoning."[167] This was still quite traditional. The Thomas splint became so significant because it systematized this treatment and moved it to an earlier moment, namely that of transportation from the site of injury to the dressing station. Immobilizing the thigh and calf to prevent bone fragments from injuring the patient further during transportation, and perhaps to slow toxemia, constituted an important goal for the device; warmth was the other.

The Shock Committee was instrumental in rapidly convincing the British Expeditionary Forces of the significance of the splint's blanket system: by the spring of 1917, the Thomas splint had been put to "universal application," with great significance being applied to the blankets. At the end of that year, Cowell treated the universal introduction of the splint as an achievement: it "has provided increased comfort to the patient and immobilization of the fragments of the bone, [and] the incidence of secondary shock has been greatly diminished."[168] The committee's instructions to maintain a hot-air chamber under the patient when he lay stable at a dressing or clearing station also aimed to maximize the value of the blankets. According to subsequent reports, the application of the splint and blankets contributed to a substantial drop of femur wound–related deaths from wound shock. The drop in the death

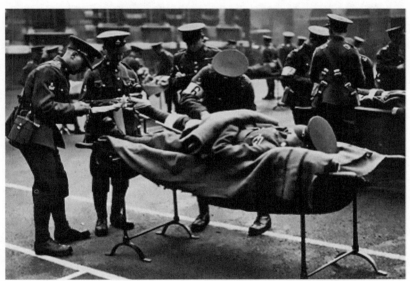

FIG. 2.3. "Thomas's splint" drill, with two layers of blankets folded on top of the "patient" and four layers beneath him; the remaining two layers are hanging on the sides until the drill is nearly complete. Wellcome Library, London.

rate would continue to be celebrated; some fifty years later, Cowell was still praising the splint as follows: "As the result of the widespread adoption and practice of the method, the mortality of gunshot wounds of the femur was reduced by over 30% in forward areas. French gunshot femurs still arrived at the forward hospitals, dead from bleeding or wound shock. This was one of the great advances in front-line surgical treatment."[169]

The matter of warmth did not end with the blankets, nor with the Shock Committee—partly because of the priority, in nursing practice, of helping patients warm up. Historian of nursing Christine Hallett has emphasized the widespread use of water bottles, hot-air hoses, and *réchauffement* devices such as the cage.[170] Wallace noted that "in more severe cases we employ a 'light bath' of electric lamps beneath a cradle, or else a 'hot-air bath' extemporized by leading under the bedclothes a pipe connected with a primus stove."[171] Meanwhile, French medical ser-

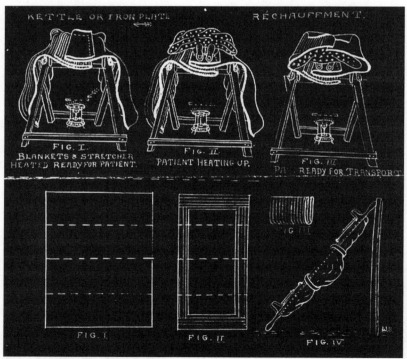

FIG. 2.4. "'Blanket packet': folding three blankets to give the effect of four. Method of carrying an army blanket folded in a waterproof sheet. This packet is the same size as a stretcher pillow." Cowell Archive, GC 116/5, Wellcome Library, London.

vices separated patients "in a state of collapse" into a *déchocage* ("de-shocking") room where their wounds could be disinfected and their arterial pressure constantly monitored. English and American physicians remained convinced that the French efforts were quite meek.

Attempts at a remedy became central to the Shock Committee's effort in England, in particular through the recognition that "recovery from wound shock has been shown to be associated with a rise in blood volume."[172] Experiments concentrated on blood volume—on finding an alternative to the administration of blood transfusions and saline solutions. Saline solutions had an only temporary delaying effect on shock.[173] Blood transfusion was highly complicated because of the difficulties of stabilizing blood at the front, and it was still a matter of debate among the British, French, and New Zealander forces.[174]

Bayliss was already working on "transfusion fluids" in 1917, when he developed the gum acacia solution in Hampstead.[175] He tested the solution by sending various concentrations of sodium bicarbonate ($NaHCO_3$) and gum to the front to try out on shock patients. His emphasis on gum arabic and sodium bicarbonate relied on the function of gum arabic as a colloid that helped maintain, and even raise, blood pressure.[176] Although Bayliss doubted Cannon's emphasis on the drop of alkali in the blood and was not convinced that bicarbonate in the solution was essential,[177] he nevertheless included it for some time in the solution, seeking to tackle multiple problems at once.[178] Under the cover of caring for the wounded, he manufactured and shipped different amounts of gum and sodium bicarbonate and waited for the results.

Bayliss and Dale probably experimented with gum acacia on cats, but the definitive—or rather, definitively indecisive—reports came from its use at the front, by surgeons affiliated with the Shock Committee, notably Cowell, who offered criticisms of particular concentrations of gum.[179] Clinical experience and surgery combined anew with therapeutically minded experimentation. Cannon reported, "I saw magical effects from these injections in cases which had reached the stage of air hunger."[180] Others were not nearly as convinced,[181] and after the war, thanks to gradual improvements in blood transfusions, the generalized use of the solution declined because it was considered insufficient and ineffective.[182] For the time being, though, Bayliss, Wallace, and the Shock Committee managed to keep the faith of the RAMC and to present the solution as a partial and effective response to shock,[183] despite the repeated failures that sparked the debate that led to the contentious Boulogne conference four days after the Armistice.

The Shock Committee's more detailed theory did not satisfy competing research labs. Carrel dismissed gum arabic as "inefficient; if in large amounts, toxic."[184] Meanwhile, Crile, at a surgeons' conference only a week after the Boulogne conference of November 1918, scorned physiology's dedication to gum, contending that it was out of touch with surgical reality. In his diary account of the event, he decried physiology's trust in gum arabic, concluding his own *la théorie c'est bon* with a quip that amounts to an indictment of aiding and abetting: "The physiologists were represented by . . . Cannon. They had their day first. They sang the hymn of praise! Then the practical surgeons of the Argonne fight sang the swan song! Gum was condemned! Exit gum! Exeunt physiologists! *Der Tag!*"[185]

Precedents and History

On the basis of its experimental and clinical work, the Shock Committee compiled a theoretical and therapeutic canon. Those excluded from it were relegated to prehistory. Crile, despite being a contemporary, was excluded on the premise that nervous exhaustion of the vasomotor center could not be demonstrated to be the cause of shock.[186] Cannon tersely dismissed Porter as harmful to patients because of his claim that fractures were the primary originator of shock and fat embolism its predominant cause.[187] Other theories were also lobbed off as part of the confusion that had marked the starting point for the committee's eventually successful research.[188] At stake in this parsing was more than exactitude—the functional-disease status of shock required a rejection of monocausal explanations. Anglo-American experimental physiologists were actively distancing their work from the nineteenth-century German-originating psychophysiology of kinetics, energy, and fatigue— including the human-as-motor model that Anson Rabinbach has placed at the foundation of modern work science.[189] The new physiology was quick to identify brittleness in an integrative, multisystem body, on the basis of a condition that performed physiology's own experimental and research capacities. (This is perhaps all the more pronounced in that the issue of heat could have easily led researchers back to thermodynamics.)

The actual history of shock, far more complex than the prehistory quickly dispensed with by the Shock Committee, reveals a curious disjunction.[190] First, the committee's labored revision of the understanding of shock brought about extremely limited advances in treatment. Second, although this was not especially visible given clinical urgencies in World War I, the therapeutic outlines and broad sense of shock as a whole-body condition predated the Shock Committee's work by half a century. In this section we follow two paths. Beginning with a broad outline of the history of shock, we foreground its preexisting status (as a clinical condition that defied categorization) in order to close the aperture on the meaning of shock in 1916–1919. Then we trace the conceptual frailty of wound shock after World War I.

The diagnosis of shock dates, in general terms, to the eighteenth century, but shock and collapse became important concerns during and after the US Civil War and the subsequent nosologies of railway spine and surgical shock.[191] This identification of shock with the 1860s was moderately common in World War I: the French physician A. Lapointe hinted at a history going back to Hunter Holmes McGuire, and Cannon identified both the beginnings of shock as "reflex paralysis" and the difficulty of distinguishing between its different kinds in the Civil War.[192] In civil life, Crile dominated the post-1890 scene through his theory of anoci-association and surgical shock; his experimental studies, aimed at reproducing the effects of shock and railway spine in canines, were later expanded on by Porter and Henderson. Meanwhile, shock reappeared in the Boer War and the Russo-Japanese War.[193] A 1907 article on military surgery in *The Lancet* cautioned that the Russo-Japanese War demanded the classification of shock in a trifecta of dangers, along with hemorrhage and sepsis: "Early operative procedures on the severely wounded are dangerous beyond the degree which modern surgery would lead us to expect," and the medical officer who renders first aid had "to combat shock with warmth, stimulants (unless contra-indicated), and opium, the last being particularly valuable."[194]

The British surgeon Edwin Morris deserves particular attention. In an 1867 treatise Morris declared the symptoms of shock to be "sudden vital depression," "great nervous depression," "final sinking vitality," "nervous shock," "violent mental emotion." He proposed two distinct classes of the phenomenon known as shock: "*shock* following surgical operations and injuries," and "*shock* arising from Mental Causes."[195] In his treatise, he limited his commentary to the first class, using recorded cases and those under his own care to examine "constitutional alarm or shock" and to indicate a disturbance to the whole—not a disturbance limited the "wound or injury producing it," or to "any vital part":[196] "The loss of a limb is more than most can bear. It would seem if simple irritation in a part was capable of affecting the whole of the nervous system," it would lead to a distribution of "general sympathy over the entire body."[197]

Morris's account is important for several reasons. First, in 1867 already, we find language that would later denote bodily behaviors that, while recognizable and consistently recorded, were nevertheless difficult to explain—shock, constitutional alarm, disintegration. Second, and this is perhaps easiest to overlook, Morris's concern with clinical cases and individuality contrasted sharply with Crile's, Porter's, and Henderson's experimental work on railway spine and shock.[198] Third, Morris traced a longer prehistory, pointing beyond the US Civil War: he credited George Guthrie's 1827 *Treatise on Gunshot Wounds* on the importance of "constitutional alarm" and John Hennen's 1818 *Principles of Military Surgery*, for asking whether shock from surgery may put the patient at greater risk than the wound.[199] Fourth, the medical officer was able to recognize that shock had legal ramifications and required not only intuition but also proof—things that "tax the diagnostic skill of the surgeon."[200] Morris crafted this "work of medical jurisprudence" to help in "unraveling those intricate cases in which there is reason to believe the symptoms are simulated" and "which have caused confusion amongst medical men." This legal concern with simulation was crucial for whole-body constitution because it raised the question of whether wound localization and simulation were the only options.[201]

Morris also criticized John Eric Erichsen's 1866 *On Concussion of the Spine, Nervous Shock, and Other Obscure Injuries of the Nervous System*,[202] which has become regarded as the classic account of railway spine. In Morris's reading, Erichsen showed a fundamental bias toward spinal injury that ignored what Morris considered his own central contribution, namely that injuries were less significant than the systemic disintegration and constitutional alarm they produced through a local wound, bodily jar, sharp vibration, or "ébranlement."[203] Morris called for definitions of shock and railway spine to include injuries to the nervous system more broadly defined.[204] Finally, Morris was keen to describe and compare the success of methods of "the treatment, or rather management, of *shock*": opium, brandy, mustard cataplasms, flannels soaked in hot water, and spirits of turpentine applied to the abdomen. All seemed to help, but most effective, and again prefiguring the ostensible World War I–era "discovery" in the case of the Thomas splint, was "warmth applied by means of blankets."[205]

Looking back to Morris helps us appreciate the lack of novelty of wound shock in World War I—as a condition, a diagnosis, and especially a framework for care. Morris, Guthrie before him, Crile after, and later Cheyne, Henderson, and the surgeons surveying the Russo-Japanese War were all well aware of the condition, its vagueness, and its danger. What the Shock Committee did was to provide a physiologically detailed reconstruction of a preexisting condition. Morris also shows that despite the similarity of World War I–era wound shock to its predecessors, something crucial did change. Until the London Shock Committee's intervention, surgeons and experimenters had offered or relied on theories of general disarrangement. Now, that generic holism became integrationism: through their codependence, the different systems of the body were brought together, conjugated into a set of coefficient behaviors. Taking a step back to look at the longer history allows us to see that the concepts changed, and all the more so the meaning of what was essentially the same description. What had seemed possible, the body as a frail whole, was a now well-articulated reality, a differential and disintegrating system of systems. In 1916, shock had become an epistemic enabler par excellence, something far more than a difficult side condition: it confirmed new concepts and hypostatized new ways of knowing.

If the war offered this new paradigm, it did not for all that end the instability of shock as a condition. No new shocked soldiers entered the wards after the Boulogne Conference of November 1918, and intense doubts took their beds. Many researchers equivocated and developed misgivings about whether the condition even existed as anything more than a misshapen medical hermeneutic caused by the poverty of old medical categories. To begin with, shock still raised the nominalist specter, especially once translation issues and French categories came into play. Its staggered relations to surgical shock, railway spine or spinal shock, shock in "diaschisis" in aphasia, nervous or shell shock, and other functional disorders became an issue again. The memory that shock had struck in situations where surgeons faced a concurrence of uncontrollable, morbid conditions—sepsis, hemorrhage, and so on—did not improve stability. Nor did the long (if less intense) aftermath of the original acute episode of shock in each patient. The ostensible return to health was far from complete, the patient often suffered from a disordered heart ("soldier's heart"), and the long-term condition habitually caused misdiagnoses, most frequently with syphilis. The conflict of interpretations, like the profusion of symptoms and the long-term difficulty of correlating these to a particular kind of shock, did not give wound shock a continuous story. Like other functional diseases, shock garnered its significance within a clinical-therapeutic environment concerning each patient, but it lacked the continuity and stability in an experimental and therapeutic context that physiology required.

This confusion generated considerable fatigue with shock in general. Crile, who was criticized as overreliant on a too-generic category of exhaustion, and whose surgical priorities were in contrast to the trend toward physiological theorization, resisted the expansion of shock. Others shared his frustration toward generalization. Columbia University physiologist F. H. Pike, known for his studies of "spinal shock," wrote to Crile in 1916: "My patience with the current theories of spinal shock is about as thoroughly exhausted as yours is with current conceptions of surgical shock. With reference to this point at least, we are on common ground."[206] Pike's use of *exhausted*, with all its self-referential implications, is almost endearing here, not least because Pike went ahead with a full-throated critique of Crile's own theory anyway. A similar fatigue with the term *shock* was evident among researchers of shell shock, such as Myers, who rejected the term as early as 1916 as "singularly ill-chosen."[207]

The London Shock Committee physiologists also complained after the war about the long-term unsatisfactory standing of wound shock. The same scientists who had offered a detailed physiological picture were neither succeeding in providing a satisfactory or well-accepted therapy nor really in agreement any more. As the war ended, their individual theoretical priorities returned, and what in the Shock Committee reports had resembled capaciousness, leeway, and complexity now smacked of incommensurability. Dale uttered a striking complaint at the close of his Harvey Lecture in New York in November 1919 after expounding his toxemic theory: "The word shock . . . is used, by surgeons and experimental workers alike, for a number of different conditions, having probably different origins but presenting certain features in common. Shock as yet is not so much a clinical entity as a dumping ground. Gradually we ought to be able to rescue from this scrap heap one well-defined clinical or pathological entity after another. I have put before you one such effort at salvage."[208]

If Dale could not fasten on the exact nature or cause of wound shock, he would lay blame on the symptomatic confusion and nominalist dumping ground. It was the fault of shock if his toxemia theory, with its focus on histamine or "a histamine-like structure," was not proving satisfactory. Cannon, meanwhile, worked periodically on shock until 1923 but then decided to call this work a "parenthesis" from his experimental physiological work on bodily regulation and the physiological basis of the emotions.[209] This was perhaps because his own reliance on acidosis had also failed to gain traction. We have seen Crile's contemptuous reaction, but acidosis was criticized closer to home, too: in 1919, in that same Harvey Lecture, Dale offered a backhanded compliment by praising the "enthusiasm and enterprise" with which Cannon had advocated the acidotic theory that Dale was belittling as an unsatisfactory counterpoint to his toxemia explanation.[210] Despite the reference to a parenthesis, Cannon made shock central to his later thinking on homeostasis, offering it as decisive evidence for the operations of the "fluid matrix" and the regulation of the blood in his *The Wisdom of the Body* (1932).[211] Dale did so as well, in his thinking on autopharmacology and the role of histamine in the body.

If wound shock was exhausting, this was for structural reasons. What made it distinct also unmade it: it was a non-organic-lesion, whole-body condition that affected multiple organ systems in dovetailed succession. As a clinical reality it was already difficult, thanks to its temporal unfolding in the individual and the considerable difference between cases, and it became far less frequent in peacetime. Corroboration, follow-up, extended testing, and systematization posed obvious problems. What changed considerably with the London Shock Committee physiologists' involvement was not the treatment of wound shock or the broad sense of the disease. Rather, it became possible to offer a minute, detailed account of the various processes that occur in shock, an account that was as clinical as it was physiological, that offered a mechanistic yet integrative protocol to explain the different kinds of suffering experienced by shock patients and to show how their bodies responded to the external intrusion.

The value of wound shock was fundamentally conceptual. What shifted with the London Shock Committee's work was the understanding of what shock was for the body, what it made the body do. With this changed the concept of the body, the logic of its functioning and integration, the sense of its response to the environment's intrusions. The body could be conceptualized as a unit that responds to external aggression or intrusion systemically, a unit that could nevertheless be examined in detail for this systemic response. If the basic diagnostic and therapeutic sense remained largely the same as they had been before the war, the ontological status of organism, intrusion, and disease had been transformed.

The London Shock Committee's set of interconnected, occasionally competing accounts offered a medical imprimatur for the condition and a physiological, clinical, and experimental logic for it. Unsatisfactory as this logic may have been, even to those who had pronounced it, it remained perhaps the crucial moment in the conceptualization of the body as an integrated being—indeed, as a being that worked because of the ways its systems were integrated and, as a result, were always ready to collapse.

What Is a Wound?

What is a wound? Medical dictionaries from the 1910s propose that a wound is a "solution of continuity in an internal or external surface of the body,"[212] or "an injury or traumatism, usually with a solution of continuity."[213] The most extensive lexical consideration of wounds, William Whitla's *Dictionary of Treatment* (first published in 1892 and in its sixth edition by 1920) avoided a general definition altogether, focusing on incisions made by the surgeon ("operation wounds") and on procedures of sterilization and strategies to avoid infection.[214] The ambiguity in the phrasing of "solution of continuity" is telling: a wound may be said to exist if there is some sort of a physical break, if a body's surface has been rendered penetrable. A wound is identifiable with an injury: the surgeon is called to repair the break, to dress the wound, to avoid doing further harm, and to see the patient through the process of healing. This was a peacetime definition: Whitla treated the surgeon as the frequent or main originator of the incision or wound, and he declared the surgeon's main objective to be "to obtain primary union in all wounds which he necessarily inflicts during a cutting operation."

A famous passage from Bowlby's 1915 lecture on shell injuries offers us a clear sense of the distance of "wounds in war" from the lexicographer's orderly "solution of continuity":

> These shells vary in weight, . . . and they consist of a thick iron case containing in a central cavity a violent explosive charge. The latter is, in the case of German shells, trinitro-toluene, and may contain as much as 200 lb. of this explosive. Such shells are burst upon percussion by a detonator, which acts by the impact of the shell upon the ground or on some other object. . . . The injury they do is in chief part by the jagged fragments into which they are split by the explosion, and also to some extent by the impact of portions of buildings, such as stones or bricks, which are scattered with immense force by the violence of the explosion. The fragments of the shells are always very rough and jagged and of every variety of size and shape. For example, the base of a 17-inch shell may weigh 150 lb., and if it struck the body of a man [it] would completely destroy it. Other fragments may weigh a few pounds and may tear off a limb or crush it to pulp, while in the smaller shells there may be scores of fragments about the size of the end of the finger or much smaller. . . . The mere explosive force of the gases of a large shell exercises great

> powers of destruction. The expansion of the gases is sufficient to kill, and in the only case in my experience in which an autopsy has been made the brain was the seat of very numerous petechial haemorrhages.[215]

Much should be said about Bowlby's explosion semiotics; still, having listed splitting of limbs, tearing, pulping, and so on, Bowlby paused on the effects of the gases of a shell, hinting at an altogether different kind of wound: the very nearly invisible wound, the nonlesion wound, the damage visible only postmortem. This category was particularly applicable to forms of shock, at times including cases of wound shock, but all the more so those of shell shock, a condition that after the war became, on the backs of its sufferers, persistent, identifiable, and famous. In wound shock, what mattered was not the physical tear recognizable on the fabric of the body. The injury was a complex object, not necessarily of the order of a limited wound, but also not accounted for in a separate order.[216] Wound and shell shock, like other war neuroses, posed as forceful a challenge to the conventional understanding of the wound as lesion, specifically thematizing the apparent lack of a localized "solution of continuity" and raising the question of what kind of wound applied to the kind of violence that seemed to leave no physical trace behind.[217] From 1915 onward, invisibility became a central trope in writing on wound and shell shock: in wound shock, the blood was invisible, vanished; in shell shock the wound itself was invisible. For the remainder of this chapter, we focus on the invisible wound—shell shock—in its relation to the body and specifically to wound shock.

Shell shock is commonly historicized from the perspective of psychic trauma: supposedly, it was first a diagnosis that erroneously aimed to locate psychic disturbance on a physical register, until physicians recognized the absence of lesions and turned toward nonphysiological psychiatry and psychoanalysis.[218] This historicization relies on the success of psychoanalysis, the psychiatric treatment of trauma, and the post-1975 advent of posttraumatic stress disorder, and depends on the idea that Myers was generally right, that Frederick W. Mott (who sought microlesions in the nervous system) was basically wrong, and that eventually psychiatric care triumphed over retrograde, reductive biology.

Yet this postulates the wrong conceptual frame, that is, a harsh mind-body distinction between psychology and physiology, with psychiatric versus biological sorts of evidence and theoretical and therapeutic claims residing in a no-man's-land between the two disciplines. Shell shock was not, however, a psychological category; nor was it simply salvaged from overeager physiologists by discerning psychologists like Myers or W. H. R. Rivers. It was a category circumscribed by physical forms of shock, and specifically wound shock. We seek to radicalize here a point made by Tracey Loughran: that with regard to shell shock, "there was no straightforward transition to a psychological understanding of the war neuroses: psychology, physiology, and biology were all inseparably blended in many theories."[219] Only with the instability and especially the therapeutic failure of physiological explanations did the category of shell shock become liberated, and it was liberated not from physiological symptoms in general but from localized symptoms. Its functional dimension remained and was not in question. As opposed to the image of a condition emerging out of the illusion in medicine and neurophysiology that they would discover physical lesions, we want to show how shell shock was conceived on the model or in the order of wound shock, from which it was at times distinguished only because it lacked an original intrusion, a visible injury.

The conceptual difficulty was evident from the beginning of the war. Recall that Crile, in his early depiction of the wounded at the hospital, highlighted as a symptom of exhaustion what would come to be called traumatic nightmares.[220] Crile similarly reported on discussions about the capacity of high-explosive shells that "caused death without physical contact" and often without leaving external marks because "the rarefaction and condensation of the air caused such violent changes in the gaseous tension in the blood as to rupture blood vessels in the central nervous system thereby producing an injury in a vital part and causing sudden death. The process is in a measure comparable to 'caisson disease' or 'bends.'"[221] This even concerned wound shock: in 1934 Cannon recalled "two instances of terrifying experiences which occurred during the War, in which the symptoms of [wound] shock appeared, followed by death . . . [but] a postmortem examination revealed no gross lesion."[222] As the war ground on, invisible injury became more widely known, exhaustion became incorporated into the often political category of neurasthenia, and "war neuroses" and the limits of different forms of shock became matters of broader debate. Physically invisible wounds came to be folded into the new categorizations of functional disease, especially surrounding shell shock.

To physiologically inclined medical theorists like Mott, shock may have been invisible, but its sites and consequences could be discovered. Mott claimed that "punctiform *hemorrhages*" and "organic lesions" were the principal contributors to shell shock.[223] In publishing images of stained neurons appearing deformed due to shock or concussion, Mott at first argued that shell shock was a direct, exact neurophysiological result of the physical shock itself. The stem of the brain and the spinal cord were immobile, "prevented from oscillating" by cerebrospinal fluid: "A sudden shock of great intensity would be transmitted through this incompressible fluid, and, seeing that it not merely surrounds the central nervous system but fills up the ventricles and central canal and all the interstices of the tissues, serving as it does the function of lymph, it follows that a shock communicated to the fluid of sufficient intensity would make itself felt on all the neurons."[224] Mott was writing amid a slew of publications concurring that nervous damage was behind shell shock and identifying shell explosions, sound waves, ear damage, commotion, and nervous injury. Nervous injuries were of different varieties and often unclear in their effects, but for Mott the injury could be

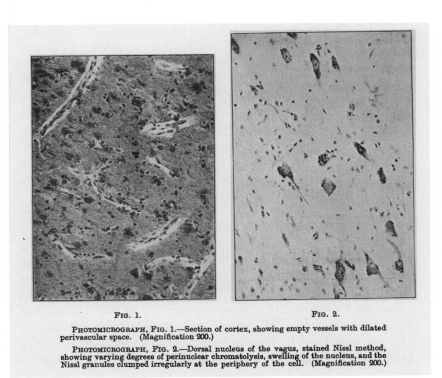

FIG. 1.   FIG. 2.

PHOTOMICROGRAPH, FIG. 1.—Section of cortex, showing empty vessels with dilated perivascular space. (Magnification 200.)

PHOTOMICROGRAPH, FIG. 2.—Dorsal nucleus of the vagus, stained Nissl method, showing varying degrees of perinuclear chromatolysis, swelling of the nucleus, and the Nissl granules clumped irregularly at the periphery of the cell. (Magnification 200.)

FIG. 2.5. Wound-shock case discussed in Mott, "Microscopic Examination of the Brain in Cases of 'Surgical Shock,'" in *Proceedings of the Royal Society of Medicine* 15 (January 1922), 25–26. *Proceedings of the Royal Society of Medicine* (London: Longmans, Green, and Co., Paternoster Row, 1922) (public domain).

microlocalized through a postmortem. The microlesion was crucial and its effect systemic.[225] For Mott the effect of shell shock was akin to that of wound shock, except it assaulted the nervous rather than the cardiovascular system. Indeed, he found some confirmation in physiologists' references to wound shock's effects on the nervous system.[226] As the war progressed, Mott complicated his original interpretation of the injury without stepping back from the physiological, "punctiform" basis of the trauma. By the end of the war, he was quoting at length from Crile, Cannon, and Bayliss, having come to accept that the onset of shell shock on the nervous system was due far more to a mixture of exhaustion and functional effects resembling the shock of an open wound.[227] Even as he continued to search for microfissures, Mott explicitly derived shell shock from wound shock–like conditions:

> There is a great difficulty in differentiating the symptoms of emotional from commotional shock. The absolutely sudden onset of emotional shock by a horrifying sight, apart from war conditions, can only be explained by a sudden fall of blood-pressure, by arrest of function of the vasomotor centre. This fall of pressure is followed by cortical anaemia causing loss of consciousness and cortical dissociation, which dissociation in a neuropath may persist in whole or in part, causing psychic deafness, blindness, mutism and amnesia. If we accept Crile's theory we can understand why a normal neuro-potentially sound individual may from prolonged stress of war become so run down in kinetic reserve energy that an emotional shock suffices eventually to produce collapse.[228]

Mott published his research findings in the London Shock Committee's March 14, 1919, report on toxemia and continued in the early 1920s to seek physiological disruptions in the brain, basing them on surgical shock.[229] His endorsement of an emotional dimension to shock's microlocalizable aspect came with neither a refusal of its physiological quality nor a newfound trust in some autonomy of the emotions, but with a recognition that it was a functional, whole-body disorder at least akin to wound shock. Its physical effects certainly existed. They may or may not be visible under the microscope, but they straddled, like wound shock, the physical-psychic barrier.

If Mott represented a physiologistic tendency, Myers embodied the antiphysiologistic alternative, emphasizing the absence of visible lesions.

Myers, too, began with the assumption that shock was a functional disease, with principally emotional effects.

Myers had begun his career in Rivers's psychophysically oriented Cambridge lab in the 1890s, and he joined the Cambridge Torres Strait expedition of 1898, on which, alongside Rivers, he carried out psychological and perceptual tests on the way indigenous inhabitants of Murphy Island experienced sound.[230] Regarding shell shock, Myers argued that neither ear pressure, whose pathologies he had studied in divers at Murphy Island, nor localized physiological symptoms were needed to understand the condition. This propelled him toward functional disease, and the hypnotic and psychiatric treatments he recommended were specifically aimed at affecting this functional psychophysiological level.[231]

To Myers, the novelty of shell shock was that it was a nonlesion phenomenon. This is also what rendered the understanding of its causality so profoundly troubled. Myers's causation model, occasionally outlined until it was described extensively in his 1940 book *Shell Shock in France, 1914–1918*, is itself difficult, if not contradictory. It describes his conversion to the idea that no essential role need be attributed to the shell explosion itself, that shock was functional in the sense of being "only mental."[232] Myers nevertheless had great difficulty in dissociating the nonphysical, psychic trauma from the actual explosions, which he referenced extensively.[233] The functional condition he described was then a matter of classification and therapeutics. As a result, treatment and rehabilitation—not research into the invisible—became the principal issues, and here the distinction from wound shock became gradually more evident, not least as shell shock exhibited a very different temporality. Shell shock did not develop in a matter of minutes or hours; rather, it demanded a long-term management as it mixed protracted daytime exhaustion with traumatic nightmares that repeatedly pulled the patient back to the original scene and moment. Invisibility guided toward the psychiatric sense of the term *functional*, with the whole body's reaction determining the mental effects.

Despite the later, crucial physical-mental dichotomy, shell shock was not simply liberated from the physical wound; rather, it was treated as an organism-wide disturbance that might or might not require a visible wound. The alternatives all assumed functional disease and the response model of wound shock, but they called it something else. It was not so much that physicians recognized the inexistence of a physical wound, but more that they began by assuming a functional disturbance whose psychological and physiological components were unclear.

Coda

To close this chapter, let us return to the conceptual problems that were associated with this most devastating and most fragile of conditions. As a concept, wound shock was complicated not only by the naming process and unstable referent but also by the implications of the term and by the kind of causalities with which it was associated. The widely shared dissatisfaction with the term *shock* was apposite when we consider the term's various implications.

All assemblages of shock seemed to share the general picture that an organism that had been pricked responded, and that this response was significant. But in the particular details proposed by different approaches, the picture changed considerably. Crile's emphasis on surgical shock attached shock to the exhaustion in the organism generated by a scalpel's or shard's intervention. The vasomotor center became exhausted, and the organism also became exhausted under the scalpel. Surgical intervention was an intrusion from without that threatened collapse by guiding an already-existing fatigue. For the psychophysical, phylogenetically stabilized organism that Crile regarded, the formula was simple: exhaustion, mediated by the intrusion, but without internal compensation, caused the rapid decline.

By comparison, wound shock indicated that the wound itself was at the root of the shock; the response was associated with the wound or intrusion. Thus wound shock posed the question of how the organism was shocked in the wound's aftermath rather than adequately responding. Because of the perforation, shock subjects the organism to a kind of ecstatic displacement of itself and turns the response to the intruding other into internal self-destruction. This image of shock—very different from Crile's—became central: the outside intrudes and ruptures skin, flesh, and tissue; the body reacts to the rupture with a massive force that overwhelms the organism and, in attempting to regain equilibrium, further endangers and destroys itself.

This inward turn is radicalized in other versions of the condition. Dale's toxemic model went in an altogether different direction when it shifted from considering toxemia as a factor in shock to looking only at histamine shock—that is, only to internally induced shock. In his argument emerged an internal or patient-based causality, with the organism *creating* shock rather than collapsing because of it. Exaemia went even further in ignoring the external cause and focusing on the organism itself being drained—drained as if from the inside toward an outside that simply didn't exist. Finally, the shock in shell shock had a different sense attached to it in that the shock was ascribed to the response to the shell rather than to the organism alone; the organism was then understood to absorb the shock. The sense of having been blasted from one's position by the shell, coupled by percussive theories and the invisibility of the wound, produced a very different agency in the term *shock*. In each case, a metaphysics prevailed, along with a particular conception of the body. These metaphysics are the subjects of the upcoming chapters, not least because the shock transposes onto the body, onto one's body and self, a logic inherent in aggressive total war: the anguish of being pierced through by the enemy and crumbling as a result of one's own highly complex, integrated organization.

3

The Visible and the Invisible: The Rise and Operationalization of Case Studies, 1915–1919

No matter how metaphysical or symbolic it may be, make no mistake about it, it will cut through flesh like a thousand razor blades.
—Haruki Murakami, *Kafka on the Shore*

Halfway through the war, the experience of soldier-patients whose nosologies exceeded existing diagnostic categories had become a primary concern for physicians throughout Europe. An unsigned editorial published in *The Lancet* in March 1916 described borderline cases in startling geopolitical terms:

> It is . . . in the study of border-line cases that true progress can be effected. Frontiers in medicine, as in geography, are artificial boundaries established in a more or less arbitrary fashion. Different names are assigned to the nations on either side of the border, but so far as the people are concerned these demarcations may mean very little. In medicine there is a neutral zone, a no-man's land, a *regnum protisticum* which really defies definition. This nebulous zone shelters many among the sad examples of nervous trouble sent home from the front, and it is only by accumulation of facts that we can hope for new definitions of permanent value.[1]

The description begins hopefully: borderline cases can be treated anew; "true progress can be effected" because it is possible to rethink these "frontiers." But the analogy of medicine and political geography that follows undermines this objective: *The Lancet*'s editor describes geopolitical demarcations in analytically sober terms that also seem strikingly naïve. Were these boundaries permeable because demarcations meant little to those on either side, or because the two sides were now at war? Were nervous disease patients content to be relegated to either side or caught by a medicine that, like political authority, imposed its rules with little regard for the suffering of patients? Just as names are assigned to countries, so too are nosological entities given new designations on the basis of symptomatic consistency: the protists care as little for these names as the patients. If anything, the analogy becomes tendentious: the editor treats the kingdom of simplest organisms that traverses the kingdom of humans as the site for borderline cases, the one site at which "true progress can be effected."

But what of the patients whose suffering is not categorized—who lay indeed in this biomedical no-man's-land? Though the war was not mentioned in *The Lancet* editorial, it populated and mapped new territories of patients as *cases*. The needs of documentation and classification brought forth various types of case reporting mired in conceptual tensions regarding the individual patient. Cases ceased to be peripheral to the work of medicine or auxiliary to the sciences of the body: they became determinant of every condition in this medical neutral ground. In this chapter we attend to the force of the "case study" or "case history" during and after the Great War as a narrative, analytical, administrative, and hermeneutic device, particularly within the spectrum of war neuroses, neurological damage, and physiological disorder.[2] Case histories, already a psychiatric, criminological, and medical device by the fin de siècle, proliferated during the war.

This mobilization of the case study form had far-reaching consequences: cases formed the basis of new nosological understandings at the limits of soma and psyche; they shaped the administrative logic of transportation, treatment, and rehabilitation; they informed the moral and social dimensions of disease and disorder. They produced a tension between the individual patient and an aggregate of patients, based on criteria of injury and possible treatments. Perhaps most striking, cases became a technology through which value could be given to or deprived from the individual who suffered. In short, cases mattered because they illustrated a certain individuality of the patient while subsuming that patient—that is, subsuming a version of that patient's individuality—within a medical knowledge suddenly unsure of its classifications.

We focus on how different case forms emerge within American, English, and German contexts. We first look at the work of the English psychophysicist and Royal Army Medical Corps (RAMC) psychiatrist Charles S. Myers, who first aligned the use of case histories with the study of shell shock. Next we ask how cases came to be operationalized toward different and sometimes contradictory ends. We look in more detail at the role played by the staging of cases, and the *croisement* between the case and the problem of visibility. Furthering the problem of bureaucracy and epistemology, we engage the American psychologist E. E. Southard and the RAMC physician Arthur Hurst, then compare the uses of cases in maxillofacial surgery to Adhémar Gelb and Kurt Goldstein's work at the Frankfurt Neurological Institute, W. H. R. Rivers's efforts at Craiglockhart Hospital in Edinburgh, and surgeon René Leriche's studies of the "surgery of pain." This logic of visibility, we argue, was intimately bound with the therapeutic impulse and the problem of individuality as this was reconstructed in pathological situations that defied categorization.

The Urgency of Categorization

In the previous chapter, we considered the designation *functional* as it was applied to physical and mental cases involving nonlesion injuries, diseases, and conditions. Nowhere was the need to resolve the functional more urgent than in conditions where injury was invisible. So-called nervous diseases, including most notably, and most vaguely, the conditions of neurasthenia and shell shock, were exemplary: they showed no obvious wound and they seemed to be experienced in each case differently. Some (Frederick W. Mott, for example) responded to the problem of invisibility by reaching for the visible: microscopic lesions that could be identified only posthumously. By rendering the invisible visible, Mott managed to reinsert a statutory conception of the body ruled by physiological laws, except that these laws were more complex than had been recognized thus far.

But to those who thought that borderline cases constituted genuine medical problems and likely medical failures, thinking in cases offered a very different logic: rather than physiological laws, patient history held sway. Each problem required aggressive refinement that could be made relevant for particular patients—and not a refinement of the patient's symptomatology so that it could fit into existing categories. It was also a pressing concern because a correlate effect of the turn to patients at the Western Front was that, in a specific sense, the patients became paradoxically similar, in their gender, also their race, and certainly as regarded their anatomical body: they were male, mostly white, relatively young, comparatively fit. They did not match the bodies from which doctors had generalized before, men and women of different ages, strengths, abilities, looks, even shapes, who grew and aged differently and were treated at different points in time. Indeed, it would be difficult to overstate the intensity with which the young male body became the model of patienthood and the criterion for treatment. In other words, suddenly the patients *should* be the same, and yet they were anything but.

Myers has been credited with studying shell shock in France from the early months of the war and with disrupting the early conviction that shell shock had a necessarily organic basis.[3] From the time of Myers's Torres Strait studies until 1914, when he went to France with the RAMC, he had remained convinced of the importance of physiological and psychological laws. Things changed; as Emily Martin has shown, he gradually undid the focus on standardizing the experimental subject that had been key to Wundtian psychology.[4] In December 1914 he set out to study trench foot and "collated 150 cases and published [them] in the *British Medical Journal*."[5] In the previous month he had already seen his first case of "'functional' mental and nervous disorder,"[6] and this brought forth a very different description:

> This patient was a man near whom several shells had burst. . . . Immediately after one of the shells had burst in front of him, his sight, he said, became blurred. Another shell which then burst behind him gave a greater shock—"like a punch on the head, without a pain after it." The shell in front cut his haversack clean away and bruised his side, and apparently it burned his fingers. This man was found to be suffering from "functionally" contracted fields of vision, loss of smell and taste, and slight impairment of visual acuity. Two other cases, respectively due to a shell blowing in a trench, and blowing the patient off the wall, were characterized by similar "functional" symptoms and by well-marked loss of memory. The second of these three cases was the first I attempted successfully to restore the patient's memory by means of slight hypnosis.[7]

Myers published these cases starting in February 1915, and in one of these articles, he wrote later, he "must have been one of the first to use the term 'shell shock.'"[8] There was a mutable power, not of cause and effect, but of the blended sequelae of injury, in the clinical cases he encountered. Myers notes how the "functional" dimensions of cases exhibiting a psychological dimension were important: "I was careful to point out the 'close' relations of these cases to those of 'hysteria'; and did not suppose, as Lieutenant-Colonel Frederick Mott was then attempting to show, that they arose from the effects of minute cerebral haemorrhages or other microscopically visible lesions."[9]

This may be overstating the point: Myers's focus was on mental "repression" and "dissociation," but never swerved too far from the physicality of the wound.[10] Cases were not opportunities to demonstrate precision, parsimony, or categorical assuredness; while the ones Myers encountered had features of dissociation, he was not convinced that "all cases of 'functional dissociation' arose solely from mental causes."[11] Cases clouded more than they clarified, yet seemed to demand hard and fast assertions about causality and scope.

For Myers, the clinical and experimental knowledge of the case, and the ability of clinicians to think through and alongside the observed injury, had to be largely sacrificed in the face of bureaucratic expedience. Drawing a distinction between the physical and the psychical, "where the tolerable or controllable limits of horror, fear, anxiety, etc., are overstepped,"[12] Myers insisted on the difference between shell concussion and spinal concussion, and forms of neurasthenia and hysteria. "A shell, then, may play no part whatever in the causation of 'shell shock'; excessive emotion, e.g. sudden horror or fear—indeed any 'psychical trauma' or 'inadjustable experience'—is sufficient."[13] The administrative result was clearer than the diagnostic and therapeutic one: depending on whether there was a visibly recognizable wound, Myers issued reports with the letter W (wound) for cases with shell concussion or other battle wounds, and S (sick) for those without one.[14] Through the rabble of triage, the two designations came to determine the priority of transportation and possible convalescence in England.

Cases at this point operationalized a distinction as viewed from the military perspective of hospitalization and rehabilitation. The condition was at once apparent and invisible, and the case systematized just that: W identified a site or lesion, often at the expense of the complexity of the experience; S identified the need for a narrative that was perhaps absent. Myers's cases were stories about patients and assessments of their memories, which tested the administrative limits of sorting soldiers.[15] Did patients meet critical criteria, or were they simply malingering? In Myers's administrative criteria for moving soldiers into appropriate hospitals and aid stations, he showed that his task was as much about managing an inventory of clinical variation as it was about clarifying the nosology of disorder. The case remains caught between two modalities of knowledge: therapy as getting wounded soldiers channeled to the right clinic, and organization as stretching categories of S and W to accommodate the breadth of injury and disorder Myers witnessed.

Myers's worry about a definition was endemic to the condition. No sooner had he uttered "shell shock" than he expressed dissatisfaction with that term. Shell shock offered a false pathogeny by pointing at the wrong vector, it indicated unity where different conditions could be involved, and it constructed an inexistent single object of disease. This helps us appreciate the weight and complexity of the undertaking to refine categories of disorder. At the bureaucratic and administrative levels it was massive enough, involving the rehabilitation and often re-enlistment of thousands of soldiers. At the epistemological level—which cannot be distinguished from the bureaucratic—the worry was that what seemed necessary and nearly impossible was to acquire a vast, if not encyclopedic, appreciation of current research, which was being produced across disciplines at impressive speed, and then to successfully systematize that new knowledge as a matter of course. The case thus emerged as a demonstration of the complexity of a condition and as a variable to be categorized without a clear or definite sense of what the category would be.

History and Theory

Despite the ubiquity and variation of "the case", there is no general history of the medical case study or case history as a narrative and hermeneutic technology in the modern period.[16] Although it is not our purpose here to construct such a history,[17] establishing the historical and theoretical stakes of the case and its emergence is essential for understanding its import.

Cases appear as early as the Hippocratic *Epidemics*,[18] and their logic was partly echoed in Aristotle's reference to "infinitely various" individuals.[19] The historian Owsei Temkin locates the beginnings of modern case recording in the seventeenth-century medical thought of Thomas Sydenham, "for whom disease represented an abstract, self-contained natural object from which the case or *historia morbi* could be derived"—not an account of individual illness but a report on the disease process in time and word.[20] By the 1830s, cases had become significant in psychiatry. Philippe Huneman reflects on "such a simple thing" as the clinical case in Philippe Pinel's early nineteenth-century psychiatry, a story (historiette) that "reports how one has fallen ill, and with what illness, what has happened to that patient once he or she became ill, how the patient is being treated, and finally if the person recovers."[21] The case history, Huneman explains, was not a demonstration but an attempt to interrogate an ontological limit—and at the same time to demonstrate ontological insecurity on the part of the patient and the physician.[22]

By the late nineteenth century, and particularly in psychiatry,[23] the case had acquired what Steven Shapin has called "epistemological decorum," that is, common acceptance as a form of the acquisition and dissemination of a kind of knowledge that is necessary for an understanding of the case's veridical claims.[24] Around 1900, the case became commonly accepted in medical pedagogy as well: advocating its expanded use in medical schools,[25] Walter Cannon argued that the case study "correlates scientific and clinical sides of medicine" by questioning and testing physicians' knowledge and diagnosis of particular conditions, forcing attention to the particular individual patient, the specificity of the condition, and the need to judge the data.[26] In a moment of bravado, Cannon later gave himself sole credit for the introduction of cases into medical teaching.[27] (Christopher Lawrence's explanation of the late-Victorian and Edwardian belief that "incommunicable knowledge" should be attributed to clinical care also has to be seen as a necessary pedagogical component of case writing.[28])

By 1900, the case was also a device valuable for the production and critique of scientific knowledge about the individual patient. Tracing the rise of the device in the 1930s, the historian of medicine Arturo Castiglioni remarked: "Every physician may recall cases in which, without any apparent logical reason, he has had the sudden perception of a hidden truth, in which a difficult diagnosis became in a moment clear and transparent. This kind of eminently personal and subjective judgment, which may become dangerous if not carefully controlled, constitutes that part of a physician's activity which makes him an artist in the full sense of the word."[29] Cases joined a host of other scientific practices that tightened the aperture on individuality in medicine and psychology, including the growth of the statistical paradigm of normality,[30] the vastly expanded aggregation of examples, the individual-to-whole relationship in cell theory, and the theorization of the relationship of individual to type, which was central to the establishment of knowledge in psychology.

In her study of psychologists Adolf Meyer and E. B. Titchener, Ruth Leys spells out the stakes and difficulties involved in Meyer's design of a "life chart" and of "composite" photographs:

If by Meyer's own account the greatest challenge then facing psychiatry was how to establish a science of the individual presumed to be unique, there could have been no more vital methodological contribution to meeting that challenge than the recognition that the requirement of absolute singularity implicit in the *ontological* concept of individuality and the notion of belonging to a norm or type inherent in the *relational* concept of individuality are not incompatible, for being an individual simply means belonging to a type—a type of one.[31]

Meyer's reliance on the pragmatism of William James and John Dewey led him, Leys argues, to a fundamentally paradoxical notion, wherein the contradiction between individual and type is no contradiction at all. Similar efforts to relate individual and type in the work of Alfred Binet also played management and specificity against each other in the establishment of knowledge itself and in the treatment of patients.

Jean-Luc Nancy imagines the status of the case in medicine and law in ontological terms when he writes, "Bodies: their space is juridical, just as the space of the law is a space of bodies configured according to cases. The body and the case fit each other. There is a jurisdiction proper to each body: 'hoc est enim.'"[32] Cases in medicine came to closely resemble—and not only for pedagogical purposes—the slightly earlier rise of case law. As opposed to statutory law, which is enacted through the deliberations of a legislative body or court system, case law is based on precedent, building on cases that come before to guide present decisions and forecast future arguments. So too does the medical use of case evidence stand in contrast to experimentally derived reliance on physiological laws: knowledge that is argued and established and that becomes the basis of present understanding, an approach debated as to its value in teaching at medical schools around 1900.[33] The case in medicine also performs a function distinct from what is commonly understood in relation to the legal use of cases, which are aimed not at producing new evidence but at establishing precedent.[34]

In medicine, moreover, the case study is meant to produce insight into collective phenomena through the individual example of various expressions of disorder or of the course injury takes over time. In some circumstances, an exceptional example is best, as contradictory as that sounds. To echo Roy Porter, the case study opens and closes "medical history from below" insofar as it offers what Cannon called "the perspective of symptoms,"[35] premised on the interpretation of signs as symptoms, all the while eschewing precisely the categories by which this would be done a priori.[36] It plays a role because of medicine's unrelenting obsession with management—with managing patients, certainly, but also with that administrative logic concerned with segregating, culling, and ultimately arriving at indexical categories through evidence offered. The case is worked on not as a dialogue through self-reporting and conversation, but as an increasingly depersonalized encounter, which, as L. Stephen Jacyna and Stephen T. Casper point out, has taken the patient as a category of thought further out of its historical frame.[37]

Michel Wieviorka observes that in medicine a case is "good" when it can demonstrate something rare yet diagnosable, defined by its occurrence in the individual patient and yet independent of the patient.[38] And as Angela Creager, Elizabeth Lunbeck, and M. Norton Wise note: "Since [the late nineteenth century]—in the human sciences and in law and medicine—'the case' has functioned much like a model system. Cases in these disciplines are valued for their specific material reality, their individuality, and at the same time their typicality. Cases, it is assumed, capture individuals in all their complex uniqueness while at the same time rendering them in a generically analyzable form."[39] Kathleen Frederickson similarly notes, "Narrative case history promises not to lose sight of the specificity of the lived instance in the face of statistical aggregation."[40]

One disruptive element of the case study that goes beyond Creager, Lunbeck, and Wise's argument that cases are valued for disrupting well-ordered constructions was particularly evident during World War I.[41] This is that the individual case study underwrites something known *as it comes into being known*; its value is, in Lauren Berlant's formulation, that "the case demands judgment on something that is as yet not fully established or normative."[42] If you think the body follows these rules, then what about this largely unclear piece of evidence that seems relevant and yet not rule bound? Carlo Ginzburg explains that for this reason, cases impose an "unsuppressible speculative margin":

> The group of disciplines which we have called evidential and conjectural (medicine included) are totally unrelated to the scientific criteria that can be claimed for the Galilean paradigm. In fact, they are highly qualitative disciplines, in which the object is the study of individual cases, situations, and documents, precisely because they are individual, and for this reason get results that have an unsuppressible speculative margin: just think of the importance of conjecture (the term itself originates in divination) in medicine or in philology, and in divining. Galilean science, which could have taken as its own the Scholastic motto

> *Individuum est inefabile* ("We cannot speak about what is individual"), is endowed with totally different characteristics.[43]

Medicine sought to operate like a "Galilean science" in Ginzburg's description: it was intended to be a general body of knowledge that could be used as a body of laws. We have seen a version of this problem in the physiological engagement with wound shock, and especially in the persistence of references to wound shock as mysterious or obscure. And yet cases both grounded this desire to be a science and undid such grounding. In the face of physiological and statistical thinking aimed at laws uniting the body with knowledge, the individual case history offered a counterpoint to generalizability and aggregate claims. Case thinking restored a sense of the fragility of extant categories by analyzing cases that were not easily attributed on either side of a divide and by showing conditions, problems, questions as they were emerging, that is, before they had been fully established.

The case blurs individual and collective into taxonomical possibility, as a tension between individual practice and interactions and the science of phyla. When Julia Epstein argues that the case history implies that the human body can be known through an "etiological narrative" commensurate with its physiological reconstruction, we wonder whether this approach does not conflate the case and the example, something that did occur, for example, with Cannon's advocacy of case-based pedagogy.[44] There remains a difference here between simply demonstrating the already known and recording the expression of disorder as an experimental task to test the limits of that definition and expression. If all case history did was put into narrative form an etiology of established disorder, then the case history would be redundant because the possibility to establish abstract, unassailable parameters of that disorder would already exist. As an experimental technology, the case provides not only the object but also the means to test, observe, and record phenomena: the case directs and defines, without committing.

For these reasons, the case is unstable and fundamentally paradoxical in form. It aspires to exemplarity, yet in the same gesture denies it; a case that would be simply representative of a condition would not be sufficient to produce the type of experimental knowledge, with therapeutic aims, that physician-researchers like Myers and Goldstein sought. By comparison to the highly standardized forms of fitness and aptitude testing to which different armies—the French in particular—were committed in preparing, judging, and testing their soldiers, the case remained quasi singular, inherently paradoxical. The paradox is amplified in the genre of collections of cases, such as those compiled by Southard, which permit categorization through operations of aggregation that admit of approximation and speculation—operations that do not amount to laws but merely reveal similarities. In this ambiguity, cases constituted a true science of the individual in Michel Foucault's sense.[45]

We invoke the critical and not merely affirmative dimension of case reporting in the establishment of scientific knowledge to insist on the powerful operationalization of cases during World War I. Cases proliferated during the war in precisely a fashion that systematically, relentlessly utilized their ambiguous and epistemologically troubled status. There is undoubtedly a tension between the individual and the collective: the problem of how an individual patient offers insight into the possibility of intervention and the scope of therapeutic options is bound up with the prerogative of helping others who are similarly afflicted. But the problem resides in the intentional use of this binary throughout the war. Case studies serve several functions, and they were adopted in ways that accepted their double status as both experimental and established, both individual and aggregated. The capacity of some soldiers to withstand while others succumbed forced physician-researchers to consider just how unique these capacities were when they drew conclusions from similar phenomena.

Most problematic here was the staged opposition between the qualitative and narrative forms of the case, and the case as it appears in aggregate, quantitative terms.[46] Both of those forms were used to refine and upend a body of knowledge within which individuals and wounds would become functional. In what follows, we attend to several types of work, two (Hurst's and Southard's) for the connection between administration and scientific delimitation, and three (Goldstein and Gelb's, Rivers's, Leriche's) for the connection between therapeutics and epistemology. We present each of them separately to identify tensions in their writing that pertain to the status of the individual war neurotic; it is through these tensions that we inquire into the questions of experimentation and therapeutics. These thinkers, these cases, these tensions built the new theaters of subjectivity.

Administrative Epistemology

The nexus of the delimitation and valuation of case knowledge was engaged by virtually every author who used case histories during and after the war. Hurst and Southard offered two of the more radical solutions to the epistemological and administrative problem posed by cases. Hurst's solution centered on the depiction of a composite patient.[47] In *War Neuroses* (1917), a twenty-seven-minute film that aimed to show that shell shock was easily and quickly treatable,[48] Hurst, the founder of the British Gastroenterology Society and known for his studies of constipation,[49] documented shell-shocked patients upon their return from France.[50] Hurst worked largely independently of other shell shock researchers, but he was no stranger to physiological psychiatry, and his interest in the various neurasthenias and cardiac conditions brought about by exhaustion and trauma stemmed from his time in France working under the tutelage of Jean-Martin Charcot's former students— Fulgence Raymond, Joseph Babinski, Joseph Jules Déjerine, and Pierre Janet.[51] Hurst's film contrasted scenes from before and after treatment to create a visual measure of the therapeutic trajectory. (Similar before-and-after films of soldiers with functional disorders were made in Germany, notably by the neurologist Max Nonne at Eppendorf Hospital, in Hamburg.)[52]

Hurst was not preoccupied with the etiological and nosological elements of shell shock. He asserted a simple notion of general causality and individual fitness, blaming, in 1917, "neurasthenia due to prolonged physical strain, [or] the utter nervous exhaustion caused by pure physical fatigue."[53] Convinced that neurosis in soldiers was "no different from symptoms in civilian life," Hurst prescribed "rest, and exposure to cured patients" as the primary means of intervention.[54] His film and its viewing were unambiguously purported to hold therapeutic value: they showed the recovery of fellow soldiers, who thus modeled behaviors others could practice. Hurst's terminology took up the character of learning; we find him using the terms *re-education* and *retraining* interchangeably: "re-education is all that is required to cure most cases of hysterical paralysis."[55]

Hurst offered some solutions that could be generalized to all patients under treatment (no visitors, consistency of nursing staff, bed rest, isolation, no discussion of experiences), and some solutions that were specific to a particular trauma (for example, "wool pulled over the ears at night during storms" for those suffering nightmares of artillery barrage, "perpetuated by war dreams, and [where] in severe cases the mind is absorbed by day as well as by night by pictures of the horrors which the individual has witnessed; every sound reminds him of shells, and every moment suggests the approach of danger").[56] Individual constitution would guide Hurst's collective picture of disorder, and the amelioration of trauma hinged in part on predisposition.[57] But this essential character remained unpredictable, primitive, and secondary to the aims of recovery as Hurst saw them.

The cases presented in the *War Neuroses* film announce their own illustrative value. In the "before" section, patients shake violently as they walk, and in the "after" section, markedly less so—but because Hurst (like Nonne) did not record the treatment sessions, cure alone was evidence, and this accordingly limited the analytical value of the film. As Edgar Jones has shown in a remarkable study, besides the tricks and reenactments that overstated after-treatment successes, a second performance was in play in the film: the before-treatment images were reenacted after the "cure."[58] Numerous cases were involved in each case of Hurst's film: the latter were composites, built around the idea that active psychotherapy merely contributed to an insistence on rapid improvement in organic nervous diseases; Hurst held to the notion that hysterical symptoms were often grafted onto symptoms caused by organic disease. Each patient-actor doubles as a synthetic war neurotic. His condition becomes synonymous with war hysteria, and his treatment is reduced to a demonstration of healing prowess.

The composite patient used by Hurst stood at one end of a hyperdifferentiated and hyperclassified individual case. At the other end was Southard's patient. Southard, director in 1917–1918 of the US Army Neuropsychiatric Training School in Boston, NOT, had been engaged in the diagnosis of syphilis and the administration in psychiatric settings of prophylaxis for syphilitic patients and their families immediately before the war, and at the war's end he published a large volume on neurosyphilis.[59] His wartime and postwar publications demonstrated a career of curating case studies, most famously in his nine-hundred-page *Shell-Shock and Other Neuropsychiatric Problems* (1919), in which he presented 589 case histories from World War I literature.[60] Southard bookended his cases with a preface and a ninety-page "epicrisis," in which he established his clinical and epistemological priorities.

Like Myers, Southard expressed dissatisfaction with the term *shell shock*, and remarked on its imprecise ordinary use. "Shell-shock denotes, to say the least, shocks and shells—yet we know Shell-shock sans any shock and sans any shell, nay sans either shell or shock."[61] The problem of precision was a problem of "diagnostic delimitation."[62] Working under the umbrella of "the newly combined fields now collectively termed *neuropsychiatry*," he argued, gave him the chance to narrow the field of misrecognition and to, step-by-step, distinguish different conditions.[63]

Unlike Myers, who sought to distinguish among his own shell-shocked patients, or at least those under his administrative charge, and unlike Hurst, who agglomerated their images regardless of divergence, Southard designed a very different empirical task for himself. He used published scientific papers—not charts, case records, or military reports—to create an inventory of evidence. He then organized his 589 cases in a dual effort: he sorted similar cases into piles, then aggressively refined the dimensions of diagnostic categories by carefully distinguishing different kinds of disorders.[64] Southard thus created clear lines of division across others' work, massaging some two hundred reported cases into eleven categories: syphilopsychoses and hypophrenoses (feeble-mindedness and imbecility), epileptoses and pharmacopsychoses (alcohol, morphine), encephalopsychoses (focal brain lesions), somatopsychoses and geriopsychoses (senility, a null class), schizophrenoses, cycothymoses, psychoneuroses, and psychopathoses. Clinical accounts of single patients offered him a way to show how very different sets of disorder (for example, meningitis, a gunshot wound to the head, and a syphilitic brain abscess) might all fall under a particular grouping—in this case, focal brain lesions and not shell shock. Then, like a jurist comparing legal precedents to choose the relevant ones among them for his purposes, Southard turned to cases of true shell shock and broached the subject of its nature and causes.

Southard attempted to "delimit"—that is, standardize[65]—shell shock first by showing all what it was not, then eventually by locating it among the psychoneuroses.[66] He drew from leading researchers in the 1914–1918 period: cases published by Otto Binswanger, Leriche, Mott, Myers, Babinski, James Parkinson, Jean Alexandre Barré, and Rivers are represented, as are researchers from France, Belgium, England, Germany, and Russia. Asking about the coexistence of different functional phenomena—syphilis and wound shock, for instance), Southard wrote:[67]

> When so much theoretical doubt concerning organic and functional neuropathy holds sway, the practical doubts in the individual case under the varying conditions of civilian practice and in the upheavals of military practice must be still more in evidence. Case after case described in the literature of every belligerent has passed from pillar to post and from post to pillar before diagnostic resolution and therapeutic success. Colleagues meeting, for example, at the Paris Neurological Society, find

themselves reporting the same case from different standpoints, the one announcing a semi-miraculous cure of a case which another had months before claimed only as a diagnostic curiosity.⁶⁸

Southard writes of his method, "I would call the collection not so much a poesy of other men's flowers as a handful of their *seeds*."⁶⁹ His reported cases were soft composites of individual patients holding to categorical distinctions to provide taxonomic order amid experiential chaos. And he looked for evidence where he could find it: "By the way, just as I found John Milton had said things that fitted neurosyphilis, so also Dante is observed in the chosen mottoes to have had inklings even of Shell-shock."⁷⁰ On the basis of classification and delimitation, one could draw the boundaries necessary for further engagement. Shell shock was not syphilis, but it could coexist with syphilis; it was not an organic condition based on a focal lesion, but it could involve a blend of functional and organic elements. It was fundamentally but not exclusively emotional, in the sense that there was no organic cause and no preexisting or hereditary functional disease: "Purely psychogenic war cases exist."⁷¹

Southard had two unrealized projects: a collection of the literature on railway spine ("shell-shock's congener") from the previous century, and a collection of cases on "the effects of suggestion and psychotherapy."⁷² He collected one hundred case studies to detail what he considered the great therapeutic potential of cultivating a close relationship between social work and psychiatry. His friend the physician Richard Cabot praised Southard for having the sympathies of a social worker and the insight of a philosopher in dividing the "*science* of psychology from the *art* of psychiatry."⁷³ Cases served the administration of both, and through them he attempted to create a philosophical foundation of care.

Mobilization for Therapeutics

If there was an administrative imperative to divide cases for reasons of bureaucratic expediency and experimental exactitude, then there existed an equally strong impulse to manage the cases toward curative ends. These therapeutic efforts were sometimes at odds with administrative priorities. Often there was little agreement between the therapeutic logics that operated in psychiatry, neurology, and surgery particularly regarding the nature and purpose of the case. Was the aim to produce generalizable results? To lay a restorative path for similar cases in the future? To outline a singular experience of suffering? The case as a type was at odds with the individual clinical case that expressed individual features of dysfunction and disorder. A second, more pronounced conflict existed between natural, clinical phenomena and the phenomena of experimental manufacture.[74] Three thinkers in particular help us to understand the stakes of the case for individualized therapeutics and its successes and limitations: the British anthropologist and psychiatrist Rivers, the French surgeon Leriche, and the German aphasiologist and neuropsychiatrist Goldstein.

Rivers began his career as a psychologist and in 1898 became one of the core participants in the Torres Strait expedition. After working further in neurology, psychology, and ethnography in the 1900s, he became interested in the social function of medicine and psychiatry and in the structure of care within society. "Medicine is a social institution," he later declared on the basis of his ethnographic work in the Solomon Islands.[75] Rivers gradually became more interested in psychoanalysis, and he began psychiatric work for the RAMC in 1915. It matters that he arrived late to psychiatry, the unconscious, and the individualized care of patients, and that he did not get there via the psychology to which he had devoted the better part of his earlier research. Nor did his anthropological work produce the conditions that made a study of psychiatric care and the unconscious possible.[76] His emphasis on individual case studies would have been hard to anticipate given the psychological and ethnographic methods that he, Myers, William McDougall, and Charles Seligman used on Murray Island, in Kiwai, and in Maguiab:

> We had taken out with us the equipment of a small psychological laboratory, and the disused missionary house in which we lived was fortunately large enough to enable us to fit up the more complicated apparatus, especially that for reaction-times, in one room, while other parts of the house and verandah were used for different purposes. . . . The subjects, which were investigated, included visual acuity and sensibility to light difference; colour vision, including testing for colour-blindness, colour nomenclature, the thresholds for different colours, after-images, contrast, and the colour vision of the peripheral retina; binocular vision; line-dividing; visual illusions, some of which were investigated quantitatively; acuity and range of hearing; discrimination of tone-difference; rhythm; smell and taste; tactile acuity and localization; sensibility to pain; temperature spots; discrimination of weight and illusions of weight; reaction-time,

including auditory and visual simple reaction-time and choice-time; estimation of intervals of time; memory; mental fatigue and practice; muscular power and motor accuracy; drawing and writing; blood pressure changes under various conditions, etc.[77]

They had lugged from England an arsenal of psychophysical technologies and as long a list of targets for comparative and standardized testing. Indeed, the Torres Strait expedition had a distinct anti-Spencerian effect, in that it became clear to the Cambridge researchers that, for example, poor hearing was a specific bodily response to the islanders' commitment to pearl-shell diving, not a sign of evolutionary backwardness and atavistic physical prowess.[78] But their attention to observable, objective elements typical of psychophysical research was sidelined in the new century. Rivers began to set up nonpsychological kinship studies for the Todas and other peoples, and he began to look at individual experience.

In an extended obituary for Rivers, Myers wrote, "Rivers's interests did not lie in the collection of masses of heterogeneous data; . . . they lay throughout his varied career in studying and analysing individual mental differences, in getting to know the individual in his relation to his environment."[79] The story of Rivers's psychiatric work at Craiglockhart Hospital in Edinburgh has been told many times, notably with reference to his ostensible empathy and his most famous patient, Siegfried Sassoon.[80] Despite his broad concerns with the physiology of the nervous system and the problems of nervous response to trauma, Rivers directed his efforts at Craiglockhart not toward the generalizable aspects of battle fatigue and war neuroses but toward the individual expressions of shock in his patients.[81] He developed a logic of visibility that could square the conflicting demands of rehabilitation: to return soldiers to active duty was to render their psychic injuries once again visible to them, sufferable, and also invisible to others, unimposing. For this reason, Rivers sought to make soldiers remember, accept, and visualize scenes that they were "actively repressing," but unlike psychoanalytic practice proper, he made them dwell not on deeply unconscious elements but directly on their practice of repression. What served well the image of Rivers as a heroic psychiatrist caring deeply for individual patients was the way he came to regard each case with its own successful or unsuccessful resolution—a practice that for him involved the reignition of patients' instinct of "self-preservation" and even the restoration of subjective autonomy.[82]

The images of Rivers offered by Sassoon and Frederic Bartlett hinge on this identification phrased through a project of autonomy. Here is Bartlett on Rivers' care: "There is really no word for this. Sympathy is not good enough. It was a sort of power of getting into another man's life and treating it as if it were his own. And yet all the time he made you feel that your life was your own to guide, and above everything else that you could if you cared make something important out of it."[83] Rivers effected this identification and love through his understanding of repression, which was at stake throughout his treatment, because, as he wrote, "many of the most trying and distressing symptoms from which the subjects of war neurosis suffer are not the necessary result of the strains and shocks to which they have been exposed in warfare, but are due to the attempt to banish from the mind distressing memories of warfare or painful affective states which have come into being as the result of their war experience."[84] A one-by-one treatment was thus essential if the particular episodes repressed were to be released of the practice itself.

Rivers returned time and again to the patient: one patient improved despite reacting grudgingly to having to "gradually accustom himself to thinking of, and hearing about, war experience." Another came to accept his doubts about further military service, abandoned his neurotic cheerfulness, and escaped his suicidal thoughts. Rivers wrote: "I advised him that it was one which it was best to face, and that it was of no avail to pretend that it did not exist. I pointed out that this procedure might produce some discomfort and unhappiness, but that it was far better to suffer so than continue in a course whereby painful thoughts were pushed into hidden recesses of his mind only to accumulate such force as to make them well up and produce attacks of depression so severe as to put his life in danger from suicide."[85] On a third occasion, Rivers proposed "helping the patient adjust himself to the situation."[86] If adjustment runs through the psychiatrist's pursuit here, autonomy from unconscious or particularly painful impulses becomes the trophy of recovery.

Commenting that "the physical conception of war-neurosis has been gradually replaced by one according to which the vast majority of cases depend on a process of causation in which the factors are essentially mental," Rivers agreed with Myers (in opposition to Mott and Hurst) that a way should be found for these mental factors to be sidestepped.[87] Unlike Myers, though, Rivers turned the cases into a narrative form: each case was that of a total individual who needed to be restored to him- or herself—or rather, who needed help to reacquire the capacity to restore him- or herself. In "On the Repression of War Experience," delivered before the Section of Psychiatry of the Royal Society of Medicine on December 4, 1917 Rivers narrated a number of his treatments. Each of the narratives was a few paragraphs in length, "to illustrate by a few sample cases some of the effects which may be produced by repression," and he took care to emphasize the goals and criteria for success.[88]

Repression, for Rivers, was an active process, an active removal from consciousness—and by the same token from narration—of deeply painful experiences. Returning these patients to a state wherein they could readopt memories was the opposite of the "evil influence" of efforts like Hurst's that pushed them to actively forget.[89] In a passage that shows how key the rhetoric of restoring of the patient to himself or herself was for the value of the case, Rivers concluded: "We must not be content merely to advise our patients to give up repression; we must help them by every means in our power to give up repression. . . . We must show them how to overcome the difficulties which are put in their way by enfeebled volition and by the distortion of experience when it has long been seen exclusively from some one point of view. It is often only by a process of prolonged re-education that it becomes possible for the patient to give up the practice of repressing war experience."[90] If psychoanalysts might object that Rivers was removing transference from the picture and falsely assuming that patients were simply doing on their own what he told them to do, Rivers was at once well aware of suggestion and committed to removing it (whether he was successful is another matter).[91] In cases of "active repression" (by which he meant "active forgetting," as he used "suppression" to denote Freudian repression), he insisted that as he restored the patient's will by directing it away from actively removing experience from consciousness and toward reclaiming it, his own role became marginal. In regaining their will as their own—in becoming active subjects again—the patients surpassed transference and changed their status from an ongoing case toward a successful one.

The use of cases—each one individual, each one therapeutically oriented—had a second purpose, to negotiate the contours of the concepts Rivers was using. Rivers defined *repression, suppression*, and *dissociation* at the beginning of his lecture, in a gesture that worked specifically against the Freudian use: "It is necessary at the outset to consider an ambiguity in the term 'repression,' as it is now used by writers on the pathology of the mind and nervous system." Later in the lecture he noted, "In the cases I have just narrated there was no evidence that the process of repression had produced the state of suppression or dissociation."[92] Repression versus suppression, repression versus dissociation: the meaning of these concepts, like the differences between them, was to be worked out and sustained on the basis of the evidence provided by the cases themselves. This was the second value of the case study for Rivers: in the use of the case, the very practice of psychiatry could be articulated, defended, and turned into a science. When Rivers urged the reform of medical practice, particularly so that it could take into account the psychological problems occurring in shock patients without visible lesions, he was using his patients to explain the insufficiency of existing physiological practice and making repression into a core concept.

Cases for Rivers were narratives in which patients might be returned to themselves through the involvement of the psychiatrist. At the same time, the psychiatrist might gain precisely the basis for confirming which parts of psychiatric practice were worth retaining, which forms of drawing out the traumatic narrative were deserving of pursuit, and in which manner the psychiatrist might sufficiently withdraw so as not to influence the patient's regaining of narrative and subjective autonomy. Therapeutics and the production of knowledge could potentially be resolved in the same gesture. But where in the body that gesture was to be directed remained a question.

"Deserters from the Army of the Upright"

When the French surgeon Leriche began his career, he was caught in the apparently inescapable lure of the stomach.[93] In his massive 1906 thesis on gastrectomy's potential to cure stomach cancer, he outlined the great promises of surgery, contrasting it to the slow improvements over the previous thirty years. Later advances owed greatly to wartime surgery. Indeed, Leriche's technical and analytical insights were reshaped almost entirely by his experience during the war. His 1937 book *The Surgery of Pain* credits his World War I–era work on the vasomotor apparatus and causalgia: "On the occasion of this present study, I would pay sincere tribute to those pariahs of pain, my causalgic patients of the Great War."[94] Yet the promise of surgery was muted for Leriche by what he saw as a dangerous departure from reason and experience. Following closely the perspective of his mentor surgeon Antonin Poncet, and drawing a line back to Claude Bernard, Leriche advocated for the importance of surgery as an experimental art. That art was threatened by a downgrading of clinical experience.[95]

In *The Surgery of Pain* we find the clearest expression of his priorities concerning clinical experience with surgery and the experiential dimensions of pain. Leriche's study was based not on a preponderance of secondary sources but on his own observations over twenty-five years, starting with the Great War.[96] By his own assessment, his study could not be exhaustive: it relied on numerous cases from which he built his own experience-based perspective and technical insights.[97] Leriche flaunted his outright contempt for an experimental method in physiology that conflated therapeutic practice with laboratory observation and the development of laws of the body. So much so that Archibald Young, the translator of the 1939 English edition, contemplated that in his "new doctrine"—the modern development of the surgery of pain that originated in the Great War—Leriche "obtains his results from procedures based on the conception at which he has arrived, and if these results appear to be incapable of adequate explanation on the basis of classical physiological doctrine, so much worse for the latter."[98]

Leriche's commitment to establishing "equilibrium of sensory harmony" in his patients was not a metaphysical or antimechanistic endeavor. He was forceful when he stated that vasoconstriction, one of the chief causes of pain, could not escape the whole physiological action of the organism.[99] And yet mechanistic theoretical priorities—the reality of a shared structure and function—did not erase the individual because it did not erase his pain. Leriche wrote: "There is something else in the origin of pain beyond the crude bringing into play of a mechanism indifferent to life, and alike in everyone. *And we understand physical pain aright only when we realise the part played by the individual in a system morphologically the same for everyone, when we have analysed what I have termed earlier—in order to strike the imagination—living pain.*"[100] He then asserted: "Everything in pain is subjective."[101] Pain follows a contour made by the individual and is shaped within the body; it is not shaped solely by stimuli and predictable reactions.

Discreet, descriptive research thus cannot hold a candle to the full scope of appraisal after surgical intervention and measured recovery on each specific patient. Pain in the laboratory is not pain at all, but rather painful sensations, nerve impulses, and anesthesias "tested on the healthy man." Experimental pain is enacted, recorded, and contained, always within parameters, always with an anticipated end.[102] By contrast, "in the suffering patient, the pain is like a storm which hardly admits assessment once it is over."[103] Leriche spent page after page marking and amplifying the distinction between painful sensations and pain proper, which in his view were wholly different objects for the physiologist and the surgeon, and moreover, are distributed differently throughout the body of the individual.[104] Crucially, pain does not bend to the will of the surgeon:

> It would be possible on these lines to formulate a purely experimental plan, the working out of which would doubtless put new aspect on the problem of pain, and yet would merely follow out this correlation of the study of nervous and hormonal functions—an old, yet everpresent, problem of physiology. But surgery justifies itself mainly by reason of its application to treatment.[105]

Even in something as straightforward as his 1918 manual *The Treatment of Fractures*, Leriche used pain to emphasize the empirical, "established" (that is, clinical, naturally observed) features of the labor leading to his conclusions.[106] His frustration is raw. In his notes on the basic care of a wound and the avoidance of infection, he chided those wishing to manufacture experimental knowledge where a clinical example exists: "I cannot understand how a fact so well established by experience can still be discussed at the present time: it is neither paradoxical nor revolutionary; it is compatible with the traditional teachings of classical physiology; it can be verified at leisure every day. Why meet so obvious a proposition with specious arguments when direct verification is always possible?"[107] It was not for reasons of professional vanity that Leriche made such harsh assessments. Experimental priorities, in his view, erased the individual and substituted parsimony for understanding. He was equally ungenerous with those who would supplant the features of a patient's experience with a gross moral impulse for meaning; he lashed out against any poets and philosophers who might assert that "pain is a blessing in the moral order" and made it clear that they lack "any real appreciation of human suffering":[108] "I am convinced that, *almost always, those who suffer suffer quite as much as they say they do, and that, if we assess their pain with the degree of attention which it deserves, they suffer indeed, more than we would imagine.* There is only one pain that is easy to bear: it is the pain of others."[109] It is the causalgia—the pain from a damaged nerve from the bullet wound and the destruction of the tissue surrounding the wound—that retools the world, and that world comes crashing down on the individual in pain in ways that enfold the milieu as much as the body. In some cases, even "the slightest sounds make the pain worse."[110] The lesion might look the same, but "the most striking feature of the disease is the *radical change in the disposition of the individual.*"[111]

The force of Leriche's argument was attached to the invisibility of pain. This was perhaps clearest in his discussion of vasomotor phenomena of the face. Few subjects in surgery during 1914–1918 have been discussed as much as maxillofacial surgery, and famous practitioners like Henry Tonks, Harold Gillies, and Johannes Esser mobilized new techniques of reconstruction to restore facial disfiguration that prevented patients' reintegration into postwar society.[112] Facial wounds were particularly cruel because of the dehumanizing visibility with which they urged as much the repair of the lesion as the restoration of the individual to society. But in Leriche's cases, pain is hidden, moving in or under the flesh. Whereas Myers had sought a narrative in the absence of an identifiable wound, and whereas Goldstein would look for total behavioral changes at the expense of visible, comparable details, Leriche faced the invisible. His surgery of pain represented the opposite of the surgery of facial reconstruction, and yet questions of vitality, recovery, and the taming of phantasms both seen and felt occurred in the shadow of pain. In both cases the individual could not easily follow the grain of living, but forced instead across it. In the surgery of pain, the individual case was all there was.

Whether discussing vasomotor dysfunction and the spread of pain, or the disfigurement of facial injury, we return again and again to the problem of the individual among the collective. As Frederic Manning offers bluntly in his memoir of the Somme, "They had nothing; not even their own bodies, which had become mere implements of warfare."[113] During the war and its aftermath, the threat of erasure was real. The most visible example was that of facial wounds.[114] And yet it is the overwhelming problem of invisible injury that probes the character of priority over other types of recovery and that heightens the demand for visibility. Hence the reception of Hurst's films over Leriche's accounts of highly individualized pain.

: : :

Perhaps no thinker valued patient case studies as much as the German neurologist Kurt Goldstein. Goldstein's career dates to the 1900s; what defined him as a neurologist and a psychiatrist was his work at the Frankfurt Neurological Institute, which led to therapeutic and experimental clinical case studies of brain-injured soldiers (many of them coauthored with Adhémar Gelb, and sometimes with other colleagues from the Neurological Institute), and the development of an approach to therapeutics. Only 4 percent to 10 percent of brain-injured soldiers survived in the 1914–1917 period; Goldstein later claimed that he had carried out a "systematic" study of two thousand patients and had conducted highly detailed research on some "90 of them regularly over the following decade."[115] During the 1920s, Goldstein wrote at least two books and a plethora of papers on the forms and effects of aphasia and on the treatment and rehabilitation of brain-injured soldiers.[116] In both the experimental and the therapeutic dimensions of his work, Goldstein insisted on the priority of the individual patient. He went so far as to note: "Important though it may be to seek repeated confirmation of our findings through new case material, such confirmation adds nothing essential to our knowledge."[117] The specific goal was to "the total picture of the patient," which facilitated longer-term care and enabled him to predict "with relatively great certainty how he will behave in any situation."[118]

Goldstein's argument in *The Organism*, his 1934 work systematizing twenty years of neurophysiological and psychiatric work, included a refusal of atomistic symptomatology—the study of symptoms as separate and adequate indicators of damage.[119] In his theorization of diagnostic and therapeutic efforts, Goldstein charged that atomism went hand in hand with a thin comparative analysis that centered on lesions rather than on the organism's attempt to comprehend and compensate for the loss of functions. He also insisted that atomism did not offer a way in which such compensation might be turned to the patient's advantage.[120] Atomism presumed that damages resulted in the same symptoms and disturbances, that individuals were affected in the same way, and that comparative analysis only replicated itself accordingly.[121] Goldstein refused to see particular disturbances merely as facts; his inquiry came to be guided instead by the question, What kind of a fact does an observed phenomenon represent?[122] While words may have been the therapeutic currency for the patient and the therapist, Goldstein occupied a wholly different set of signifiers: not symptoms as such, but first and foremost the observed phenomena he referred to as performances.[123]

Goldstein founded his theoretical and therapeutic practice on the detailed study of "total behavior" or the "total picture of the organism" based on patient performances that he elicited and observed in therapeutic and experimental clinical studies of brain-injured soldiers. He recorded extensive accounts of the range and consequences of injury, most famously in the case history of a twenty-four-year-old former miner named Johann Schneider. On June 4, 1915, Schneider was struck in the occipital lobes by two shards of mine shrapnel. One wound was above the left ear; the second was in the "middle of the back of the head, ... penetrating into the exposed brain."[124] As Anne Harrington has proposed, Schneider became a veritable Anna O for Gelb and Goldstein, as well as for philosophers interested in their work, including Ernst Cassirer and Maurice Merleau-Ponty.[125] In a 140-page study, Gelb and Goldstein argued that even though Schneider passed the traditional tests designed to point out reduced capacities (tests ostensibly premised on atomistic symptomatology, reflex psychology, and localization), upon closer observation he was incapable of a natural performance of most of the tasks he had managed in tests. Gelb and Goldstein highlighted a series of failures that led Schneider to attempt to reconstruct the lost performances differently. Reading, for example, required "a series of minute head- and hand-movements—he 'wrote' with his hand what his eyes saw. He did not move the entire hand as if across a page, but 'wrote' the letters one over the other, meanwhile 'tracing' them by head-movements. An especially interesting aspect of the case was the patient's own ignorance of using this method. . . . If prevented from moving his head or body, the patient could read nothing whatever. . . . If required to trace a letter the 'wrong' way, he was quite at a loss to say what letter it was."[126] Simply put, here was a patient who could carry out concrete tasks, yet could handle only even basic uncontextualized mental operations in an extremely belabored way, if at all. Schneider sought to compensate in his performances for the natural attitude he had lost, an attitude that Gelb and Goldstein called abstract or categorical.[127]

In detailed clinical accounts that he published throughout the 1920s and in several research films, Goldstein recorded his observations about the limits of individual capacities under the conditions of illness and injury, as well as the crises brought about by demands the environment places on the human organism.[128] These observations were made after subjects were tested in controlled experiments either in the hospital or during periods of recovery outside clinical settings. Goldstein's filmed assessments and tests showcase his experimental priorities, particularly regarding the status of the single patient and the patient's world.

Our following description of parts of Goldstein's 1930 film *Tonus* illustrates a point of contrast between Goldstein's practice of using film to show individual features of symptom elicitation and Hurst's treatment of film as a therapy showcase, all the while providing a rare insight into clinical labor with patients performing a range of tests in order to demonstrate their loss of capacity after brain injury. We see Goldstein working in two of the three primary domains of his research, namely motor-kinetic disorders and sensory-perception disorders—the absent third being language disorders, famously aphasia. Blending experimentation and care, he directs his subject through various tests that demonstrate precise performances of disorder as follows:[129]

A thin man in a wool suit is seated in front of a white sheet. Goldstein, wearing a white laboratory coat, enters the frame to hold the patient's arms outstretched in front of his body and pointing toward the camera. When the patient closes his eyes, his arms begin to migrate outwards from their original position. Goldstein returns, turning the patient's head first to the right and then to the left. With each turn, the patient's extended right arm lurches further to the right, then the left arm moves to the left. A small stand with a horizontal metal bar is then placed in front of the patient. Goldstein marks a point along the horizontal bar. Closing his eyes, the patient raises his arm repeatedly to touch the point; with each try, his finger moves further from the original point, his arms bouncing with the pointing action; Goldstein marks the points where the patient touches along the axis of the metal bar. Next, the patient is instructed to touch the top of his right hand with his left, first with his eyes open, then closed; again the extended arms begin to migrate outwards. The patient is at first able to repeat the touching motion, but eventually misses the top of his hand as the arm sways too far from its original position.

The setting changes: now the patient is sitting at a table in an outdoors courtyard. Looking upwards, out of frame, he holds his palm open and hand extended. Goldstein touches particular points on the patient's palm with a long knitting needle. The patient attempts to repeat the action with a pointer of his own, concentrating deeply as he struggles to do so.

Back at the original setting, the patient is instructed to keep his eyes closed and hold his arms outstretched; once again they begin to rise. A young assistant in a laboratory coat stands behind the patient, playing violin; the patient smiles at the music, and his arms begin to sway in a jerky rhythm, again straying far from their original position. Next, a piece of equipment is placed on the patient's head, with cups covering the ears attached to a long hose. Goldstein, visible behind the patient, asks him to point, and then blows a small amount of air pressure through the hose and into the patient's ears. The patient leans left and his arm moves wildly leftward. . . . After another cut, the patient's eyes are covered with a mask attached to wiring. As Goldstein instructs him to move from the right to the left, the patient makes jarring movements with his arms and torso. With his eyes closed and opposite arm extended, Goldstein strokes

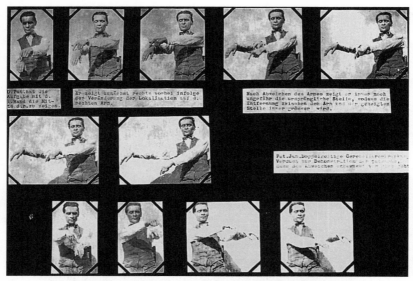

FIG. 3.1. Presentation slide involving the patient from the *Tonus* film attempting a pointing test. Box 18, Slides, Goldstein Papers. Columbia University Rare Book & Manuscript Library, Kurt Goldstein Papers (used with permission).

the patient's side. The patient's extended arm begins bobbing up and down in rhythm with the stimulation to his torso.[130]

Goldstein's experimental methods were not always of his own design: we find Gestalt psychologists, neurologists, and experimental human physiologists using similar techniques to assess kinetic and neuromotor functions. Unlike them, Goldstein was more interested in assessing the functional capacities of his individual subject than in generalizing his results. This introduces anew the tension between experiment and therapy in the case patient—between the generalizations afforded by dispassionate research science and the seemingly circumstantial insights developed in clinical care. Experimentation obliges the patient into a milieu of demands that she or he is not capable of carrying out properly or normally. Patients' loss of a "categorical attitude" forces their divergence from the norm as well as from their preferred behavior:[131] we observe a patient's failure to do as asked, but we also recognize that the tasks are unnatural to them. Throughout his research in the interwar period, Goldstein continually returned to the patient's coming to terms with the world: experimentation was his "means of discovering the capacities of the individual."[132] This is not an abstract concern. The patient functions in the world in everyday life, and everyday life often sets tasks that are impossible to carry out. The clinical case, tending through experiment toward rehabilitation and therapy, brings forth less the individuality of the laboratory than a performance demonstration of divergence—cipher by cipher, elementary performance by elementary performance—from earlier, "normal" integration.

Thus in *Tonus* the patient does not attempt to order the total spatial situation in which he has been placed but responds to specific requests that he perform repeatable movements.[133] Added stimulation causes the patient's limbs to move or even become wildly uncoordinated, making it impossible for him to perform the given task.[134] By this, Goldstein could show how a brain lesion, however localized, affected the patient's capacity to perform tasks—in the case of these tonus tests, even to steady the body in space without disequilibrium ensuing. Though the tasks in the experiment may appear not to be directly relevant to any actual situation, the performances concern elements of likely situations and are not so artificial as to be meaningless beyond the experiment; instead, Goldstein suggested, they are needed in life and are often performed out of habit.[135] The point was to totalize the picture of an individual patient in this place, at this time, with these demands of the milieu, and with these reactions forced on him specifically.

In other words, Goldstein recognized here the threshold between assessment of functional capacities under controlled conditions, and the labor and activities needed to live in the world where crises arise or the individual's impairments block self-realization. For him, as for Bernard before him, these real-world demands did not undermine the knowledge gained through experiment but could be mobilized to sustain knowledge, therapy, and individuality in collaborative tension. Therapeutics, by this token, involves a mixture of retraining, in the sense of helping the individual to regrasp for herself behaviors that are expected by biological and social milieus, and healing, in the highly specific sense of facilitating adaptation as the restoration of adequacy, given tendencies toward preferred behavior, the mostly excessive demands of the environment, and the danger of catastrophic situations.

It is not enough to say that experimentation underwrites therapeutics, though it does that too. Rather, experimentation forces patients into situations that are unnatural to them but to which their preferred behavior may immediately or one day partly adapt, both so they can experience a certain comfort and so they can relearn and manipulate their milieu. Experimentation directs individuals toward reoccupying and remaking disorder along contours of their experience. What appears to the untrained eye as a failure to carry out tasks expresses for

Goldstein the tension and collaboration between experimentation and therapeutic intent: it might allow for some movement toward a new, less impeded life.

Hence Goldstein's therapeutic dream: by demonstrating through systematic experimentation the patient's shrunken milieu and its contours, limits, and new geographies, and by always bearing in mind that experimental performance does not mirror the world outside but produces responses specific enough to become predictable, he would collaborate with his patients in a first step toward the world, one in which new norms do not await but are achieved. The case was for Goldstein an individual who, while following a general pattern of disordered behavior, did so along the individual's own path—and that was everything.

For Myers, the crude nature of treating vast numbers of people made him constantly aware of and anguished about the fact that his charge forced him to make overgeneralized distinctions between observable injuries and those buried out of view. He resented the forced generalization deeply. But this predicament was not one that the radical turn toward single cases—in Leriche or in Goldstein—could solve. Said another way, Goldstein was not the antidote for Hurst or Southard. Goldstein was untroubled by the onslaught of patients requiring gross sorting precisely because he was not in a situation with the same demands. His gold standard of the solitary case that accounts for the needs of the individual has an uneven polish. Myers was able to offer broad strokes in the clinical assessment of shell shock despite his perpetual critique of the hand making those strokes—namely his own. Leriche was forced to reconcile the technical, anatomical dimensions of surgery with the individual character of pain and its sequelae. Goldstein completely sacrificed scope and potential standardization of care in order to manage precision, rendering distinction unnecessary in the face of individualistic clinical aims. There is a real difference between the seemingly unending stream of patients that needed assessment for reasons of treatment and rehabilitation, who were judged by Myers or reported by Southard, and the individualized care that Goldstein moved toward, for the purpose of achieving ever more precise understandings of the nature of certain brain pathologies that could then be universalized on the premise that no two cases are the same.

At both extremes, physicians, surgeons, and physiologists still contended with the modalities of the case study, which required individuality as much as it could not be reduced to it, which mobilized comparison while insisting on difference, which extrapolated largely without compositing. Perhaps this is why the case study was so important, so failing, so contentious: it was such a plastic form, a technology that each of these thinkers amended toward different ends and in different measures. Perhaps the case study not only helped them think about individuality or created hypothetical knowledge; it also constituted a tool for ratcheting up a certain kind of incomplete knowledge that could maneuver imperceptibly from individualized care—of a hypothetical individual, a patient under treatment—to the operationalization of entire fields of pathology. Case studies could speak to both of those ends, and through them, into the nature of disorders in that no-man's-land that could neither be described through laws nor be neglected.

Coda

After all the labor of making cases appear, rendering their hidden qualities, domesticating categories of diagnosis to deliver order to the tangle of symptoms and lesions, we are still faced with the problem of the invisible. To make therapy possible, a disorder that was not already as visible as an open wound required visibility—categorical, etiological, and psychophysiological. Injury was brought forth so it could be worked on. But to return order to the disordered soldier, the wound must again be buried, rendered invisible to all, a scene of violence forgotten. What torments these therapeutic aspirations is the trace: the wound as it remains. The lines and divots and missing flesh, the tic, the unsteady gait—signs of an organism pushed to a "margin of tolerance."[136] Georges Canguilhem famously remarked that disease is often imagined moving through the sick person as if it were passing through a door.[137] And yet so much dis-ease in this passing was etched into the flesh and psyche of these men: the door could hardly remain shut.[138]

We are reminded of a portrait painted by Otto Dix of his friend and fellow World War I veteran, the printer Max John (1920). His friend grins happily, half-shaven. His intense eyes are blank. He does not so much look at as through the viewer. Max John is elsewhere. Dix renders this portrait through an aesthetics of trauma founded on the decipherability of subtle traces on his friend's face, on the face of Max John and countless others like him—not so much on the face as in the tissue, pushed into the clay of the visage. Max John, the war neurotic, has recovered and reintegrated, but his invisible wound (can we call it memory?) is clear. If forgetting was an active strategy of recovery during and after the war, then this activity forgot to forget traces that linger on

FIG. 3.2. Otto Dix, *Portrait of Max John* (Bildnis des Malers Max John), 1920. Museum für Neue Kunst, Freiburg im Breisgau, Germany. Otto Dix (1891–1969), Portrait of Max John, 1920, Oil on cardboard. © Otto Dix / BILD-KUNST, Bonn - SACK, Seoul, 2017; Städtische Museen Freiburg, Museum für Neue Kunst. Formerly the collection Dr. Fritz Salo Glaser, Dresden. Purchased Augustine Museum 1959. 2007 restituted to the heiress of Dr. Glazers and reappropriation The City of Freiburg with the support of the sponsors: Cultural Foundation of the Länder Ernst von Siemens Cultural Foundation Commissioner d. Federal Government for Culture and the Media.

the body and visage in the portrait of Dix's comrade three years after the war. For Kurt Goldstein, the present moment of the injured soldier is always one that pulls his past injury into new situations, and into a future where new terms for relating to the world are continually established anew while never wholly forgetting the terms that came before. For Arthur Hurst and E. E. Southard, the injured soldier finds external legibility in his ability to occupy a clinical category long after cure has been granted. And for René Leriche, the physical expression of pain is not simply found in a single moment but woven into the fabric of the organism, even years after the sounds of battle fade. In disparate forms, Dix's portrait joins so many of the medical thinkers we have discussed in its uneasy refusal of erasure.

Part Two

4

Brain Injury, Patienthood, and Nervous Integration in Sherrington, Goldstein, and Head, 1905–1934

Of all the scenes of injury dramatized in World War I memoirs, the encounters with men suffering or dying from head wounds stand apart. Brain injury was widely seen as a particularly horrific mutilation, impossible to forget, impossible to ignore, a constant anxiety for one's self's survival. Robert Graves offered one classic version of such encounters in *Good-Bye to All That* when he described a man with a head injury "making a snorting noise mixed with animal groans. At my feet lay the cap he had worn, splashed with his brains. I had never seen human brains before; I somehow regarded them as a poetical figment. . . . Even a miner can't make a joke that sounds like a joke over a man who takes three hours to die, after the top part of his head has been taken off by a bullet fired at twenty yards' range."[1] A similar scene was offered by Ernst Jünger in *Storm of Steel* that involves himself as the injured party: "I got a blow on the skull, and fell forward unconscious. When I came round, I was dangling head down over the breech of a heavy machine-gun, staring down at a pool of blood that was growing alarmingly fast on the floor of the trench. The blood was running down so unstoppably that I lost all hope. As my escort assured me he could see no brains, I took courage, picked myself up, and trotted on. That was what I got for being so foolish as to go into battle without a steel helmet."[2] In both of these texts, the brain-injured man is doomed to suffering and despair, to a grotesque, animalistic death ostensibly unmatched by other fatal injuries. Graves's narrator stages a theatrical encounter with utter viscerality, forcing a new and emotional knowledge against his attempts to dissociate himself.[3] Jünger's "growing pool of blood" brings him to singular despair, then comfort once he has deflected the real threat, as if his head made up for the missing helmet. For Graves the morbidness empties all reality of comfort and possibility.

Devastating as much to the body as to the self that inhabited it, head injury became a matter of considerable social, medical, and aesthetic distress. What memoirs absorbed in terms of the blasting apart of the self, what artists from George Grosz to Henry Tonks represented in terms of facial and social deformity, neurological thought engaged in terms of the staggering survival of a body that bore, perhaps in perpetuity, a destruction akin to social, symbolic death. What occurred in the encounter between researchers and those who escaped death only to live, profoundly transformed, with brain injury? How was this ruin of the self to be conceptualized? More than any literary, philosophical, or artistic encounter with brain injury, medical and neurophysiological thought offered approaches that asked detailed questions of the meaning of neurological damage and attempted to grasp what individuality and healing the injured patient could perhaps hope for.

The neurophysiological integration that was undone by traumatic brain injury pressured to the extreme the logics that we studied in previous chapters: that of a self-regulation compensating for shock but perversely bringing the patient's body to collapse, and that of the case history as a narrative and hermeneutic device aimed at getting a handle on the complexity of variegated classification all the while operationalizing the patient. This chapter pivots to the work of Henry Head and Kurt Goldstein, two neurologists and aphasiologists who became prominent through their work with brain-injured soldiers during and after World War I, and who influenced virtually every major contemporary neurophysiologist from Karl Lashley to August Bethe to Alexander Luria.[4] As soldiers with head trauma poured into hospitals, Head concurred with Goldstein that "the war was producing a series of cases unique in the history of the subject."[5] Working independently, they advanced a forceful critique of the existing understanding of aphasia and linked disorders. Their predecessors had, by and large, explained the loss of linguistic capacities by pointing to the cerebral localization of linguistic centers: damage to particular areas was understood to cause damage to particular ways of using language. But to Head and Goldstein, as to many contemporaries, this was nonsense at the theoretical level, nonsense that guaranteed that the medical and psychiatric experience of the patient would be stamped and signed by unceasing violence. Head famously charged aphasiologists with acting entitled "to lop and twist their cases to fit the procrustean bed of their hypothetical conceptions."[6] Like Theseus on his way to Athens, Head and Goldstein formulated new theories of how the healthy mind operates, how it is integrated, how individuals function as individuals, how they partake of language, and how they respond to loss.

By *integration* in this chapter we mean three things: first, the integration of the nervous system, which Head and Goldstein extended to an integration of the many systems of the body into a single whole; second, the integration understood, both a contrario and experientially, as the instance of health right before and around the collapse caused or precipitated by brain injury; and third, a concept of integration and disintegration that took over after the traumatic event. Each of them saw the effects of this posttraumatic integration very differently, in both philosophical and therapeutic terms; indeed, it defined their different conceptions of the organism.

: : :

With the publication of *The Integrative Action of the Nervous System* in 1906, Charles Sherrington's understanding of integration in the nervous system built the foundation of neurological and neurophysiological conceptions of integration. Arguments favoring neurological integration and opposed to mechanistic assemblies of the body had been widely available in the prewar period. They included evolutionary theory, Gestalt psychology's holism,[7] and, significantly for our purpose here, John Hughlings Jackson's approach to aphasia. In 1876 Jackson criticized the localizationist model, contending that it was necessary to consider the individual brain as a whole to understand neurological disturbance. In his Croonian Lectures of 1884 on "The Evolution and Dissolution of the Nervous System," he used a Spencerian framework to establish the nervous system as evolving from an automatic, animal basis common to all organisms that share it to a developed, specialized, voluntarist system in humans. In injured human beings, according to Jackson, this highly developed system regressed to a lower level.

But the work that rendered the language of integration into a decisive component of neurology's research program was Sherrington's. *The Integrative Action of the Nervous System* scaped the land on which neurology would situate the problem of bodily integrity, fundamentally displacing earlier neurological theories. Immediately hailed as a masterpiece, *The Integrative Action of the Nervous System* established neurology as its own ground by articulating a grammar of reflexes and reflex arcs and returning to other biological fields (like evolutionary theory).[8]

Before moving to Head and Goldstein, we need to retain two elements of Sherrington's theory. First, nervous integration is central to the body and its capacities to move, act, and respond to stimuli. It occurs initially at the level of the reflex arc: "*The unit reaction in nervous integration is the reflex*, because every reflex is an integrative reaction and no nervous action short of a reflex is a complete set of integration."[9] Reflexes initiate the most basic coordination of sensation and motor response—they provide the alphabet out of which animal organisms are phrased. Sherrington postulated the "purely abstract conception" of the *simple reflex*, a "convenient if not a probable fiction," according to which each single reflex action could be simply distinguished from the rest of the activity of the nervous system. From the basic integrative unit, *the reflex arc*, Sherrington proceeded to more complex units, discussing integration at local and regional levels.

Hence the second element we need to retain: the brain and the central nervous system are the central, dominant site for bodily integration. The central nervous system is "an organ of co-ordination in which from a concourse of multitudinous excitations there result orderly acts, reactions adapted to the needs of the organism. . . . Out of this potentiality for organizing complex integration there is evolved in the synaptic nervous system a functional grading of its reflex arcs and centers."[10] Sherrington's nervous integration concerns the body alone; there is little sense in these passages that the organism's behavior toward its world is at stake, or that anything other than experimentation can be used to engage these questions. Sherrington wrote: "The motile and consolidated individual is driven, guided, and controlled by, above all organs, its cerebrum. The integrating power of the nervous system has . . . constructed from a mere collection of organs and segments a functional unity, an individual of more perfected solidarity."[11] As we shall see, Head and Goldstein built their theories of integration very much in critical conversation with Sherrington's, with marked differences in tone, object and argumentation.

Head and Goldstein: A Parallel Introduction

Henry Head is remembered as a major figure in early twentieth-century English neurology, though as L. Stephen Jacyna has recently argued, his main care was "medical science" and the ways it provided for patients.[12] After nearly forty years of work, Head proposed in his 1926 *Aphasia and Kindred Disorders of Speech* a general approach to speech disorder, which he expanded to a theory of consciousness, normal nervous functioning, and pathological behavior. For Goldstein, his 1934 *The Organism* constituted the second intellectual event of his career, after his detailed writings on aphasia during the late 1910s. Goldstein proposed an entire theory of the organism, a radical revision of neurological and physiological theories that, he contended, underappreciated its integrated structure. He focused on the way in which an organism compensates for lost performances, and the way in which the loss of an autonomous, abstract level of thinking as a result of brain injury leads to complex, bodywide counterefforts.

Head and Goldstein did not know each other personally and did not maintain much of a correspondence.[13] Still, Head concluded his 140-page history of aphasia with a highly favorable discussion of Gelb and Goldstein's analyses, describing them as "admirable work along psychological lines" and "a notable exception" to existing approaches.[14] In the single letter from Goldstein surviving in Head's archive, Goldstein noted his appreciation for Head's praise at a time when his work lacked broad recognition in Germany. He countered by complimenting Head's book as "a standard work in our research"[15] and by promoting it.[16] This convergence of intent offers a sense of scientific practitioners working in parallel;[17] even granting minor reliance of each researcher on the other's work in the 1920s, it is astounding to see the similarities of thought in the absence of systematic interaction.

Head and Goldstein trained in neurology in the 1880s and 1890s. Head began his studies at Cambridge, but was defined by his years in Halle, Germany;[18] Goldstein trained under Carl Wernicke in Breslau. Both arrived at the problem of aphasia well before the war—Head by 1910,[19] and Goldstein by 1906, when he was still at Königsberg.[20] In 1914, Goldstein moved to Frankfurt as director of the Institute for Research into the Effects of Brain Lesions,[21] becoming a leading researcher in the treatment and rehabilitation of war neuroses involving cerebral injury—like Head did at London Hospital and, as of 1915, at Empire Hospital. Each, moreover, carried out significant research with psychologists: as we have suggested before, Goldstein's principal collaborator over two decades was the Gestalt psychologist Adhémar Gelb, who died while Goldstein was writing *The Organism*. In Head's case this was W. H. R. Rivers. In 1903 Head and Rivers carried out a major experiment in nerve regeneration in which, after Head's radial nerve was surgically spliced, Rivers carried out sensation tests on his arm over a period of two years.[22] Head's experiment became famous and was still being reported as far and late as the *New York Evening Post* nine years later.[23] By 1918, both Head and Goldstein felt uneasy in their discipline. Each emphasized his credentials across broader fields of study than neurology; each developed a philosophical and methodological bent.

Both studied at length consciousness, its neuronal scaffolding, and its relation to language. Both sought to explode the diagnostic and explanatory category of aphasia, decrying existing formulations as profoundly insufficient and criticizing their predecessors harshly.[24] They shared a number of targets, including brain localization—that is, the tendency to identify faculties such as language, sensation, and memory with particular sites in the brain. Localization was a traditional concern of psychology and had been central to anatomy and physiology since Franz Joseph Gall. Head and Goldstein denied that such faculties were separate from the organization and integration of the nervous system that underwrote and substantiated consciousness. As a result, both rejected the prevalent psychological and physiological expectations of predecessors and contemporaries as insufficient to the point of being harmful to the patients under their care; they thought that patients' highly individual reaction to brain injury was being routinely con-

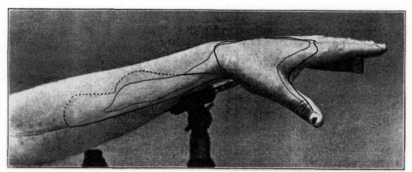

FIG. 4.1. Mapping the extent of the loss of sensation fifty days postsurgery. Image reproduced in the *New York Evening Post*, September 8, 1912 (public domain).

FIG. 4.2. Henry Head, his eyes closed, faces away, while W. H. R. Rivers uses a long pin to trace the spaces of protopathic and epicritic sensibility on his arm. PP/HEA B.9, Wellcome Library.

cealed.[25] Both found considerable inspiration in Jackson's writings on language and aphasia and regarded his work as prefiguring their own, a voice in the wilderness amid the nineteenth-century obsession with localization.[26] Yet both eschewed the evolutionary language that Jackson (following Spencer) and Sherrington relied on: their engagement with the integration of the nervous system and with aphasia would be based on neuropathological mechanisms, not tested against the *longue durée* of Darwinian evolution.[27] Finally, both described aphasia as a disorder of totalizing consequence for the organism and treated it as the single pathology that enabled an understanding of the structure of the self. For Head, at stake was not only the behavior of an organism suddenly limited in its access to language, but also neurological integration and language as fundamental wholes constitutive of social experience. Goldstein expanded from the case problem we have already looked at and from the nervous system to consider the healthy organism's capacities to abstract beyond immediate, concrete circumstances, to maintain an organismic wholeness, and to self-actualize—that is, to develop a decisive measure of autonomy vis-à-vis the environment. Head shared the path of generalization from brain injury toward subjective wholeness: for him, the brain's access to language—the ontology that makes language possible and constant—would become the central concern. Working in parallel—in the same way as Kuhn described researchers discovered energy conservation in parallel[28]—each of them established the brain-injured organism as a unit on the verge of breakdown. What follows here is a consideration of each of their approaches to aphasia and their way of handling integration. These were philosophical, therapeutic, and technical issues, a matter of how tests, patients, and revisionism contributed to two specific and different senses of integration and collapse.

Head's Orders and Disorder

Head's principal publications on aphasia span the period from 1915, when he republished Jackson's essays on aphasia, to 1926, when he assembled his two-volume magnum opus, *Aphasia and Kindred Disorders of Speech*, using expanded versions of his publications from 1919 to 1924 and a set of twenty-six case studies, each of them between nine and forty pages in length. Head drafted much of his critique of contemporary aphasiology in his introduction to a republication of Jackson's essays in the journal *Brain*, writing that "the views on aphasia and analogous disturbances of speech found in the text-books of to-day are of little help in understanding an actual case of disease."[29] Jackson had "prefigured" this,[30] and his "psychological" approach deserved to be treated as current because it undercut precisely these textbook reductions and related confusion.

A work acutely aware of its own rhetorical force, *Aphasia and Kindred Disorders of Speech* presents us with a narrative movement from harsh epistemological critique and tension toward climactic resolution. "I have attempted to blaze a track through the jungle,"[31] Head wrote, and he meant several jungles growing inside one another: the jungle created by theories of aphasia that he considered profoundly misdirected and by this point scrambled into mutually contradictory positions; the jungle experienced by each individual patient enduring a *specific* collapse of the order of language; the jungle of conceptual and clinical obstacles plaguing the physician's understanding of the patient's condition. The book is structured to clear these jungles one by one.

The first volume balances historical, theoretical, and clinical claims. Head opened not with a précis of his project, but with two histories, first an autobiographical account of his engagement with aphasia, and second a 150-page history of the theory and treatment of aphasia since the early nineteenth century. Head's history "cannot be an unprejudiced account";[32] it constantly metes out severe judgment on localizationist ideology and its obsession with determining linguistic capacities. Even recent work—which he put under the banner of "Chaos"—offers a case in which "new wine is poured into old bottles with disastrous results."[33] Only Gelb and Goldstein had recently offered a serious psychological theory, and theirs was a major attempt to resolve the violence forced onto the patient throughout this history.[34] Head distilled this terrible history of violence so as to salvage and redeploy useful concepts such as Constantin von Monakow's *diaschisis* (splitting).

Head followed up with a chapter on experimental and clinical methodology, then a clinical chapter covering five of the cases from the second volume.[35] The litany of his objections to existing aphasiology is strewn across these pages as if the battle against traditional interpretations had to be fought over and over, "a restless destruction of the false god" while the promised catharsis is repeatedly deferred.[36] There exist no brain "centers" for the use of language; we do not think in images or in words, as believed in the nineteenth century.[37] We cannot distinguish a priori between conditions of sensory or motor aphasia,[38] or between categories like ataxia, agnosia, amnesia, and aphasia—which are "descriptive terms only" and not "isolated affections of speaking, reading and writing."[39]

Using clinical material and revisionist theory, Head led his argument toward a central thesis: aphasia should be reconceived as a group of disorders of symbolic thought and expression.[40] These disorders never affect merely one element of language and symbol use, nor do they correlate with specific sites of injury: "We cannot even analyse normal speech strictly in terms of motion or sensation; most acts depend on both these factors for perfect execution. For the use of language is based on *integrated functions*, standing higher in the neural hierarchy than motion or sensation, and, when it is disturbed, the clinical manifestations appear in terms of these complex psychical processes; they cannot be classed under any physiological categories, motor or sensory, nor even under such headings as visual and auditory."[41] Aphasic disorders correlate some brain injury with a "want of capacity to use language with freedom"; they are not "static" conditions but dynamic "disorders of function."[42] Because integration plays a key role, they disturb several elements at once (naming, syntax, speaking, comprehension, and "internal speech"), albeit not to the same degree, with the result that what was needed was a schema for understanding how injury affects particular patients and groupings of patients. Such new categories, or "designations" of group difficulties affecting the patient, would remain "empirical; they have no value as definitions and must not be employed to limit the extent of that loss of function to which they are assigned."[43]

Let's reconsider this set of critiques by way of the affirmative argument Head then developed regarding aphasia and integration. Tucked halfway into the book is its definitive expression, out of which Head spirals toward his narrative and experimental climax:

> Every case of aphasia is *the response of an individual* to some want of power to employ language and *represents a personal reaction to mechanical difficulties in speech*. No two patients exhibit the same signs and symptoms; still less do the phenomena of recovery follow exactly the same course. But by comparing on broad lines *the manner in which function is restored* in cases of each particular variety of aphasia, it is obvious that some acts of speech are recovered sooner than others, and we thereby obtain an insight into the essential nature of the clinical phenomena.[44]

Head's explicit goals involved the restoration of lost functions, the patient's recovery, and improved understanding of clinical data. But it is the first sentence that offers the striking formulation: "every case is the response of an individual," a personal reaction. Head did not simply write about different cases; he crafted individuality itself in terms of each patient's dynamic response to loss, lack, or difficulty. In the above formulation, the brain-injured patient's self can be found in the organism's compensatory and transformative effort at recovery—at the reconstruction of a relationship to language that has been broken, and with it, the relationship to other people. The locus of individuality is not to be found in the brain, or even in the healthy individual, but in the pathological and recuperative relation to symbolic formulation and expression, a relation that is individualized because it is not "free."

That the response should be individual is decisive for several reasons that deserve close attention: for Head, the response was a dynamic process with a social component, an integrative process that concerned language far more than any component of humanity's biological substratum, a process that exposes for us the different strata of the somatic, conscious, and linguistic integrations that form selfhood. First, aphasia is a functional and dynamic, as opposed to a static, condition.[45] In the aftermath of a wound, the patient undergoes an ongoing experience, not merely a stagnation, and this experience is at times one of recovery and at times one of further loss and decline, generally depending on the existence of any progressive disease of the brain and also the success of socialization.[46] Second, what replaces the actual site of injury as Head's focus is the integrative way in which the organism attempts to respond and compensate, which he again defines as the site and totality of the new self. Survival and welfare thus depend on a new, recovered integration, on the way this integration compensates for the original, normal, now lost integration: it is "a total reaction to the new situation." The patient is not some stable being that is tied either to the period prior to his injury or to the injury itself. Instead, he is a being imagined in continuing resistance and response to the loss; he resides in "all the powers," the "abnormal response" that is this "fresh integration."[47]

Third, Head understood aphasia as primarily a language disorder and concomitantly a psychological disorder (insofar as language has an "emotional" and not only a "formal" component). Only then was it also a neurological and biological disorder, in that the disruption of normal neuronal integration conditions the loss, and the organism's response to it needs to be grounded biologically. "Evidently there must exist a group of functions indispensable for language in its widest sense," and even if these functions are not equally essential, they are "affected by the destruction of parts of the brain."[48] A lesion is not altogether unimportant: it disturbs cerebral and physiological processes, resulting in suffering. Head kept copious notes and diagrams of sites of injury and extent of neurological damage, but he does not seem to have found significant use for them. He instead tried to work his way back to the damage from patients' failure to sustain for themselves a linguistically articulated order, to understand the meaning of the words or sentences he was uttering, to copy texts and images, and to identify regular and frequently used "symbolic expressions." Individualized responses could be approached not through physiological damage, but only through symbolic damage, and as a result diagnosis and treatment remained tied precisely to language.

Fourth, the conception of language proposed by Head allowed for concurrent processes of generalization, categorization, and precision. Language-based categorizations and terms like *word-blindness* and *word-deafness* had proliferated since Charlton Bastian published on the subject in the late nineteenth century.[49] Head—who had studied under Bastian—duly dismissed earlier reductions of language to a mere dimension of sensory or motor systems injured by aphasia. The designations he eventually decided on—with all his caveats on the limited value of categories still in place—were four: verbal, syntactical, nominal, and semantic.[50] Verbal aphasia "consists mainly in a defective power of forming words, whether for external or internal use."[51] Syntactical aphasia is a "disorder of symbolic formulation and expression" that "consists essentially in lack of that perfect balance and rhythm necessary to make the sounds uttered by the speaker easily comprehensible to the audience."[52] Nominal aphasia "is a disorder of language" that involves an "inability to designate an object in words and to appreciate verbal meaning."[53] Finally, semantic aphasia involves cases in which "comprehension of the significance of words and phrases as a whole . . . is primarily disturbed."[54]

Having established the quadripartite argument regarding the disruptions of symbolic thought and expression, Head continued: "We do not think in words." In explicitly neogrammarian fashion, he was asserting that words are merely a small part of a language as symbolic system,

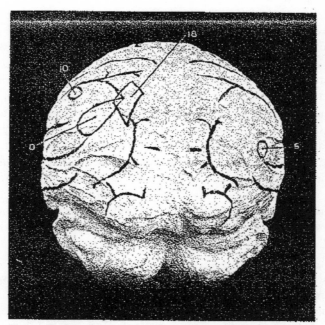

Fig. 13. To show the site of the injury to the brain, seen from behind, in No. 8, No. 10 and No. 18 to the left and No. 5 to the right of the middle line. All these patients suffered from Semantic Aphasia.

FIG. 4.3. Henry Head's efforts at localization, reported in *Aphasia and Kindred Disorders of Speech*. Cases of semantic aphasia. *AKD* 1: 456. Henry Head, *Aphasia and Kindred Disorders of Speech*, vol. 1 (Cambridge: Cambridge University Press, 1926) (public domain).

that they have multiple meanings, and that language must be understood as a dual system of its own,[55] at once logical and psychological, and one that we normally engage as a whole. The aphasic, Head claimed, is "like a man talking a foreign language": "debarred by his disability from the use of verbal forms natural to him," the aphasic "seeks some other and less difficult mode of conveying his meaning."[56]

Head's case 7 illustrates nominal aphasia (the difficulty of naming objects). His tests, his frustrations with the patient, and his empirical efforts are evident in the following passage about the case, as is his argument that patients often improved dramatically:

> A case of Nominal Aphasia, due to gun-shot injury within the limits of the left angular gyrus. There were no abnormal physical signs. Six weeks after he was wounded he was so grossly aphasic that it was impossible to obtain from him any coherent information. He could not express himself either in speech or in writing and had obvious difficulty in discovering a method of formulating his meaning. He rapidly regained sufficient use of language to permit of more complete examination, he then failed to say his name and address, or the days of the week and the months correctly. He could not name familiar objects and colors and was unable to tell the time. Repetition was gravely affected, though the sounds he uttered usually bore some remote resemblance to the words said by me. He showed obvious defects in comprehending the significance of spoken words and phrases. Oral commands were badly executed; he chose common objects or colors after great hesitation, and frequently gave up the attempt altogether. He had considerable difficulty in understanding the meaning of single words put before him in print and selected familiar objects and colors slowly with obvious effort. He failed grossly to execute more complex printed commands. With great effort he succeeded in writing his name imperfectly, but could not add his address and failed entirely to compose a letter. He was unable to write down the name of any of the colors shown to him and transcribed the time badly, although he employed numbers only. . . . He found it almost impossible to state the relative value of any two coins placed before him. Yet in spite of his confused replies, he undoubtedly recognized their monetary value, for he was able to put together the equivalent of any one of them from amongst a heap of money on the table. Drawing to order was grossly affected and he failed to

produce a ground-plan of the ward in which he lay, although he could indicate the position of the various objects visible from his bed. . . . Four years and nine months after he was wounded, his power to use all forms of language had greatly improved; yet although he could speak, read and write, he remained as definite an example of Nominal Aphasia as before. He managed to execute correctly all the serial tasks in which he had previously failed. But his powers of speech were obviously defective and he confessed that he had difficulty in finding names and that this confused him. . . . Above all he found difficulty in writing spontaneously, but he wrote well to dictation and copied excellently. He counted slowly but correctly and solved all but one of the simple arithmetical problems. He could name coins and state their relative value after some hesitation. He had recovered his power of drawing to command but could not construct a ground-plan of the room in which he worked. Orientation was not affected. He could play draughts and billiards, but not card games.[57]

Head's focus on symbolic systems and his linkage of linguistic or symbolic capacities to internal psychological losses and responses guides his conclusion that the patient was aiming directly at compensation. This linguistic and psychological approach to aphasia is not, nevertheless, the final word. Although he supported Goldstein's "psychological" approach, Head did not describe his own approach in those terms.[58] To complete his account, Head moved from linguistic pathology and case-based psychology to a neurobiology of levels of integration. Pathology, we have seen, shows disturbance in both internal interpretation and language participation. But the central term in this conclusion is *vigilance*, which complements the argument on response-based individuation. Sherrington's writings and experiments underwrite this concept, as Head attempts to explain how an activity that is apparently as fundamental to the organism's functioning as language (or, more precisely, "symbolic thought and expression") can be so fundamentally disrupted not because of the loss of a part of the brain but because of the loss of a certain integration.

To establish vigilance, Head then proceeded from neurophysiological examination, using the example of a human being whose spine had been severed alongside Sherrington's cat decortication experiments. Removing a cat's brain or handling patients who had lost the use of their legs revealed for Head a fundamental propensity of the organism to carry out certain acts beneath any possible conscious level. Here, the theory of protopathic and epicritic sensibility that he and Rivers had developed in the earlier experiment on nerve regeneration became of value once again:[59] as sensation returns to the affected area, it is first the "deeper," more intense, undifferentiated sensation that returns, with the higher-end epicritic sensation returning and allowing for a higher-level integrated consciousness only later.[60]

Vigilance refers to integrated activity carried out at the barest neurological level. The word indicates for Head the persistence of function during a sleeplike period: the capacity to hold vigil over the body. Disencumbered of the will—indeed, even in the absence of a cerebrum—the body lurks over itself, asserts eminent domain over itself without a conscious agency surviving to do so. It maintains a set of very basic postures, responds to certain forceful excitations (such as temperature), and can even have certain localized reflexes, regardless of apparent paralysis. Vigilance is "high-grade physiological efficiency," exhibiting "purposive adaptation" to external stimuli and giving the illusion of conscious behavior—acts we know to be impossible but that the body manages to maintain.[61] An integration that occurs at the purely neurophysiological level, vigilance underwrites any epicritic sensation and conscious act, it exceeds and merges local, neurological-level integrations. It allows Head to posit that from muscle tone to conscious actions a series of integrations occur of which the body is not aware. And it explains the loss of language in terms that finally tie lesions to a symbolic order on the basis of effects on the general "neural potency" of the organism:

> Every automatic act is an exercise in physiological memory and can be disturbed by vital states which have nothing to do with consciousness. What wonder that the complex powers demanded by speech, reading, and writing, can be affected by a lesion that diminishes neural vitality. Vigilance is lowered and the specific mental aptitudes die out as an electric lamp is extinguished when the voltage falls below the necessary level. The centres involved in those automatic processes, which form an essential part of the conscious act, may continue to live on at a lower vital level, as when under the influence of chloroform; they do not cease entirely to function, but the vigilance necessary for the performance of their high-grade activities has been abolished by the fall of neural potency.[62]

Head can then finally conclude the book in a manner that radicalizes this holism, positing that each mind acts as a whole, that each "conscious act is a vital process directed toward a definite end and the consequences of its disintegration appear as abnormal modes of behavior."[63] Consciousness does not normally recognize the limits of its wholeness: "the continuity of consciousness . . . is produced by habitually ignoring the gaps."[64] Only when consciousness has been broken can its facts and functioning be recognized. Indeed, not one but a whole "series of integrations" takes place, mostly "on a purely physiological level,"[65] that enable both high levels of abstraction and categorical thinking,[66] that collapse in tandem with injury, and that the surviving parts of the organism seek to reestablish and reconstruct. Individual patients, having lost the original wholesome integration, attempt through these powers to produce new conscious integrations—in the process failing in different ways to overcome verbal, syntactical, nominal, or semantic losses or disorders. In linguistic unfreedom, the aphasic becomes pathologically individualized.

Goldstein: Aphasia, Tonus, Organism

Goldstein's major work in aphasia also began during the war. It culminated in the publication of a string of detailed cases he had studied with Gelb, including the case of Schneider that we discussed in the previous chapter. Rather than focus further on the case studies or on Goldstein's dismantling of localization and atomism in his *On Aphasia* (1926),[67] we proceed directly to *The Organism* (1934), which published his research on the sensorimotor consequences of neuronal damage and attempted a neuropathologically grounded theory of organismic functioning. We then return to Goldstein's largely overlooked studies on tonus, which ground his shift from the study of aphasia to that of the whole organism.

Goldstein's argument in *The Organism* generalized his dismissal and replacement of atomistic symptomatology and cerebral localizationism.[68] As we have seen, the two approaches to aphasia denied the tremendous complexity of pathological performances exhibited by each organism, and offered each patient little by way of therapy because capturing the precise restriction of abilities was well beyond their reach. With *The Organism*, Goldstein aimed these criticisms straight at Sherrington's *Integrative Action of the Nervous System*, obviously attempting to replace it as the neurophysiological standard.[69] For Goldstein, reflex theory retained an atomistic epistemology, subdividing real-life behaviors until it could pretend that their components were low-level arcs and reflexes. It further posited that responses to identical stimuli were themselves going to be identical. But reflexes, Goldstein argued, are themselves milieu dependent and not of equal consequence in different contexts.[70] Thus, he charged, neurophysiology had become fundamentally reductive, with researchers testing local reflexes or reflex arcs rather than remembering that what hurt the patient were the complex lost performances.[71] He did not see disease or disturbance as the sole origins of a disruption of behavior; rather, they forced the organism to an adaptive, conservative adjustment in which its disordered performance tended toward compensation or readjustment that would generate a new order, a new normality.[72]

The Organism's sculpting of a broad alternative approach leads to a thorough rethinking of central biomedical notions, such as health, norms, disease, and cure. The core of his argument involves *order* and *disorder*. In injury, the organism's original "ordered behavior" is disrupted, and the organism finds itself seeking, in a disordered fashion, to restore itself to a new order:

> In an *ordered* situation, responses appear to be constant, correct, adequate to the organism to which they belong, adequate to the species and to the individuality of the organism, as well as to the respective circumstances. The individual himself experiences them with a feeling of smooth functioning, unconstraint, well-being, adjustment to the world, and satisfaction, i.e. the course of behavior has a definite order, a total pattern in which all involved organismic factors—the mental and the somatic down to the physico-chemical processes—participate in a fashion appropriate to the performance in question. And that, in fact, is the criterion of a normal condition of the organism.[73]

In this passage Goldstein deploys three core concepts: wholeness, performance, and order. First, the totality of the organism is involved—all processes, not solely psychological or bodily ones. Second, Goldstein insists on the organism's performances; normal performances define health and order and are affected by disorder. Third, order is fundamental: when the organism finds itself "dis-ordered," it attempts to compensate, to achieve order anew. Certain kinds of disorder can result in a catastrophic reaction in which the organism fails to handle any of the environment's requirements.

In a passage where he intentionally conflated a disorder within and the presence of disorder around the patient, Goldstein wrote: "The principal demands which "disorder" makes upon them are: choice of alternatives, change of attitude, and rapid transition from one behavior to another. But this is exactly what is difficult or impossible for them to do. If they are confronted with tasks which make this demand, then catastrophic reactions, catastrophic shocks, and anxiety inevitably ensue. To avoid this anxiety the patient clings tenaciously to the order which is adequate for him, but which appears abnormally primitive, rigid and compulsive to normal people."[74] Compensation, anxiety, and order versus disorder are central points in the effort to overcome Sherrington's *Integrative Action*. Specifically with regard to central cortical injury, Goldstein offered a further point: that its effect was to destroy patients' capability for abstract thought and to leave them able to respond only to immediate concrete circumstances. Compensation for disease and disturbance causes the organism to lose its well-ordered *abstract* or *categorical* capacity:

> Whenever the patient must transcend concrete (immediate) experience in order to act—whenever he must refer to things in an imaginary way—he fails. On the other hand, whenever the result can be achieved by manipulation of concrete and tangible material, he performs successfully. Each problem that forces him beyond the sphere of immediate reality to that of the "possible," or to the sphere of representation, ensures his failure. . . . The patient acts, perceives, thinks, has the right impulses of will, feels like others, calculates, pays attention, retains, etc., as long as he is provided with the opportunity to handle objects concretely and directly. He fails when this is impossible. . . . He can manipulate numbers in a practical manner, but has no concept of their value. . . . He is incapable of representation of direction and localities in objective space, nor can he estimate distances; but he can find his way about very well, and can execute actions that are dependent upon perception of distance and size. . . . The patient has lost the capacity to deal with *that which is not real—with the possible*.[75]

Abstract behavior, Goldstein insists later in the book, involves "the ability of voluntary shifting, of reasoning discursively, oriented on self-chosen frames of reference, of free decision for action, of isolating parts from a whole, of disjoining given wholes, as well as of establishing connections."[76] By contrast to this grasp of order—this wholeness and normality of experience—the concrete attitude is "realistic" and does not imply "conscious activity in the sense of reasoning, awareness, or a self-account of one's doing."[77]

The loss of the abstract attitude in brain injury was central to the elimination of the old order and to the reduction of patients to a disorderly situation in which all they could cope with was the immediate, concrete situation surrounding them. Cortically injured patients face "systemic disintegration" rather than only the loss of particular performances: these merely evince and complement the organism's fragmentation.[78] Faced with abstract demands, patients begin to sense intense anxiety over their inability to fulfill them and seek emphatically to return to some order. These "catastrophic" situations may involve a further reduction in the capacity to cope, and the organism, operating without abstract or discursive thought, pushes blindly a *restitutio ad integrum*—however restricted that *integrum* might be.[79]

Thus Goldstein argued "from the pathological to the normal"; indeed, this was the basis of his thought.[80] Whereas Head focused on the new integrations actively carried out by the injured patient and compared those to "normal" language use, Goldstein proceeded directly from his patients' performances to a theory of the structure of the organism.

Goldstein further distinguished himself from Head by attending to the motor capacities and posture of the organism, in particular the problem of tonic musculature evident in films like *Tonus*. At every moment of our lives, the tonus of our musculature is regulated, and without this constant, unconscious adjustment we could neither walk nor stand nor even sit. Goldstein first studied the correlation between brain lesions and modifications in head posture, and he would later be credited with the discovery that tonic neck reflexes "disclosed the existence of an extended motor relationship between different parts of the body."[81]

In several studies written between 1924 and 1930 and recounted in *The Organism*, Goldstein described in detail a series of specific tests that pertain to the body's tonic musculature and that are aimed at the tendency to compensate for injury so as to achieve some optimization, however difficult or tenuous.[82] Scholars have prioritized the optical tests, but it is in the studies of tonus that the point about performances and the extrapolation of neurological concerns toward organismic wholeness and patient care becomes clear.[83]

In addition to grasping and pointing tests, which allowed Goldstein to observe movement and neuromuscular response, he began incorporating tests for assessing equilibrium and sensory response with the aid of instruments as simple as a set of keys, a tuning fork, or a violin.[84] In

FIG. 4.4. Use of air puffs in *Tonus* (1930): The patient is instructed to hold his arms steady, and then, with closed eyes, to repeatedly touch the top of his left hand with his right. With pressure applied in the inner ear, the left arm begins to veer leftward and the right one fails over and over to touch it. Box 18, Kurt Goldstein Papers, Columbia University Rare Book & Manuscript Library.

Tonus, Goldstein uses air puffs connected to a tube to create pressure —a common test for lesions of the inner ear used to assess percussion injuries and their effects, especially on equilibrium and dizziness.[85]

Although the failure to react to ear stimulation was most associated with cerebral tract lesions,[86] for Goldstein its physiological and psychological relevance lay elsewhere as well: in the patient's experience of the environment. He used the air puffs to assess the performance and meaning of disequilibrium and vertigo,[87] not despite, but rather because of similarities in performance related to inner-ear damage, vertigo, disequilibrium, and brain lesions. "Similar-appearing symptoms can be of essentially different origins," Goldstein wrote, and "only by knowing that latter can one avoid inadequate treatment and achieve better results."[88] To examine and understand the patient's loss of order and abstract attitude, it became imperative to test symptoms in groups and seek the correlations of performative failures.

Thus Goldstein rejected views that dispensed with the environment as mere background while the organism's malfunction remained internal, its abstract attitude independent. The psychologist Kurt Koffka, for example, insisted that "there is no behavioral environment" for tonus,[89] and Head thought that neurobiological vigilance supported a model of the organism based on a reflexlike protopathic response to the environment. Goldstein, in contrast, saw changes in tonus as modifications occurring on the total neuromotor level; tonus marked a deeper ingraining of the environment into the structure of a disordered or partially reordered organism. Such modifications *incorporated* the environment. He could thus ask what a disordered modification even in a single area of functioning meant to a specific individual given a particular demand.[90]

This incorporation of the environment called for compensation and optimization, and Goldstein discussed these at length in relation to organismic responses to new contexts. They resulted in a new—and to the naked eye, distorted—"preferred behavior":

> The adaptation to an irreparable defect takes essentially opposite directions. *Either* the organism adapts itself to the defect, . . . yields to it, . . . resigns itself to that somewhat defective but still passable performance which can still be realized, and resigns itself to certain changes of the milieu which correspond to the defective performances; *or* the organism faces the defect, readjusts itself in such a way that the defect, in its consequences, is kept in check. . . . In patients with one-sided cerebellar lesion, we often find a "tonus pull" towards the diseased side. All stimuli which are applied to this side are met with abnormal intensity, with abnormal "turning to the stimulus." This leads to deviation in walking, to a predisposition to falling, to past pointing, etc., all towards the diseased side. Usually the patients display simultaneously an abnormality of posture in the form of a tilting of the body, especially of the head. As long as the patient remains in this abnormal posture he feels relatively at ease, has less subjective disturbances of equilibrium, less vertigo, etc. His objective performances, such as walking, pointing, etc., are better. Deviations may disappear completely. However, the subjective, as well as the objective, disturbances immediately reappear as soon as the patient reassumes the old, normal position of the body. Apparently the abnormality of posture has become the prerequisite for better performances, has become the *new preferred situation*.[91]

The subjective-objective division in this passage matters: in "objectively" abnormal behavior, we notice that the modified, deviant responses represent a profound discomfort at the supposedly normal, objective demands. A new normal has arisen, specific to the patient's restricted perspective, and these deviations indicate the reconstructed and preferred—and in many respects, healthier—perspective. The patient has found a way to compensate for losses or changes in functions; put another way, the patient bends in a particular direction because the new bearing that she or he adopts is the one she or he experiences as least limiting. Only in a potentially catastrophic crisis arising from the inability to meet environmental demands does the postinjury state become not simply objectively pathological but subjectively felt as such. In experimental work on tonic movement, Goldstein could demonstrate modifications in the capacity of the body to perform motor tasks with or without such loss, giving some clue as to how the environment of capacities had been narrowed and how it was being manipulated anew by the modified individual. In tonus pull, we find not only the apparent defect but also the sense in which the individual has compensated and reordered the world.

In patients reported as cases or appearing in Goldstein's research films, we can see this preferred behavior expressed through groups of symptoms exhibited when "normal" actions are required of the patient, who then shows unilateral disequilibrium, "spontaneous inclination" (leaning and swaying) of the body, a passive turn of the head, further aggravated body and arm deviation, and so on. Together the results of these tests produced facts not about all of the patients under examination but about the capacities of each single patient, and the difference between this patient's preferred behavior and the "normal" one.

In his *Philosophy of Symbolic Forms*, Ernst Cassirer (Goldstein's cousin) provided a summary of the clinical scene at the Frankfurt Neurological Institute, yet he retained only the pejorative sense of disorder and did not emphasize the productivity of compensation:

> One psychic blindness patient could not, with his eyes shut, orient himself in regard to the position of his head or another part of his body. If, . . . for example, his right arm was raised sideways into a horizontal position, he could not *immediately* make any statement regarding the posture of the arm, although by an arduous detour, by executing certain pendulum-like movements of his whole body, he could ultimately come to certain conclusions about the arm's position. . . . Gelb and Goldstein believed that when his eyes were shut, the patient possessed no spatial representations at all.[92]

This was not quite the endpoint of Goldstein's interpretation. Goldstein emphasized the organism's fundamental tendency toward preferred behavior, which is "determined by the total attitude of the performing person,"[93] and which would generate the greatest comfort.[94] The body, struggling against the demands of even minimal tasks, generated new norms in posture and spatial identification: a subjective normalcy at odds with our sense of normal behavior during the accomplishment of such tasks.

Integrations: Similar and Diametrically Opposed

Head and Goldstein's divergences in argument and method culminated in distinctly different conceptual arrangements of the body. What brought them together—the shared, dominant imperative in their work, the syntax behind their similarities and divergences—was a logic binding integration to disequilibrium. Their distillation of order, integration, catastrophe, vigilance, and related concepts out of observations of their patients displays the centrality of this logic to their thought. Both authors ask how an organism exists as a single whole and tends again and again toward wholeness during and after injury; both establish collapse and wholeness as a single problem, and interestingly, they do so across different registers. Head's turn to vigilance, his focus on linguistic deficiencies, and his different levels of integration—reflexive, neurological, linguistic—all suture an organism quite different from Goldstein's holistic-psychological organism that constructs a new preferred behavior after losing its abstract capacity. These different bodies and selves played out across other registers too: of humans' engagement with language, of mind-body monism and dualism, and of the facts and concepts of disease and recuperation. In the remaining pages, we consider their models of the integrated body and its collapse by sculpting each in the other's relief.

Head's conception of integration occurs within a two-tiered system. On the first tier, a basic neural and corporeal integration is abstracted not from clinical work but from Sherringtonian neurology; elaborated in the argument on vigilance, it is reminiscent of protopathic sensibility. We might call this Head's conception of the *integratedness* of the organism: integration is a given, takes place deep beneath consciousness, and endures even in decorticated mammals. The action and reaction of impulses, including reflexes, and vigilance more broadly are essential for any psychoneural integrated movement of the organism. Here this integratedness at the physiological and unconscious levels simply exists and is limited only by the loss imposed by particular lesions. For anything to happen in the organism, this bottom-up integration remains essential; it is sensed but unaccounted for.

On the second tier, in both his theoretical and his clinical descriptions Head makes much of active integrative forces, occurring especially in the aftermath of injury. Injury is followed by "a fresh integration carried out by all available portions of the central nervous system."[95] (In comparison, Goldstein's *The Organism* describes the patient as lacking freedom, as having altogether lost abstract capacity, as struggling between disorder and a new, restricted, highly precarious order that is constantly undulating.)[96] For Head, active integrations help the injured patient recover, usually to a considerable degree.[97] Consciousness integrates actively. Head formulates the organism as an arsenal and often tracks its integration in terms of militant or military activity: "Faced with a new situation, the organism puts out all its powers, conscious, subconscious and purely physiological, in order to produce an adequate response directed towards its welfare as a whole."[98]

Goldstein's theorization involves an integration described along two nontiered schemas, arranged concentrically around the patient. First, the abstract versus concrete schema: brain injury annihilates the abstract capacity of the organism and results in concretization, an inability to plan, and an inability to extract oneself from immediate circumstances.[99] Goldstein minces no words: the loss of abstract attitude is devastating. It imprisons the patient into the immediate situation. Although some parallel could be drawn to Head's model on the basis of loss of epicritic sensation, unlike Head, Goldstein sees the "destruction of the abstract capacity" as irreversible.[100] Language and performance are apiece with abstraction, and true integration resides only in the whole, healthy, normal organism, not in neurological or active conscious work. For this argument, Goldstein's focus on performances—on the natural execution and continuous, uninterrupted, untroubled performance of particular tasks, often not recognized as tasks—is essential. In mooting Sherrington's integrative action and Head's identification of vigilance as the basic neurobiological regulator, in focusing instead on the reparation of the subject's interwovenness with his milieu, Goldstein depicts normal life as constantly threatened with disintegration.

Second, Goldstein uses the order-disorder-catastrophe triplet to establish what sort of rehabilitation the physician may hope for in a patient. Integration follows from Goldstein's argument on norms: the healthy organism is ordered to the point of never recognizing or sensing the existence of this order; she is normative, free from order, freely integrated, autonomous, driven to "self-actualization."[101] By contrast, freedom in Head is freedom to use language without major constraints. For Goldstein, the disorder that occurs in disease involves at once the patient's and the physician's recognition that an order has been lost. Although the management of disorder may lead to new orders, it also submits the patient to the purview of norms and to a depreciation vis-à-vis those norms: "A life that affirms itself against the milieu is a life already threatened," in Georges Canguilhem's words.[102] Autonomy is dramatically lessened, and only through the scene that the physician makes possible can a hampered, stuttering self-actualization perhaps provide comfort within these norms. Catastrophe involves not only immediate collapse but also a cascade of restrictions by the milieu; amid this "systemic disintegration,"[103] the organism thrashes at any restricted stability it can get.[104]

Goldstein's two schemas imply each other without being coextensive: order does not concern merely circumstances of brain integration, and some order can be provisionally attained regardless of abstract disintegration. Goldstein focuses the two schemas toward a concept of self-actualization, which can occur in an injured organism as well as in a healthy one. It is a value and ideal for the healthy to grasp as well as a purpose to restore to the injured or sick man. In the silence of wholeness, the capacity to self-actualize emerges. Life is a relation to the production of oneself as nonthreatened, autonomous, and normative: one is an individual in the sense of being in-divisible. Life, even animal life, "points . . . to an individual organization, on which basis alone it becomes intelligible as the expression of the tendency to actualize itself according to the circumstances."[105] Systemic disintegration is the gravest danger for this self-actualizing being.[106]

Head and Goldstein thus set the stage very differently. The course of disease, and especially of consciousness in disease, like the site of active integration in the diseased or disordered subject, are different. The degree zero of integration resided for Goldstein in the automatic production of an alternative tonus and its lesser, abstractionless biological and social norms; for Head it lay in neural vigilance and the constant effort at linguistic restoration, however partial. Whereas Goldstein saw no grand restitution of wholeness once the abstract attitude was lost, and looked at alternative preferred behaviors, however seemingly deformed, Head identified no integration in the interaction of consciousness with its milieu. Instead he saw a neural integrity beneath all consciousness, and he tracked the patient's gradual improvement into language, treating it as a fresh, active integration.

Head's attention to language, founded on neogrammarian theory, accentuates the very difference of his subject from Goldstein's. Head posited that each of the multiple levels of integration bears profoundly on language and its use.[107] Vigilance constituted the hinge between them: any substantial reduction in its "neural potency" amounted to a destruction of the joints between the realm of consciousness, speech (interior and exterior), and language. At this point, "the mind/body problem does not exist so long as we are examining the consequences of functional disintegration." In other words, the organism became one in its attempt to create a "fresh integration" and in its use of "all available portions of the central nervous system" to achieve it.[108] Language and symbolic expression were only partly unavailable to the patient, who remained broken in terms of "normal" access to the outside world. Even this broken self pursued a new self in language. Goldstein's attention to tonus and its partial restoration of order and self-actualization beneath abstract wholeness offered little regarding the patient's suspension from the order of language.

This raises the difficult problem of monism and dualism—so central to psychology, physiology, and philosophy—which Head and Goldstein negotiated in quite different ways. Goldstein saw monism in self-actualization, in the fully healthy individual seeking life without interruption by the environment. For him, this monistic pursuit in the individual was undermined by brain injury, disorder, and the loss of the abstract attitude.[109] Having escaped death only to lose their selves, Goldstein's patients remained shattered into pieces: beneath consciousness and the lost chance of wholeness, their bodies engaged different norms to subsist in a new preferred manner.

Head proposed, instead, that what was at stake were the particular ways in which the linguistic dimension of the self collapsed—and with it all direct access to language and society—so the self would break, too. Losing the normal form that allowed its singularity, the coexistence and coextension of integrations, this broken self began its vigilant pursuit of an internal reconstruction; it sought a new wholeness without normal selfhood being quite available. Head's patients broke and then reconstructed a certain wholeness that crisscrossed language, a wholeness that kept them in perpetual tension with the demands of the linguistic order. This wholeness could not be easily or perhaps even adequately understood from the physician's linguistic perspective, and patients could be treated only to aid their movement toward a new and reduced harmony with the world. In other words, their pathological self-reconstruction was profoundly monistic, the self becoming one out of the fragments as it sought the new integration. Goldstein accorded his patients no such trend toward unity.

This means that Head and Goldstein developed very different conceptions of individuality. For Goldstein, only the healthy and free individual was truly individual. For Head, by contrast, individuality resided in pathological reintegration. It was precisely because each patient was a case unlike any other, with only barely any capacity for categorization into language and society, and because each patient was carrying out reintegrating processes while remaining unfree, imprisoned by injury, that any patient could become truly, differentially, pathologically individual.

Or, as Virginia Woolf put it in *On Being Ill*, "But with the hook of life still in us still we must wriggle."[110]

5

Physiology Incorporates the Psyche: Digestion, Emotions, and Homeostasis in Walter Cannon, 1898–1932

Macbeth's advice that "good digestion wait on appetite and health on both" is now well-founded physiology.
—**Walter B. Cannon**[1]

In 1898 Walter Cannon, then a medical student at Harvard, published his first article on the movements of the stomach, based on experiments he carried out on cats using the recent technology of Röntgen rays (X-rays).[2] Cannon's publication effected a drastic shift in the study of internal organs and the functioning of the digestive system because his method made it possible to visualize the internal organs of a living being while it was active and basically undisturbed.[3] Cannon devoted the subsequent decade to studying the digestive system—from the alimentary canal to the colon, from chewing to defecation, from peristaltic waves to tonus. His work was heralded as the complement to Ivan Pavlov's research on conditional reflexes in the digestive system of canines, as well as to William Bayliss and Ernest Starling's research on hormones.[4] By 1904, he was lecturing at the British Physiological Society. His audience included Charles Myers, Henry Head, W. H. R. Rivers, and Charles Sherrington.[5] Seven years later Cannon compiled his studies in a paradigm-setting book, *The Mechanical Factors of*

Digestion.⁶ Meanwhile, intrigued by the effects of anger and fear on the animals on which he was experimenting, Cannon turned his attention to the adrenal glands, the factors influencing their action, and their "emotional" effects on digestion.

Cannon's book on states of great excitement affecting the whole body, titled *Bodily Changes in Pain, Hunger, Fear, and Rage*, appeared in 1915. In it he named these states "major emotions" and concentrated on their physiology and their meaning. Focusing on the confluence of the sympathetic nervous system and the secretions of the adrenal glands, Cannon presented the neurological and hormonal basis for the actions of animals and human beings in situations of physiological derangement. He then left for Europe to participate in the study of wound shock. Returning to America in 1919, Cannon turned to the thyroid's involvement in internal regulation, and once again to the physiological basis of emotional activity. In 1929 he published an expanded version of *Bodily Changes*, harshly criticizing the James-Lange theory of emotions (which is still influential today), and witing further on the coordination of the "sympathico-adrenal" system and the emotions.⁷ *The Wisdom of the Body*, Cannon's 1932 crowning work on homeostasis, guided strands from his earlier research toward a synthesis of the devices through which human organisms control their own stability.⁸ By this point he was widely recognized as the pioneer of American physiology. He published his last books, *Digestion and Health* and, with Arturo Rosenblueth, *Autonomic Neuro-Effector Systems*, five years later.⁹

The purpose of this chapter is to show how Cannon moved from identifying experimentally the mechanisms on which the digestive system was built to theorizing the neurophysiological and pathological arrangement of the emotions, and eventually to presenting the body as an integrated system that he called a "fluid matrix." The digestive machine had given way to a hormonal and emotional whole and, later, to a homeostatic self. More than any of our other protagonists, Cannon exemplifies the shift to a concept of the individual as brittle and constantly imperiled. "Our bodies," he pronounced over and over in the 1930s, are "made of extraordinarily unstable material;" only organism-wide stabilization processes hold us together.¹⁰

Shifts: Evidence, Object, Epistemology

Cannon often recounted his research as a linear advance of the light of experimental physiology through landscapes untraveled, its lens eventually tightening around a complete theory of the body and the self.¹¹

At first sight, his plethora of publications on digestion and the emotions confirms the claim of continuity, which Cannon, perfectly content to assume the mantle of the canonical figure, accentuated. Since the beginning of his work, Cannon's texts were made taut by tensions and compromises in their experimental and evidentiary structure, as well as in the epistemology they articulated. The interesting thing about Cannon is not that he made intellectual choices and then papered them over; rather, just as his choices mirrored and spearheaded a broader shift in the concept of the body, retroactive continuity placed the cloaks of legitimacy and even majesty on the shoulders of this new concept.

To tell Cannon's story—the story of this body he described and redescribed—is to look for signs of the puzzle being reordered. In this section, we first follow Cannon's permutation of types of experiments, the evidentiary support for an organism suspended between physiology and pathology, between laws of the body as a given structure and forces turning it into a plastic, self-regulating being.[12] Second, we follow the change of object in his research, from the mechanical digestive system to a less static, less ruly unit, different in gradation and complexity. Third, we look at the shifts in epistemological priorities, the kinds of knowledge he deemed necessary for his argument to hold. These included major theoretical assumptions that Cannon gradually abandoned or came to treat as auxiliary: notable among them is the sidelining of the mechanist basis of the organism, the effacement of Darwinism as an explanatory tool, and his movement from tacit acceptance to sustained critique of the James-Lange theory of the emotions.

In his 1898 "Movements of the Stomach Studied by Means of the Roentgen Rays," Cannon wrote, intentionally, as if no real research had taken place before: he made only a single citation—a disparaging one about earlier failures to see actual stomach movements.[13] He also opened his 1902 essay "The Movements of the Intestines Studied by Means of the Roentgen Rays" with a disdain for his peers' work that bordered on stylized pity: "The investigation of intestinal movements has been beset by the same difficulties that characterized the investigation of the gastric mechanism. Pathological subjects or animals subjected to the disturbing action of drugs and anesthetics and of serious operations have been the only sources of our knowledge. A considerable difference of opinion as to the nature of the normal movements in the intestines has resulted from observations made under these necessarily abnormal conditions."[14] Cannon proceeded to cast doubt on the results of research based on those premises: despite elsewhere advocating the pedagogical value of case studies, here he dismissed the use of "pathological subjects" and similarly refused as affected any data gathered from anesthetized patients. He again accounted for recent work as though describing obsolete, archaeological findings. X-rays provided the necessary authority where other experimenters relied on guesswork. In "The Movements of the Intestines" Cannon equated proof with seeing: "I have repeatedly seen"; "Only twice have I seen"; "I once saw"; "I have never seen."[15] Because X-Rays did not disturb the operations of the body, they constituted the first true method of physiological observation since the one-off event in which William Beaumont had famously studied "the obscure process of digestion" by peering through the open wound of a gunshot survivor.[16] It was a triumph for positivist mechanism: in technologically enhanced *seeing* would lie the definitive criterion for the sciences of life.

In Cannon's subsequent work, X-rays continued to provide evidence and detail, but they were no longer the only proof: the seeing-is-knowing rhetoric all but disappeared after 1902, as did his aversion to deriving the normal from pathological cases. Having drawn as much from X-rays as possible, Cannon began citing his earlier results to validate other kinds of research, and he had the necessary authority to make claims over experiments in which X-rays were not useful. Animal experimentation and clinical pathology, which he had recently mocked, became essential.

Animal experimentation deserves a note: it was standard in physiology labs, and by 1905, with anti-vivisection controversies raging on both sides of the Atlantic, it became politically identified with physiology itself. Part of Cannon's evidentiary shift is no doubt due to these controversies. Like Bayliss and Starling, who were personally embroiled, Cannon participated vocally, publishing pamphlet after article in vivisection's defense.[17] The controversy united physiologists in defense of the value of their science, reauthorizing common experimental protocols, offering the sense of a community resisting valiantly in the name of science.

At this point Cannon was using a complex of protocols that included varieties he had recently disregarded. Compare, for example, his 1902 passage disparaging experiments predating his X-ray studies of the intestines to his 1905 essay "Gastro-Enterostomy and Pyloroplasty: An Experimental Study by Means of the Röntgen Rays," which compared two methods of surgery on the pylorus (the lower end of the stomach).

Despite the article's subtitle, the bulk of his evidence came "from clinical sources, from recent physiological investigation, and from observations on animals on which gastro-enterostomy or pyloroplasty has been performed."[18] Clinical sources and case studies gained in value once Cannon began to discuss the "Influence of Emotional States on Functions of the Alimentary Canal" (1909). For the first time he cited cases as examples, something that was possible precisely because he had covered the "normal functions" and was able to recognize that emotional influence played a role on such functions.[19] The early aversion to disturbing the body was nowhere to be found in "The Nature of Gastric Peristalsis" (1911), where Cannon reported surgically removing the entire stomach. He did the same in 1914 with the adrenal glands.[20] Hormonal tampering and surgical manipulation of organs no longer posed an evidentiary or epistemological problem; they all had become complementary to the new standard. On the one hand this seemed a step back from the assurance of vision; on the other, it changed the questions Cannon could ask.

Such questions now included how emotions were felt related to bodily functioning. The new claim that "stomach contractions . . . cause hunger," and not the other way around, would become a staple of his midcareer research, which relied on this shift to an anything-goes set of protocols. In "An Explanation of Hunger," written with his student A. L. Washburn in 1912, Cannon offered three kinds of evidence: first, observations of hungry people reported in recent research; second, self-observation and introspective psychology (What does one refer to when one calls oneself hungry? A "dull ache"? A "gnawing sensation"?);[21] and third an account of self-experimentation, in which Washburn "accustomed himself to the presence of a rubber balloon in his stomach and a small tube in his oesophagus."[22] Washburn swallowed a balloon connected through the tube to a manometer that recorded stomach contractions. This allowed Cannon and Washburn to determine that contractions preceded the awareness of hunger and that hunger "is normally the signal that the stomach is contracted for action."[23] Thus Cannon could argue that the hunger sensation was not the physiological result of an emotion imposed on the stomach, but the whole-body effect of a local, determinable, physiological cause.

The availability of all these techniques surely played a role in the shift in evidence, and so did reasons inherent in institutions and the politics of science.[24] The course of Cannon's focus of research, his objects, and the epistemological questions attached to these objects were also important in guiding the shift. By 1909, his search for causes turned to

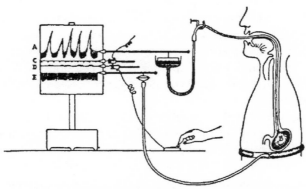

Fig. 13. Diagram showing the method used to record the gastric hunger contractions. A, Kymograph record of the increase and decrease of volume of the gastric balloon, B. C, time record in minutes. D, record of the subjective experience of hunger pangs. E, record of the pneumograph placed about the waist; this record proves that the hunger contractions do not result from action of the muscles of the abdominal wall. (From Cannon, Article "Hunger and Thirst," "Foundations of Experimental Psychology." Courtesy of the Clark University Press.)

FIG. 5.1. From Walter B. Cannon, *The Wisdom of the Body* (1932; New York: Norton, 1963), p. 71 (public domain).

the adrenal glands and the nervous system, as well as to the organism's experience of external and internal disarrangement.

: : :

Let's turn, second, to Cannon's shifting intellectual priorities, the scope of his research, and the parts and systems of the body that now served as his objects. In his 1898 essay on "Movements of the Stomach," Cannon indicated the effects of restlessness, anxiety, and petting on a cat's stomach movement.[25] Similar references to the destabilizing effects of disquiet and anxiety on the digestive system appear in "The Movements of the Intestines," published in 1902: "a tranquil mood on the part of the animal was found to be quite as necessary for seeing movements of the intestine as it was for securing normal activity in the stomach."[26] Cannon reprinted this particular passage repeatedly and sometimes verbatim, all the way to the very last pages of his 1911 *The Mechanical Factors of Digestion*.[27] He later interpreted this and other passages he recycled to mean that he had been attentive to the emotions since the very beginning of his research.[28] But that is misleading.[29] His "difficulty" at this early point was that certain conditions disable or disarrange the researcher's tools, and he insisted on the "importance of avoiding so far as possible the states of worry and anxiety, and of . . . grief and anger and other violent emotions."[30] Anxiety and intense emotional reactions were not objects of research, but inhibitors to be mastered.

The study of digestion, in other words, necessitated control over the emotions. This sense of the emotions as disturbances persisted in *The Mechanical Factors of Digestion* even though Cannon had moved on.[31] Cannon's first published text on the emotions and their effects on the alimentary canal came out in 1909. By 1911 he was consistently privileging adrenal secretions and the sympathetic nervous system—which directs adrenal secretion, as had just been recognized—as the crucial sites for studying meaningful emotional influence.[32] In the years 1911–1915 the majority of his publications not dealing with the vivisection controversy focused on adrenal stimulation and what Cannon called the "major emotions." His language for dealing with these emotions had moved away from treating them as destabilizing factors to recognizing that they enabled research on pathology and regulation. The object that had emerged from frustration with destabilizing factors was now a productive ensemble: it linked the sympathetic nervous system (the part of the autonomic nervous system that regulates internal organs) to the secretions of the adrenal glands, to the forces that stimulate or are represented in emotions, and to the body's way of handling stimulation and disarrangement from the desired norms.

Why the adrenal glands, and why the emotions? Cannon's work here was very much in tune with that of other researchers: after Starling discovered and named hormones, they were identified as having profound effects. Yet these effects were unclear: adrenaline, for example, confused experimenters in that it could both constrict and dilate the arteries.[33] Some adrenal secretions seemed to affect both the performance of muscular work and the fatigue that hampered it.[34] And, as was already recognized, it was precisely because adrenal secretions were linked to the sympathetic nervous system that they enabled the organism's capacity to take in food, to transform it into energy, and to react to extreme conditions.

Cannon moved to conceiving of the emotions as devices whose considerable value for self-preservation was due to their regulatory ties to every other major system of the body. He later summarized the shift in his thinking: "Digestive organs do not follow an isolated system; they are one of a company of related systems. They serve the other systems and are in turn served by them."[35] He was reaching this kind of understanding around 1911–1913, whereas earlier "I did not understand the organization of the sympathetic system and the arrangements for a diffuse distribution of its influences."[36]

Disturbances of function affected "viscera hidden deeply in the body."[37] The naked eye could not see them, nor could X-rays. More complex still, they involved a "violence of excitement"[38] that brought about a cessation or inhibition of other functions. The body responded in an elevated and holistic manner. So Cannon did not pursue emotions as such, but the sources of and reasons for bodily destabilization and equilibration. In naming these changes emotions, Cannon attempted at once to establish a psychological theory on the back of his physiology, and to initiate an explanation of organism-wide disturbances that resulted from environmental disturbances. In this manner he incorporated the psyche into his physiology.

: : :

Third, we pivot to the epistemological shift. What does it mean to introduce the activity of the adrenal glands into the study of mechanistic bodily operations? The introductory answer is straightforward: the experimental and research shift entails a shift in the treatment and conceptualization of the organism. If you use X-rays to determine the movements of the stomach and the rest of the digestive canal, dismissing earlier research as having failed to see, you subscribe to a sense of the body as a group of systems and organs, each of which you study independently—as if each X-ray were a moment in a chronophotographic series that lets you envision its continuous operation. All digestive canals are the same; pathologies are inessential to qualifying the norm and cannot suggest a plurality of norms. But once you begin to combine induction from animal to human physiology, clinical case studies and pathology, introspective psychology, and surgical and hormonal modification; once you treat the adrenals as the theater in which emotional influence is born, in which they mimic and combine with the sympathetic nervous system; once you begin developing a general theory of emotion that explains these effects—then the body begins to look like a very different thing, one that revolves around the diffuse systems through which it manages external stimuli, or through which it continually attempts to gain the initiative.

And, to look ahead momentarily to homeostasis, once you think of regulatory mechanisms as a force field that regulates and organizes the body, once you speak of the latter as an "internal economy" or a "fluid matrix,"[39] once you begin prying into the systems that explain the remarkable stability of the organism in the face of external changes and intrusions, the body again morphs, this time into a being that can be called cybernetic as easily as self-curing—a body that looks neither mechanistic nor holistic in the traditional senses of those words.

The Invention and Meaning of the Emotions

By 1914, Cannon was consistently interspersing two kinds of evidence—physiological (X-Rays, blood pressure, stomach pressure readings) and pathological (case studies, surgically and hormonally induced modifications). Each compounded the other to describe a body for which physiological or biological laws could comprise the emotions, including their pathological and psychological components. The emotions were traditionally more psychological, in the evolutionary sense of behavioral responses to the environment, and also in the behavioral-moral sense inherited from utilitarianism.[40] Cannon dismissed this focus and sought evidence in the pathological effect of disturbances that occur as the organism is roused in extreme circumstances, as well as in literary, anthropological, even anecdotal examples: a full chapter titled "The Energizing Influence of Emotional Excitement" depends entirely on ethnographic and anecdotal research.[41] The social life of the emotions had become as plainly essential as the evolutionary development of the organism and physiological processes that explain pathological deviations from the norm. The strategy of *Bodily Changes* was to unite this complex universe into a specifically physiological one.

Cannon retained only two of the traditional emotions, rage and fear, and added to them the conditions of hunger and pain, and later thirst.[42] They shared one thing: they were whole-body destabilizations that could be observed when the organism found itself reacting in a forceful manner to extreme occasions in the outside environment. Cannon's exemplar was a cat strapped in a holder and either terrified or enraged by a barking dog. Emotions, he contended, do not cause bodily changes. Rather, they are states, experiences to be identified with these bodily changes even though the emotion and the bodily change may appear to be in conflict. In emotional states, the body is influenced to such a degree that its normal functions are disrupted, particularly the functions associated with digestion. Such disturbances are caused not by a physical intrusion or pressure, but by elaborate physiological processes.

Other than a few words on "feelings," Cannon did not mention "emotions which are usually mild—such as joy and sorrow and disgust" until page 277 of the first edition. Even then, he suggested that they deserved study only "when they become sufficiently intense."[43]

Thus the object of the book was bodily changes due to the correlations obliged by environmental stimuli, with pain and hunger being as significant as the two environmentally induced dominant emotions. As we have seen, Cannon understood the feeling of hunger as an outcome of stomach contractions. Fear and rage were the two whole-body conditions that had absolute effects: they were violent reactions to pressures from the environment and fundamentally distortive internal processes—deployments of the sympathetic nervous system and, through it, of adrenal secretions. Pain, meanwhile, had already been identified by "the advocates of the theory of organic evolution" as highly similar to the "major emotions"; Cannon saw it as underlying the trigger into fear and rage, the fight-or-flight response.[44]

Above the fireplace of Cannon's office at the Harvard Medical School hung two portraits—one of Charles Darwin, the other of Claude Bernard.[45] These portraits, epistemic things par excellence,[46] all but conditioned the dual evolutionary and pathological logic that Cannon deployed in *Bodily Changes*.[47] His experimental evidence would fall in the physiological category, but evolutionary logic subtended its function and meaning.

In the evolutionary framework, emotions were ancient patterns of response to the environment.[48] Cannon opened *Bodily Changes* on a specifically Darwinian note: "The doctrine of human development from subhuman antecedents has done much to unravel the complex nature of man." He continued: evolution "has proven applicable . . . in accounting for functional peculiarities,"[49] and "more and more it is appearing that in men of all races and in most of the higher animals, the springs of action are to be found in the influence of certain emotions."[50] Yet Cannon shifted the site of the emotions in the evolutionary argument. Darwin had described the external, especially facial *expressions* of emotions as mechanisms of adaptation necessary for an organism's survival because they maximized the organism's capacity to respond to stimuli received from friends and foes. By contrast, Cannon emphasized that emotions were internal reactions. Functioning beneath the level of consciousness, interpersonal recognition, and will, they allowed for the necessary physiological adjustment (in Allan Young's term, *mobilization*) through which the organism managed to respond to challenges.[51] For Cannon, the facial expression of emotions was relatively insignificant, insofar as expressions could be negotiated consciously and feigned,[52] whereas emotional states "are not willed movements, indeed they are often distressingly beyond the control of the will."[53] Cannon dismissed the other emotions on which Darwin had concentrated (disgust, joy, grief, and so on), sticking to the four particular bodily changes of his title.[54]

Cannon's revised evolutionary argument emphasized that over "myriads of years of racial experience" and over "innumerable injuries in the course of evolution," emotions as bodily changes became physiologically useful.[55] Emotional states had an adaptive rationale even if their scene of operations was the sympathico-adrenal system and not the face. This rationale guaranteed the value of drawing conclusions about humans from observations of animals: what was applicable to lower animals was by default also applicable to humans.

The adaptive rationale played another far-reaching role. Already in 1911, Cannon had not shied away from suggesting that the problem of emotions, as bodily changes, included "natural" and "modern" components.[56] In "the wild state," emotional regulation involved the storage of adrenal secretions as reserves for later emotional exertion. In other words, at moments of peace, "wild" humans preparing for a hunt could store up power that their bodies could then expend during the attack. But in "strenuous" and "exciting" modern life, protracted emotional exertion induced pathological states.[57] This remained a core theme of Cannon's Darwinism. The "ancient pattern of response" is ready to be triggered; but at the same time, modern stress and anxiety involved particular effects of the successful or unsuccessful adaptive response. In 1936 he would describe this as follows:

> Under the conditions which prevail in civilized society the emotional disturbances of digestion commonly occur when the traditional dangers in the long history of our race are not present or urgent. Only occasionally are we greatly excited by circumstances which require supreme muscular effort. On the other hand, we may become intensely perturbed while watching a stock ticker, or seeing a race, or facing an examination, or on being wheeled to an operating room, i.e., when no effort is called for. . . . Deep down in our nervous organization, however, is the ancient pattern of adaptive emotional response. The digestive processes, therefore, may be profoundly affected by inert and idle excitement, almost as much as if the utmost physical exertion were anticipated.[58]

As in this later quote, so in *Bodily Changes*: the evolutionary demonstration rationalized emotional expenditure all the way to contemporary examples. Thus Cannon's numerous ethnographic and anecdotal references, which illustrated the social life of the emotions, conversely affirmed that the semantics of the organism's action in an environment derived from evolutionary, phylogenetic factors. As examples from "wild life" of one kind or other (ethnographies of indigenous populations, autobiographical accounts of fleeing or fighting, theories of war), they performed the "ancient pattern of response" and set the stage for a physiological explanation.[59]

The eminence of this evolutionary logic should not blind us to the independence and significance of a second, pathophysiological argument, which established a quite different way of understanding the body's interconnectedness. Despite occasional convergences, this second logic was differently argued and had different epistemological implications. Where the evolutionary argument amounted to a genetic explanation, it did not cross into the space of the structural explanation that physiology provided—the space where the question eventually boils down to "how does this work for this body here?" Cannon was careful to distinguish the two: even after tracing the evolutionary explanation and citing Herbert Spencer, William McDougall, George Crile, and Darwin himself, he nevertheless accorded it no value for understanding particular internal changes, then moved to a physiological explanation of causes and meaning. What physiology offered was a structural rationale, a hermeneutics of the body: an answer to how these bodily changes are useful, how they work, how the organism functions. It is here that he could deploy evidence from physiological experiments, case studies, and pathological states.

Above all, emotions remained disturbances from a norm. Cannon claimed that unlike psychologists, he was able to treat them as productive forces. His research "revealed a number of unsuspected ways in which muscular action is made more efficient because of emotional disturbances of the viscera."[60] Such disturbances included the following: "The cessation of processes in the alimentary canal (thus freeing the energy supply for other parts); the shifting of blood from the abdominal organs to the organs immediately essential to muscular exertion; the increased vigor of contraction of the heart; the discharge of extra blood corpuscles from the spleen; the deeper respiration; the dilation of the bronchioles; the quick abolition of the effects of muscular fatigue; the mobilizing of sugar in the circulation—these changes are *directly serviceable in making the organism more effective in the violent display of energy which fear or rage or pain may involve.*"[61] The gain from the evolutionary interpretation—that emotions are useful—was to be reconceived in terms of physiological efficacy. This gave a different sense to "disturbance": the different internal processes aimed at achieving maximum efficiency in a given environment.

How were internal processes effective and efficient? Cannon established that the sympathetic nervous system, which affects glands and the smooth muscle of the viscera without having recourse to conscious behavior, constituted the primary and the fastest response mechanism to external stimuli, especially aggressive ones. The sympathetic system "is arranged for diffuse discharge, [and] is likely to be brought into activity as a whole";[62] also, it was nonspecific, meaning that it affected the viscera in general rather than through particular neural pathways. Emotions affected the body because discharge from the sympathetic system stimulated the adrenal glands to rapidly release adrenaline into the bloodstream, which resulted in a crucial and massive compounding of the nervous discharge: "[Adrenaline] cooperates with sympathetic nerve impulses in calling forth stored carbohydrate from the liver, thus flooding the blood with sugar; it helps in distributing the blood to the heart, lungs, central nervous system and limbs, while taking it away from the inhibited organs of the abdomen; it quickly abolishes the effects of muscular fatigue; and it renders the blood more readily coagulable. These remarkable facts are, furthermore, associated with some of the most primitive experiences in the life of higher organisms, experiences common to man and beast."[63] The sympathico-adrenal system spearheaded the body's unified reaction through the major emotions.[64] From the perspective of the digestive tract alone, emotions were disruptions, changes that disarranged the normal, well-functioning, healthy machine. But that disarrangement occurred precisely in the name of whole-body efficiency in the particular response to the environment. That efficiency was urgent: adrenaline made it possible to reach the point at which, faced with an external threat, the organism could react in the best manner possible. In the most extreme of situations, it could rapidly fight or flee.[65]

Fight or flight, frequently identified with *Bodily Changes* because Cannon became the concept's great promoter, barely appears in the book. Where he included it, it was so he could handle McDougall's psychology.[66] Elsewhere Cannon wrote of "fleeing or conflict," or prioritized one of the terms. Still, he never focused on fight or flight as the only possible situation. Fight or flight instead served him as the hinge between fear and rage: the *or*, like the concept itself, signified for him the moment at which the perception of an external, disruptive stimulus generated an internal disarrangement that maximized efficiency. As significantly, it was of little interest to Cannon whether one chooses fight or flight. Each serves just as well as a response to the situation in which fear and rage emerge as the reserved and all-but-interchangeable binary options that the organism feels as emotions and deploys for its survival. This radical response was the goal of the entire complex technique of the emotions.

Extrapolating from anthropological and literary examples and from laboratory experiments, Cannon theorized that the body holds "reservoirs of power" (William James's expression) through the constant presence of certain adrenal extracts and hormones.[67] He identified these reservoirs with the well-balanced secretions that always course through the organism. Thus it was not the *appearance* of adrenaline that threw off the organism from its original practices, because adrenaline flowed through it anyhow. It was the sudden rapid rise in adrenaline that resulted in imbalances, facilitated higher activity, required the shutting down of the digestive tract, and so on. In other words, the body was prepared to react to intrusions and stimulations, and often did so, on the basis of reserves of adrenaline and sugar, which, in military fashion, the body "called forth for instant service" so it could "bear the brunt of struggle."[68]

How did reserves and physiological disturbances correlate in emotional shifts? In a 1911 article titled "Emotional Glycosuria," concerning occasions of elevated sugar content in urine, Cannon laid the groundwork for his explanation of how externally induced heightened emotional states ("great grief and prolonged anxiety") could cause permanent bodily changes, including the onset of diabetes resulting from wartime bombardment and, in one man, shock from discovering his wife "in adultery."[69] Elevation of the blood sugar content, together with heightened adrenal activity, generated an arsenal of options that resembled visible physiological disturbance. An external disruption resulted in a physiological change, including, often, the cessation of important functions. This cessation was due not to the external stimulus, but to the rapid shift brought about by the body's reserves, now activated by the sympathico-adrenal system. The perception of an external stimulus or disruption at once mobilized and adjusted this preparedness: the adrenal glands now released adrenaline in considerable excess of the normal, rapidly augmenting the adrenaline that already flowed through the body, and directing it by the sympathetic system as well. Crucially, what caused the massive response was the perception of threat, *not* the actual object perceived or the seriousness of the danger. As Cannon showed on other grounds too, the perception of external threat set off an emotional response by effecting bodily changes, but the apparatus was entirely internal, all but separated from the outside world, all but independent of the stimulus.

Fear and rage were qualitatively interchangeable, their main difference being that they involved a very different display of power. Citing McDougall, James, and Darwin on the similarity of the two emotions and their imposition of different relations to the outside world, Cannon insisted on their physiological identity: "Physiological evidence indicates that differences in visceral accompaniments are not noteworthy. There is, indeed, obvious reason why the visceral changes should *not be different*, but rather, why they should *be alike*. . . . Just because the conditions which evoke them are likely to result in flight or conflict, . . . the bodily needs in either response are precisely the same."[70] In fear and rage, and in pain as well, a sudden excess of adrenaline generates a guided chemical disequilibrium that is echoed by the nervous charge. These changes are experienced not as a simple quantitative charge, but as a major qualitative one. By virtue of the "reservoirs of power," the emotional response is overwhelming: the well-prepared, overflowing organism in the process of becoming pathological is now the only organism capable of a health-preserving response to the danger. Cannon specifically linked reservoirs with the "wild state" because "the fear emotion and the anger emotion are, in wild life, likely to be followed by activities (running or fighting) which require contraction of great muscular masses in supreme and prolonged struggle."[71]

Military metaphors were central: the body, Cannon argued, normally withstood wartime violence thanks to its preparation. The reservoirs and muscular masses were part of an arsenal to be mobilized so the organism could become a warrior. What appears to us as a variance from the norm, in contemporary wild states (notably war) or in strained modern life, is the ploy of a body that is organized so it can shut down some normal operations and act—to the point of physiological and psychophysiological derangement—with the massive force necessary for survival. In this model, the body reaches a seemingly pathological state of preparedness for action in fight or flight. Insofar as it consumes reserves through everyday stress, modern life accentuates this, with frequently pathological consequences.

In the 1915 edition of *Bodily Changes*, what generated the internal, visceral response is either unknown or insignificant: the response was nonspecific, caused by the body as a whole. Cannon concluded his next-to-last chapter with a discussion of plausible rationales for the rapid onset of emotional effects in fear and rage, and for the apparent secondary status of both the viscera and the cortex in this onset. Fourteen years later, in the 1929 edition, he would modify this approach and identify the optic thalamus as the site that triggered the sympathetic nervous system's organization of emotional response. This modification relied in part on a well-known experiment with "sham rage" that Cannon carried out with Sydney Britton. Removing the cerebral cortex of cats, they found that "as soon as recovery from anaesthesia was complete, a remarkable group of activities appeared, such as are usually seen in an infuriated animal—a sort of sham rage. These quasi-emotional phenomena included lashing of the tail, arching of the trunk, thrusting and jerking of the restrained limbs, display of the claws and clawing motions, snarling and attempts to bite . . . Besides these, and more typical and more permanent, were effects on the viscera, produced by impulses discharged over the sympathetic nerve fibers."[72] "Sham rage" recalls Henry Head's "vigilance": it confirmed the nonspecificity of the emotional reaction within the organism, as well as its nonwilled status. Even without a cortex, even without an environment, rage ensued.

: : :

What, then, are the emotions? In *Principles of Psychology*, William James famously argued that we do not cry because we are sad; we are sad because we cry. In the 1929 edition of *Bodily Changes* Cannon targeted this idea (by then widely known as the James-Lange theory of the emotions). He refused both causal routes: fear and rage neither caused the bodily shift nor were caused by it—they were names for the experience of a particular goal-oriented, instrumental, efficient bodily state.[73] "Violent affective states" were "concomitants" of physiological changes: "These bodily changes are so much like those which occur in pain and fierce struggle that, as early writers on evolution suggested, the emotions may be considered as foreshadowing the suffering and intensity of actual strife."[74] Cannon's refusal of a causal explanation is critical: to accept one would prioritize among the physiological, evolutionary, and ethnographic evidence, and would also admit that an external trigger causes the emotional response. Instead, Cannon used his different kinds of evidence to opt for coextensive explanations. The stimulus-response duet was established by ethnographic anecdotes and observations of cats in the laboratory; the evolutionary explanation doubled the ethnographic evidence, without intruding into the physiological rationale, by explaining how these actions brought forth primal biological activity. The cat dove into fear or rage at the sight of a dog because long-standing processes had made the dog appear to be its "natural enemy";[75] the same went for human beings faced with external hostility, which raised "ancient patterns" as if from deep slumber.[76] But the core of this system explaining the emotional response was physiological and systemic, not in that physiology stood apart, but in that it unified the complementary explanations into a vital internal machinery coextensive with the actor.

Emotional activity was nonspecific and bodywide, carried out by an unknown (later understood as thalamic) prompter, and it was internally generated rather than reflexive. The sympathico-adrenal system operated as a whole and affected the organism not only as a whole but also beneath the reach of the will, thereby integrating bodily actions that responded to external stimuli interpreted as threats (and, to a derivative degree, also to pleasurable stimuli). What mattered physiologically was the internal excess and mobilization that made response possible, and coefficiently generated physiological imbalance—potentially with salutary but just as easily with grievous consequences, especially in modern neuroses. According to their complex physiology, then, the emotions amounted to chemically programmed deregulation and reregulation of the organism, essential for an adaptation to new situations and capable of disequilibrium and bodily damage. What made possible the activity and survival of the individual organism in extremis also made possible its deepest suffering.

Shock and the Homeostatic Conception of the Organism

One begins to see why Cannon was reluctant to abandon his research in order to participate in the war effort; why he would later call the study of wound shock a "parenthesis";[77] and yet why, given his attention to whole-body reactions such as pain and fear, he was, for the Rockefeller Institute as well as the London Shock Committee, exactly the man to work on the subject. It also becomes clearer why he prioritized the reduction of alkali reserves and a toxemic rationale such as the one proposed by Henry Dale. For him, adrenal secretions and nervous activity compounded in shock to produce an organism singularly disabled in its physiological capacity to defend itself.

Cannon folded his research on shock in the early 1920s but returned to it in his crowning work, *The Wisdom of the Body* (1932), where he used shock at length to discuss the ways in which the "fluid matrix" of the body regulated and stabilized itself, producing what he called homeostasis. Shock provided the major pathological condition that served Cannon as an example where self-regulation failed, with "very serious," often catastrophic consequences.[78]

The Wisdom of the Body is very different from Cannon's earlier books. Its object is the body as a self-regulating whole. In it Cannon all but abandoned reliance on animals and on the evolutionary, phylogenetic rationale so central to *Bodily Changes*. He clearly wrote for a nonspecialist audience: its main arguments are phrased in a studied vernacular that would have been impossible in his earlier work, and its conclusions are capped by political claims. But the shift was more fundamental still. Cannon's *The Mechanical Factors of Digestion* offers an account of the digestive tract and turns to disturbances only at the project's closure. *Bodily Changes* conceived the organism as a generally stable apparatus and identified emotions in terms of hyperactivity and pathology—disturbances of the norm. This is no longer the case in *The Wisdom of the Body*, which is about the body's capacity to self-regulate; stability had previously been taken for granted, but now its mechanisms needed to be outlined and explained.

The book's central concepts were homeostasis and the fluid matrix. Cannon defined homeostasis, to which he first referred in 1929,[79] as follows: "The coordinated physiological processes which maintain most of the steady states in the organism are so complex and so peculiar to living beings—involving . . . the brain and nerves, the heart, lungs, kidneys and spleen, all working cooperatively—that I have suggested a special designation for these states, *homeostasis*. The word does not imply something set and immobile, a stagnation. It means a condition—a condition which may vary, but which is relatively constant."[80] The fluid matrix was specifically conceived as an update of Claude Bernard's 1859–1860 concept of the inner environment (*milieu intérieur*), and we return to it in the next chapter, when we consider how the two concepts responded to a contemporary scientific concerns regarding the correct way to articulate organismic regulation.

With homeostasis, Cannon obsessed over the brittleness and the constancy of a life under continual threat. He began *The Wisdom of the Body* with the statement "Our bodies are made of extraordinarily unstable material" and returned to that rhetorical and theoretical figure:[81] "One of the most striking features of our bodily structure and chemical composition . . . is extreme natural instability. Only a brief lapse in the coordinating functions of the circulatory apparatus, and a part of the organic fabric may break down so completely as to endanger the existence of the entire bodily edifice."[82] Over and over, the body was threatened with "disaster," the effects of water loss were "grave," alterations in the chemical balance of the blood were "hazardous,"[83] internal and external conditions were "profoundly disturbing."[84]

The rhetoric of external and internal threats and constant jeopardy marked a shift from *Bodily Changes* even if it retained the clatter of war: thus Cannon christened *sentinels* those "automatic indicators," that "set corrective processes in motion at the very beginning of a disturbance."[85] But just as the problem of the emotions had come to rest on an unstable rather than simply well-prepared interior, references to war changed toward a model prioritizing the internal defense against volatility. Following Cannon's clinical and experimental work in wartime France, shock was the urgent, exemplary counter to homeostasis, its first and most radical pathology of regulation.[86] Shock couldn't be conceived without some external aggression, but it played out only thanks to the destabilization and disintegration of the internal operations of homeostatic regulation. Indeed, whereas in his earlier writings on shock Cannon had seen it as a multisystemic failure brought on by anoxemia and acidosis and then compounded by the body's efforts to regulate itself, in *The Wisdom of the Body* shock was indicative of compounding failures of regulation. It even returned in muted form in chapters where Cannon addressed very different mechanisms—for example, the reoxygenation of the blood, the development of lactic acid, and the nervous control of other systems. Bodily changes, as in rage and pain, ought to be placed in the context of responses to external threats to the organism. Subsidiary to the first three layers of response—protective reflexes, self-healing mechanisms, and defenses against bacteria—the emotions now formed only a fourth layer of response.[87]

Cannon, in other words, was inscribing a rhetoric of instability, of the constant threat of pathologization, into normal processes. Normal processes were constant efforts to avert incipient, internally generated disaster, as if disaster lived within them. An organism was integrated to such a degree that pressure, invasion, and loss of regulation would lead almost immediately to systemic collapse. Faced with anything that came its way, the body responded as a whole and suffered as a whole. Homeostasis was the name for a shared implication of multiple processes of stabilization, integration, and regulation that made the body not an agglomeration of parts, but a unit that was always under threat as one.

Once again, the evidentiary basis of Cannon's claims had shifted: his chapters in *The Wisdom of the Body* relied by and large on physiological accounts supplemented with explanations of pathologies. Anecdotes and ethnographic references, so central to *Bodily Changes*, had lost pride of place—indeed, they were altogether absent. The evolutionary rationale survived as a genetic explanation of the rise of homeostatic devices, but Cannon invoked evolution rarely—usually in passages where he cited fear and rage[88]—and its explanatory value was limited: Cannon confirmed that adapting to a hostile milieu required bodily agencies to maintain certain stabilities and standards.

In the place of earlier approaches, Cannon accounted, one by one, for different kinds of constancy in the fluid matrix (the constancy of water and sugar content in the blood, the maintenance of oxygen supply, and so on), then outlined the systems and mechanisms that act to counter particular destabilizing trends. Shock offered the strongest evidence, followed by experiments of different types—some on animals, some involving self-experimentation, some involving X-rays, many involving surgery, but all based on induced pathology. Neither physicochemical processes nor organs were at the center of his attention; instead, he was conducting an organism-wide search for particular triggers, systems, and agencies that counteracted imbalances. The simplest, sentinels, were rather straightforward switches that pressed for immediate counterbalancing. More complex devices, such as the sympathico-adrenal system and the "insular or vago-insular" system—the mechanism of the control of insulin and its regulation of blood sugar—were agencies in the full sense of the term, in a way that the cerebral cortex could never be. In this framework, the place of the emotions changed again. The emotions had already shifted from being disturbances of normal processes to being efficient, productive disturbances that spur adaptation and survival by unsheathing the primal forces of an organism forced to save itself. Once the form of the normal allowed for bodily instability to become the basis of every logic of integration and stability, disarrangement and reregulation, emotions became bare necessities for the organism's self-deployment on occasions of exertion. They aimed at the most brutal self-presentation of the body. They were apparatuses that built and used reserves to guarantee the functioning, safety, and necessary leeway for the organism's operations. For Cannon, the multiplication of sites of control and regulatory activity dispersed the organism's integrative apparatuses and established the singularity and wholeness of each body: because so many structures were needed and used, because they operated by way of reserves and advance preparation, integration was not only assured; it was total: "The integrity of the organism as a whole rests on the integrity of its individual elements, and the elements in turn are impotent and useless save as parts of the organized whole."[89]

With the advent of cybernetics, many of these agencies and regulators would come to be thought of as producers of "negative feedback," as though they were literal on-off switches, working alone or in groups. The merits of this method notwithstanding, Cannon's argument was not nearly as technologically inflected as the cybernetic reception of his work suggests. The central metaphor employed throughout the book is that of storage, overflow, and inundation.[90] In the emotion theory, storage had allowed the rapid deployment of adrenaline. Here storage and inundation helped guarantee a "margin of safety" by allowing the organism to avoid tension within certain precise limits.[91] Cannon insisted that the organism pursued economy, and the fluid matrix and its coordination or governance worked specifically toward economic aims, particularly the avoidance of loss. The organism used storage, overflow, and inundation to hold danger at bay.[92]

Margin of safety seemed a promising term for another reason. In a field still very concerned with the mechanist-vitalist distinction, the margin-of-safety argument allowed for Cannon to emphasize his credentials as a mechanist. Given the explicit holism of homeostasis, Cannon's citation of the Hippocratic tradition of a "healing power of nature," and his occasional support of phylogenetics,[93] the possible accusation of Romantic holism must have been a point of concern. The margin-of-safety argument enabled rebalancing insofar as it expanded and improved on mechanical explanations by suggesting that storage and inundation allowed for a range of the normal, thereby abandoning traditional and strict mechanist normality for something closer to statistical normality.[94] At the same time, the term *safety* carried with it all the anxieties of violence, endangerment, and brittleness that so terrified Cannon.

: : :

Twice in *The Wisdom of the Body*, including at the very beginning of his presentation of the fluid matrix, Cannon pointed out that the body was covered by a husk of dead material, the outer layer of the skin. Individual life includes and develops that layer, yet it only begins within it.[95] This figure of a nonliving container of life should be by now a familiar trope in Cannon's work, at one with the "reservoir of power" he had physiologized out William James's psychology, and at one with the images of storage and of the overflowing or inundated body. The sense that ectoderm and the liquid cover of the eye closed and enclosed a unit of life by serving as its supplement, thanks to which this life remained separated from the outside world, was an essential trope for the concept of a body that produced itself insofar as it regulated itself. Even this dead cover was generated by the internal environment, providing a minimal yet absolute barrier. Like the emotional reactions of fear and rage, which were internally generated and deployed as if without contact with the outside world, as if triggered not by an external threat but by a physiologically nonconscious and unwilled perception, followed by the encoding of an object as threatening, the container or reservoir operated only internally and regulated itself in a diffuse fashion strictly separated from the outside. That outside world began beyond it and also, biologically recalibrated as if beyond all touch, within it. Joined of course to this figure of separation was a second trope: the language of conflict and peril, of fight or flight, which cast the environment as always hostile, either because of its primal violent character or because stressful society depleted the reservoirs, leaving anguished neurasthenics in its endless wake.

It is hard not to notice that in Cannon's war, as in his peace, hardly any wound was worth mentioning, hardly any breach of this container from the outside. Endless hostility, destabilization, and deregulation left little room for actual injuries, holes, or even chinks in the armor. Life within persisted; it was destabilized then restabilized. It came undone not because of any intrusion, invasion, or pressure, but only because regulation could not always hold mechanically and the margin of safety did not always suffice. Continuity with the world was assured, but disruption mattered because it forced the systems of regulation to overload and crash. When healthy, the organism was inundated; essential nutrients and materials flooded it and overflowed. It operated in full control of itself and in disinterest toward external pressures or dangers. Yet an organism not marked by internal excess was already a threatened organism, one for which regulation had a defensive posture and purpose, one for which regulation signaled the incoming threat. (An aging organism was no better: its regulatory apparatuses were out of tune and aided—indeed, constituted—its decline.) The ostensible absoluteness of the body's ultrathin ectodermic coat reflected by way of contrast the uneasy, elaborately manufactured, and anguished equilibrium within. This barrier was the matter before which tumult reigned and behind which integration and the homeostatic body ruled, cohabited, and spoiled.

Before it the world became a world of threat; behind it, regulation reshuffled the threat and, to contain it, attempted to adjust and reregulate itself, with often catastrophic consequences.

We end this chapter, then, with the image of an internally generated, regulated, unstable life, a life that is ostensibly freedom and regulated excess, a life that, in tune with the external world of threat, is peril itself. It is this image of life that we will slowly deconstruct in subsequent chapters.

6

The Organism and Its Environment: Integration, Interiority, and Individuality around 1930

> We are always looking for metaphors in which to express our idea of life, for our language is inadequate for all its complexities. Life is a labyrinth. But the labyrinth is a static thing and life is not static. Life is a dance, a very elaborate and complex dance! The physiologist cannot consider the dance as a whole. That is beyond his experimental power. Rather, he isolates a particular corner or a particular figure. . . . The shortcoming of his method becomes fairly evident when he seeks to relate his corner to another in a far distant part of the dance. Moreover, even should he seek to treat the organism as a whole he is still almost bound to consider it as an "individual," complete and separate in itself, shut off from its environment and its history, born, as was Minerva, armed and fully equipped from the head of Jove. But in fact living beings are not so. "Individuality" comes into prominence only in their more differentiated groups.
> —Charles Singer, *A History of Biology*

Integration, Interiority, Individuality: Claude Bernard and the 1920s

Accounting for the advent of the integrated, brittle body, hardened against catastrophe, forever sinking toward ruin, this study has been reacting to the scent of "the individual": the individual human being whose digestive, endocrine, circulatory, and nervous systems, especially in their mutual influence, provided the ground for understanding the particular response this individual here has to particular external pressures; the indivisible body that, thanks to the interdependence of its systems, is endangered even by slight injuries; the patient who, perforated from the outside, finds himself gripped by internal shock; the case whose medico-psychiatric condition forces a rethinking of medical categories; the injured mind obliged to compensate for its survival.

Since the late nineteenth century, for biologists no less than for philosophers, individuality was a problem to be reframed and perhaps overcome. In most strands of biology, by 1900 it had been largely set aside in favor of studies of population- or species-wide phenomena.[1] Cell biology in particular seemed to close off human individuality. Rudolf Virchow, shortly before his death in 1902, boasted at the Medical Congress of Rome, "There are no more general diseases, but only diseases of the organs and diseases of the cells."[2] As a category the individual could be dispensed with.

Not so in physiology after Claude Bernard, and especially in the period under study here. In 1859, when Bernard defined experimental medicine and the idea of the internal environment, he attached it to the possibility of individuality and freedom. *La fixité du milieu intérieur est la condition de la vie libre et indépendante* became the motto of physiological study: "the fixity of the internal environment is the condition of free and independent life."[3] Bypassing for now the liberalism latent in the motto, we focus in this chapter on Bernard's claim that we live primarily in our internal environment and that biological life relies on internal stability for health and relative independence from its external world.

Bernard was explicit in identifying this *milieu intérieur* as "a true product of the organism." It is generated by the organism, yet it "preserves the necessary relations of exchange and equilibrium" as if the milieu, rather than the organism, were the real agent: "As the organism grows more perfect, the organic environment becomes specialized and more and more isolated . . . from the surrounding environment."[4] And elsewhere: "All the vital mechanisms, however varied they may be, have only one object, that of preserving constant the conditions of life in the internal environment."[5] Thus stability is the effect of reactions and interactions between the different inner realms of embodied life, with organs, blood, and lymph central among them. Life, as creation, becomes freer the less it is burdened by external stimuli and constraints, and the more it is the self-creation and self-regulation of a living machine.[6] It also thus creates individuals internally: essential to the science of experimental biology is the assertion that "the internal environment, created by an organism, is special to each living being. Here is the true physiological milieu."⁷ Each organism journeys down its own path, thanks to the stabilization apparatuses that produce it.⁸

This stability of the internal environment and its status as the ground of life became a standard point of reference in the 1920s.⁹ Once physiology attached the object of its research to integration and individuality, Bernard experienced a revival. He was retroactively anointed the father of the science: physiologists began to draw a line from Bernard to themselves, at times without intermediary figures, at times claiming that they were simply proving him right.

Consider Ernest Starling's "career in regulation." In his *Elements of Human Physiology* (1892) he spoke at most of an "adjustment" of organs to one another and a reflex-style adaptation of the organism to its external environment.¹⁰ The body was a "colony of amoebae," which Starling immediately compared to "colonies of men"; it could also be assimilated to a heat engine.¹¹ If there was no denying its organization, there was also no place for speaking of integration.

Sherrington introduced the language of integration in 1905, but at first the word stuck mostly in neurology. The term associated with hormones and bodily fluids was *regulation*. By 1907, in *The Fluids of the Body*, Starling had turned to a commonwealth rather than a colony metaphor to describe the body; he emphasized that bodily regulators work to provide "an average constancy of environment," but he cautioned that this constancy was relative:¹² "Constancy of any bodily condition is unattainable in the presence of the varying conditions of our environment, and is indeed not compatible with our conception of life."¹³ Starling still did not mention Bernard; nor did he treat the internal environment—or "fluid balance"¹⁴—rigidly. In 1912, rather than use Bernardian language, he portrayed the organism in relation to external threats to which it had to adapt by importing an evolutionary and environmental pressure: "The sole condition for the survival of the organism is that any act of disintegration shall result in so modifying the relation of the system to the environment that it is once more restored to the average in which assimilation can be resumed. Every phase of activity in a living being must be not only a necessary sequence of some antecedent change in its environment, but must be so adapted to this change as to tend to its neutralisation, and so to the survival of the organism."¹⁵ Starling's organism was essentially reactive.

By 1925, however, the internal environment was no longer conceived as simply composed of a system of fluids or organs. What managed it was the precise chemical arrangement, the hormonal regime on whose basis the organism remained autonomous from its surroundings. In 1929 Cannon quoted Bernard's disciple Léon Frédéricq to the effect that "each disturbing influence induces by itself the calling forth of compensatory activity to neutralize or repair the disturbance,"[16] and the French physiologist Charles Richet to argue that "the living being is stable. It must be in order not to be destroyed, dissolved, or disintegrated by the colossal forces, often adverse, which surround it. . . . The living being is subjected to all manner of impressions but resists them all; it is continually renewing itself and is always the same."[17] Others agreed. John S. Haldane declared, of Bernard's quip, that "no more pregnant sentence was ever framed by a physiologist." For Haldane, organic regulation was nothing less than "the essence of life."[18] His magnum opus *Respiration* was merely "an attempt to follow out in regard to blood reaction and oxygen supply the line which Bernard indicated."[19]

When Joseph Barcroft considered the theme, he at once complained that the Bernard motto on an organism's freedom had become so standardized as to be "grotesque."[20] Sixty pages later, he nonetheless concluded that the principle basically held: personal experiences and laboratory work confirmed that all mental activity depended on it. For the individual who experienced disorders of the internal environment, the mind was burdened, even lost.[21] For any of us to claim to be an individual, to have higher mental functions, to be free, it was essential that the internal environment be capable of sustaining this stability.

In other words, as for Bernard the precursor—at least the Bernard that had been reinvented as a precursor—so for the protagonists of this book: in the conceptual matrices they built around the integrated body, interiority and individuality went hand in hand.[22] This chapter studies the category of "the individual" as it appears in the web of concepts and categories to which these scientists fastened it, key among them "interiority" and "integration." If physiology cast itself as a science of the individual; if it discovered in Bernard a precursor worthy of extended attention; if it proclaimed regulation, balance, equilibrium, coordination, and integration to be essential components of individuality; if it raised the individual to the status of a major category—what kind of an individual was this? How did integration and interiority play into this individuality?

To anticipate, a very simple version of the answer went like this. Each body, male or female, has a different history, it is differently nourished, it is attached to its environment and to others through emotions that have physiological effects and also social links, it has faced different dangers and emerged from each of them differently, its hormonal production differs in composition and equilibration, it is faced with an environment that is singular to it: as a result, it is fundamentally individual. It enters medical or therapeutic practice as an individual, and it responds to pressures—especially injuries and pharmacological treatments—in a manner that puts its individuality on the line. By 1930, this negotiation of individuality had become the platform of integrationist physiology and neurology.

Yet to affirm this is merely to begin to broach the subject of individuality, for its announcement is by no means tantamount to its clear theorization. First, different thinkers assigned the concepts of interiority, integration, and individuality different sites and relationships. Second, while asserting the physiologically structured and integrated human being as an individual, they mostly deprived him or her of the agency that we would expect to see accorded to individuality. Instead, that agency was granted to the internal environment, the equilibrated totality of systems through which this individual responded to the outside. Although the patient or organism was called an individual, this individuality operated beneath any will, consciousness, or selfhood. Even external stimuli came to matter principally because they were endosomatically regulated. It was only as they obliged elements of the internal environment to respond, and other elements to respond to the responses, that they set up and broke down the singularity of the organism. For Haldane, Cannon, and Henry Dale, individuality had to be accorded to the body's multisystemic structure, not to the body itself. Cannon offered the now-standard term for the equilibrium of this structure: *homeostasis*. Dale coined his own term, *autopharmacology*, for the study of the release and transmission of chemicals found in bodily tissues. These endogenous substances—including histamine, endorphins, and adrenaline—generate reactions, modulate functions, and act as regulatory signals for the organism in times of need.[23] Similarly, for neurologist-psychologists like Henry Head, Kurt Goldstein, and Alexander Luria, individuality emerged as a problem with the disintegration of the organism and the attempted reorganization of the mind-body ensemble amid disorder.

To study this, we first explore the spread of other approaches to integration and whole organisms, especially in biochemistry. Then we build on Cannon's and Goldstein's examinations while introducing Dale's autopharmacology and Luria's studies of the neurophysiological grounds of behavior during disintegration. As a point of connection, we ask what the tendency to look only inside the organism did for the sense of its existence amid social norms and other organisms.

We are motivated in part by a question that Georges Canguilhem posed in his 1952 essay "The Living and Its Milieu."[24] Canguilhem described the early twentieth century as a period in which, from evolutionary biology to psychology to Taylorism, organisms had come to be conceived as dominated by their milieu. This had not been always so, he argued: from the early Enlightenment until the early nineteenth century, organisms had been conceived of as largely free of their milieus and as capable of forming them. With the advent of Romantic holism, and still in accordance with the political thought of the period, biology began to imagine forms of the organism's subservience to the milieu. By 1900, the individual was imagined across a series of disciplines as externally defined and formed, and as deprived of autonomy and thus of genuine individuality.

Canguilhem argued that the idea of the living's submission to its milieu had begun to ebb with Goldstein and Jakob von Uexküll, who postulated the organism's relative autonomy. For them, he argued, the relationship was a *débat*, an *Auseinandersetzung*, a face-to-face between the organism and the milieu, in which the milieu provided a set of demands but the organism was at once the center and the centrifugal force "accommodating" and even "dominating" this milieu.[25] Canguilhem took advantage of the etymology and Pascalian interpretation of the environment as *mi-lieu*, "middle" or "center-place": the organism is itself *mi-lieu* to the extent that is a center. The organism submitted to the demands of the milieu only when it was sick; otherwise, it remained largely free of the milieu's pressure: "A life that affirms itself against the milieu is a life already threatened. . . . A healthy life, confident in its existence, in its values, is a life of flexion, suppleness, almost softness."[26] In this formulation Canguilhem allowed for two concurrent claims, ontological as much as political. The individual living being is autonomous vis-à-vis its milieu, and the two coexist in harmony. The organism finds this autonomy in being normative, in being capable of exceeding the demands of the milieu without struggling with it. The ideal is almost Spinozist: a relationship of coextensiveness. Canguilhem quietly advocated for physiology's concerns. Yet he never hinted that physiology and related disciplines had specifically pursued an identification of the individual, healthy human being with this freedom. This is all the more surprising because he was an authority on Bernard's lectures on experimental medicine and intimately familiar with Cannon, Starling, and Haldane's work. Attributing a "shift" to Goldstein, Canguilhem overplayed his hand, indeed at a moment when the "new" physiology, with its focus on interiority and freedom, was being displaced by biochemistry, genetics, and molecular biology.

Contrasting this lineage to Canguilhem enables us to detect more clearly how tenuous and problematic the concepts of individuality and interiority were. The more firmly physiologists grasped them, the more they crumbled in their hands; between the internal environment, the borders of the body, the forms of integration and regulation, and the results of injury and destabilization, there was much to negotiate. Was individuality an epistemological category for the researcher, or a properly ontological one? How is it that both the external milieu and internal regulatory mechanisms kept arranging and pulling apart the singularity of the body?

The Competition

By 1925, *integration, interiority*, and *individuality* had awoken in the language of physiology and were routinely used against both traditional vitalism and mechanistic physicochemical reductionism. But integrationists in physiology were far from alone in postulating that the organism was a whole or a regulated being. Embryology, after Hans Driesch's research on the division of the sea-urchin embryo in the 1890s, sought mechanistic and materialist methods for managing the capacity of embryos to develop in an organized fashion. This field of research, from Sven Hörstadius to Hans Spemann to G. E. Coghill, offered one approach to the study of general organization.[27] Another arose slightly later, in the 1930s, with Ludwig von Bertalanffy's syntheses of genetic and structural forms of the living, which eventually led to his systems theory for biology.[28] Related to such tendencies—as well as to the broader crisis of mechanism, vitalism, and finalism, the main paradigms in biology—was von Uexküll's famous study of the relationship of the inner world and environment of insects and animals, whose conclusions he extrapolated to humans and which philosophers from Martin Heidegger to Giorgio Agamben have used to introduce concepts of worldhood and openness.[29]

Within physiology, a biochemical experimental approach gained ground, exemplified by the work of the Harvard physicist and physiologist Lawrence J. Henderson and other researchers, including the English biophysicist A. V. Hill and the American cell biologist Fenton Turck.[30] This came in sometimes explicit competition with the conception of individuality in the physiology of regulation. The triumph of biochemistry and molecular biology in the post–World War II period relied greatly on advances in biochemistry, cytoarchitectonics, and evolutionary biology. In what follows we focus on Henderson's approach because he was at once closest to and farthest from our integrationists, and also a most powerful figure in American academia.

Something of a Renaissance man, Henderson began as a chemist when he graduated from Harvard in 1900. In the 1910s he identified as a "biological chemist" of the cardiovascular system, befriended philosophers Josiah Royce and Alfred North Whitehead, and began thinking of the philosophical implications of his work. In 1927 he founded the Harvard Fatigue Laboratory at the Harvard Business School, where he collaborated with scholars, economists, and political thinkers from all over Harvard.[31] When he concluded his career in the late 1930s, he was mostly writing on sociology.

Henderson was particularly influential in the development of a regulation-oriented biochemistry—notably of the blood—that was mechanistic and reductionist through and through. In his 1913 article "The Fitness of the Environment," he wrote:

> We cannot imagine life which is no more complex than a sphere, or salt, or the fall of rain, and, as we know it, it is in fact a very great deal more complex than such simple things. Next, living things, still more the community of living things, are durable. But complexity and durability of mechanism are only possible if internal and external conditions are stable. Hence automatic regulations of the environment and the possibility of regulation of conditions within the organism are essential to life. . . . Certain it is from our present experience that at least rough regulation of temperature, pressure, and chemical constitution of environment and organism are really essential to life, and that there is great advantage in many other regulations and in finer regulations. Finally a living being must be active, hence its metabolism must be fed with matter and energy, and accordingly there must always be exchange of matter and energy with the environment.[32]

The difference of this passage in style and argument from the work of the physiologists we have been studying is stark. Not only did Henderson dismiss one and all critiques of mechanism and instead seek biological properties firmly under the physicochemical sun; even more fundamentally, his approach began from simple elements and basic organisms and moved up in degrees of complexity and durability. Rather than look at complex systems of organisms, humans first of all, he started off with cells and their elements and nutrients. He then postulated fundamental analogies of organized being at different levels: if organization were the fundamental principle, it could be recovered or replicated at different levels, and only the intensity of its complexity would accrue. These principles remained central to Henderson's research all the way to his later sociological writings.

The grandeur of Henderson's theory resided in this argument: the operations within the body are fundamentally physicochemical operations. Nothing—no mediator, no operational structure—links them to the whole, except the fact that the whole is organized and equilibrated by them. Thus, integrations occur at every level (molecular, cellular, intercellular, organic), but there isn't a group of interlocked systems that affect one another and onto which agency is to be displaced. The whole organism remains physicochemically integrated through the protoplasm of its cells, that "system of exquisite sensitiveness" whose molecules and ions Henderson studied in the laboratory.[33]

Henderson's main problem was not a beleaguered, pathological organism, but the experimental complexity of working on protoplasm which involved a large number of variables:

> The large number of components of protoplasm is a condition which seriously weakens our methods of attack upon the problems of general physiology. But also . . . it increases the complexity of these problems. For, in a system of n variables, each variable will in general be a function of the other $n - 1$. Thus a change impressed upon one variable will directly involve changes in all the other $n - 1$ variables. But then the change in each of these will involve secondary changes in all the others, and so on indefinitely. Hence, the number of secondary changes increases very rapidly as n increases. When, as in protoplasm, n is great, the complexity of the analysis becomes very great.[34]

This challenge obliged the use of nomographic exposition to explain the balance of possible and mutually dependent variations; the whole body owned molecules, elements, and their cooperative organization, yet these elements, in their exchanges and variations, owned the whole too. Interiority was a physicochemical quality, and the milieu was reduced to stimulation that alters the given but does so mostly within the confines of a norm determined by the complex integration of these limited variables.

Henderson followed the above passage with this analogy: "The nature of the case will be more readily appreciated by reference to another

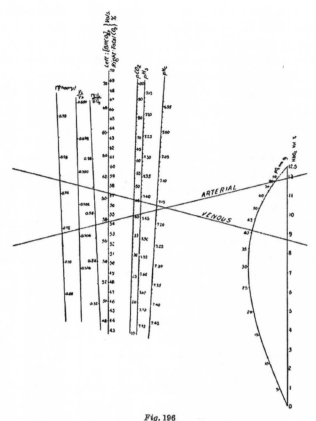

Fig. 196
Blood of Pernicious Anemia: Partial Recovery

FIG. 6.1. Henderson, *Blood*: diagrammatic organization of blood equilibrium in pernicious anemia, including details on buffers and divergence. Lawrence J. Henderson, *Blood: A Study in General Physiology* (New Haven, CT: Yale University Press, 1928) (public domain).

branch of biology in which our intuitions are better practiced. Consider a human society of n individuals and let n equal successively 2, 3, 10, and 20. When $n = 2$ or 3, a short story will sometimes suffice as a description of the society. When $n = 10$, it is doubtful if even the greatest poets or novelists have ever successfully described a single instance. When $n = 20$, a description is possible only if most of the individuals become mere puppets."[35] Clearly, Henderson had the social analogy at hand—indeed, he quoted a three-page-long passage of the sociologist Vilfredo Pareto's mathematical suppositions to support mathematical systematization and integration at all levels.[36]

Henderson's word *puppets* brings us back to his physicochemical mechanism, which he was willing to extend to society as well as to protoplasm. Though he by no means refused difference between individuals, he did deem it insignificant in biological research. To an audience of dentists in 1926, he declared: "If blood were drawn from each one of you individuals and were subject to physical analysis, . . . it is apparent that the blood would be practically comparable between all of you. Aside from slight fluctuations in the level of the carbonic acid, or something of that sort . . . , the most important variables of all are left essentially unchangeable."[37] Similarly, in a letter three years later to David L. Edsall, dean of the Harvard Medical School, Henderson described some of the goals of the Harvard Fatigue Laboratory concerning the "generalized conception of the total physiological state of individuals," emphasizing that his work concerned "stationary states" and transitions from one such state to another.[38]

Given these priorities, the term *puppets* is more apt for Henderson's account than it first appears. Divergence between organisms in Henderson's work was not constitutive of the multiplicity of minor, constantly rearranging differences, but rather was placed within a preconceived buffer zone of acceptable deviation from a statistical average. There is little indication that individuals are anything but. Henderson was nevertheless making an often implicit claim that individuality is mere quantitative difference, in biology as in sociology, and that everyone in a particular class under study was to be examined as a puppet of common elements that animated them more than difference within the organism ever could. There was little evidence of a subjecthood conjoined through the equilibration of systems or anything other than mutual coding at the cellular and molecular levels.

Henderson also attempted to guide physiology to biochemistry by rewriting its history. He was instrumental in the renewed attention to Bernard and the *milieu intérieur*. He admired Bernard profoundly, facilitating the translation of Bernard's *Introduction to the Study of Experimental Medicine* into English.[39] In *Blood*, he hearkened back: "Claude Bernard held and clearly explained that it is the task of general physiology to study the phenomena of life which are common to animals and plants. These phenomena are physical and chemical phenomena; they are to be investigated by physical and chemical methods; they are conditioned by the same physical and chemical forces that may be discovered among inorganic phenomena. But in the living being they are also always harmoniously organized and integrated, and this is their most striking characteristic."[40] Henderson distinguished physicochemical processes in the blood as the dominant factor in the fixity of the internal environment. This was correct in a literal sense: Bernard had at times equated the internal environment with the blood.[41] (Bernard had not, however, stuck to this claim throughout his writing; pace Henderson, moreover, as Canguilhem also argued, Bernard could not be consistently described as a mechanist.)[42]

Regardless, Henderson's identification with Bernard, his crafting of physiology's history so Bernard would lead directly to him, and his occasional conflicts with colleagues over his legacy offer us a classic case of the invention of precursors, and the production of authority.[43] One of Henderson's great achievements was indeed to sublate this authority into his demonstration that no agency, no sequence of forces, no subject was needed to account for the functioning and stability of blood. Like Turck, Hill, and Bertalanffy, who also looked beneath the body's systems and beneath the organ to recover the multiple elementary actors who worked with one another to achieve a functioning scenario, Henderson saw little value in foregrounding divergence as constitutive of the making and stabilization of the internal environment. This was perhaps because Henderson did not care to involve behavior and psychology in the body's integration. The result was an individual organism deprived of its individuality, thanks to the scientific activities that recaptured its physicochemical stability.

Thus, in his experimental as in his historical and institutional work, Henderson cast the subject of biological organization and the agent of free and independent life as the totality of cells and nutrients conceived as a container, full and active in constant economic exchange, and capable of creating a buffer. It was decidedly not the equilibrated and fundamentally unstable totality of systems that is traversed by elements and affected by stimuli. As with related problems of behavior, constitution, and type, Henderson considered it imperative to demur on individuality until "the facts can be usefully brought together."[44] For some contemporaries, including J. S. Haldane, such a vision of the body corroded the body's united and individual quality, and at the same time obscured— even negated—the possibility and meaningfulness of interiority too. It threw the science itself into doubt.

Haldane: Organic Unity and Reactions

When Lawrence Henderson published *Blood* and prefaced Bernard's *Introduction to the Study of Experimental Medicine*, several of his colleagues took offense—among them Haldane, and Yandell Henderson of

Yale University. Yandell Henderson went first: he wrote a febrile, vituperative piece in *Science*, attacking Lawrence in personal as well as scientific terms. His scientific critique revolved around Lawrence's ostensible failure to cite Haldane and also around his expansion of chemistry into general physiology, in what Yandell considered not only a category mistake but also a metaphysics inserted on the sly.[45] Yandell further accused Lawrence of using "weasel words," such that if an experiment had adverse results, the researcher could return to the initial claim and present it as meaning something different from how it was understood initially. The original statement could thus be shown to have foretold the result of the experiment.

Lawrence was livid. He threatened to sue *Science* for libel, and the editor responded by concurring with Yandell that Lawrence lacked taste.[46] Lawrence wrote letter after letter to inform anyone who would listen that he was astonished at the implication of intellectual dishonesty and arrogance by an ill-tempered and excitable colleague, but that he was not in the least angered by the review. By and large biochemists accepted his stance.[47] Lawrence's demand that Yandell make amends led the two into mutual recriminations that hide the fact that the scope of the discipline was what was at issue.

In the various charges, Lawrence's power and his disciplinary and personal condescension are clear, as is Yandell's perception of the threat in a resurgent, bold, and institutionally powerful biochemistry:

> Yandell (January 31, 1929): "Certainly no one who ever lectured in Yale University before talked as far down to his audience as you did. This tone of condescension seemed to me the more surprising when I came to realize that most of the new material in your book is based on the recent work of Bock and Dill, and that that work has as one of its essential and fundamental conditions the use of the 'virtual venous CO_2 pressure,' an idea produced in a Yale laboratory. Just suppose that I were to come to Harvard to talk in a condescending manner about work that was partly founded upon an idea previously produced at Harvard without ever mentioning that such was the case."

> Lawrence (February 5): "I am neither angry nor seriously annoyed because your accusations are so far from the truth that they make the same impression on me as if you had said I had stolen your pocketbook. I am past the age of being worried by such things. But you should make amends for what others

regard as an insult, even though I regard it as a mere manifestation of ill temper. There are two things in this affair which far transcend my understanding of human nature. One is that a man like you should have published such an article before saying a single word in private. The other is that after the publication you should suppose that the matter can be settled in private. It can't be done."

Yandell again (February 7): "As to 'arrogance' the jury is overwhelmingly against you, for as one of the most eminent of our colleagues, in another university and another field of science, remarked to me after reading my review: 'Everyone is afraid of L. J. Henderson.' Others cannot interpret your attitude toward less profound thinkers except as intentional."[48]

Lawrence, meanwhile, loath to insult Haldane further, included copies of his letters to Yandell when he wrote to Haldane and renounced ever having slighted him. Haldane had a sense of humor about the personal dimension of the matter. He remarked that Yandell had probably increased Lawrence's book sales, then went on to back up almost every charge Yandell had laid—"It is true you say various 'weasel words' (a delightful euphemism which I never met before)"[49]—and compounded the charges by decrying the meaning of *Blood* for physiology as a discipline. In Lawrence's account, he reckoned, the blood had suddenly been deprived of its relation to respiration, while organs and organ systems had given way to a biochemistry of protoplasm that failed to acknowledge the multisystemic complexity of the internal environment. For Lawrence it was all a matter of the stability of hydrogen ions and molecules; he didn't understand organ systems or their regulatory mechanisms.[50]

In his public response to Lawrence's book, which he first delivered before the British Physiological Society, Haldane denounced Henderson for making "very far-reaching" assumptions and, of course, for failing to understand Bernard. After delivering the lecture, he sent it to Henderson, and the two politely agreed that protocol and personal matters had been handled with the necessary tact. Haldane then published his ominous warning in both *Science* and the *Journal of Physiology*. He listed the scientific charges: failing to understand the discoveries regarding bodily fluids and organ coordination that had taken place in the seventy years since Bernard, beginning not from higher organisms but from unicellular ones and hence missing the complexity of integration, prioritizing protoplasm, and rejecting any physiology that did not submit to physicochemical reductionism.⁵¹ Henderson may have been claiming his due credit that with biochemical methods general physiology had become a mature science, but he suffered from a terrible literalism when it came to integration. For Haldane, even the cooperation of kidneys and respiratory organs was essential before Henderson should be able to turn his attention to protoplasm, yet Henderson acted as if such regulation and organization were unworthy of attention.⁵²

Haldane's aggressive tone reflects the importance of constancy, regulation, and integration in the sense emphasized by the British physiologists.⁵³ A Scottish physiologist who taught at Oxford and was renowned for his work on respiration and blood, Haldane even coined the phrase "the new physiology" in 1916 to celebrate recent British achievements. In the annus mirabilis of physiology, 1905, he demonstrated that bodily chemistry was predominantly responsible for the control of respiration, and subsequently that carbon dioxide (not oxygen) in the blood played the greatest part in this control.⁵⁴ In 1914 he embarked on studies for the Medical Research Committee on "shallow breathing," a condition in which limitations on breathing produced oxygen deprivation in the blood and uneven ventilation of the lungs, with a serious threat to life in the short term and further symptoms in the long term.⁵⁵

Having invented decompression chambers for divers and "mine-rescue apparatuses," Haldane began to build some of the first gas masks, and he tested the bodily effects of different poisonous gases on his own respiratory system. To understand how each gas actually felt, he self-experimented, repeatedly plunking himself, his collaborators, and his (grown) son in steel pressure chambers to calculate alveolar air under normal and abnormal circumstances, at times to the point of "stupefaction."⁵⁶ Comparing effects, the lore goes, allowed him to identify what gases the German army was using on the Western Front.

On the basis of the "new physiology," Haldane argued against mechanism and vitalism, and pursued, in his Silliman Lectures at Yale in 1916 as in his 1922 *Respiration*, the possibility of a language of coordination and arrangement. He identified Bernard's internal environment with the interaction between organs and organ systems, where the transport, use, and sharing of elements received from the outside world occurred.⁵⁷ During the war, Haldane began a very friendly exchange with Lawrence Henderson that immediately made their contrast perfectly clear. Henderson had written a critical review of Haldane's antimechanism in *Science* in 1914, and Haldane returned the favor, brutally and without naming his target, in a full-frontal assault on biochemistry and its mechanistic suppositions in 1916:⁵⁸

> The attempt to analyze living organisms into physical and chemical mechanism is probably the most colossal failure in the whole history of modern science. It is a failure, not, as its present defenders suggest, because the facts we know are so few, but because the facts we already know are inconsistent with the mechanistic theory. If it is defended it can only be on the metaphysical ground that in our present interpretation of the inorganic world we have reached finality and certainty, and that we are therefore bound to go on endeavoring to interpret biological phenomena in the light of this final certainty. This is thoroughly bad metaphysics and equally bad science.[59]

Haldane's later, forceful restatement of his critique could not be much of a surprise. However quaint it may seem that Haldane would go to bat over Bernard in 1929, behind the veneer of that philology lay the threat of a resurgent and mechanistic biochemistry. Haldane's fighting stance served as a dramatic way of refuting biochemistry's pretense to experimental, nonmetaphysical rigor and spared him having to defend his own holism, which he knew to be scientifically unpopular. His long description of organismic responses and regulatory functions furthers this precise direction: without mediators, Henderson's integration at the level of blood protoplasm mistook the foliage for the forest and forgot the trees too. In his letter on *Blood*, Haldane directly complained to Henderson that "the mere existence of a physico-chemical system does not give any account of" the living body's functions.[60] Henderson's effort to construct authority through a newly invented historical lineage, Haldane reckoned, fundamentally threatened the priority of pathology and of the study of the organs and their synchronization that would be imperative for a science of individual systemic wholes. Physiology itself needed to be understood in a specific *regional* sense that began with and never surrendered to objections regarding the organism, its top-down organization, and the interweaving of different reactions.

Haldane's appropriation of Bernard was obvious to him: his work more than anyone else's, he thought, was the full-grown child of Bernard's pregnant sentence. That he in turn had opened a space for a science of systemic wholes was central to Haldane's own recent work and to his 1922 magnum opus *Respiration*. Lawrence Henderson's physicochemical mechanism treated the body's organization and integration as the products of intra- and intercellular interactions that were largely indifferent to the broader systems and the individual as these systems' interconnection.[61] In comparison to Henderson, Haldane thought divergence and pathology did matter; in fact, his very concepts of integration and individuality depended precisely on the disruption—especially the reversible disruption—caused by pathological circumstances.[62] The organism was nothing if not the internal environment's constant and active reorganization around stimuli and disruptions.

Respiration was largely devoted to establishing the functioning of respiration and its effects on the organization of the blood, and Haldane found much inspiration in wartime whole-body pathologies, such as soldier's heart, shock, and shallow breathing.[63] To these he added his research on air quality in mines (which had made him a close supporter of the Miners' Union since the turn of the century and had led to his composition of several Parliamentary Papers),[64] in ships, at high altitudes, and in compressed chambers.[65] Argumentatively, *Respiration* was deeply devoted to the molecular study of abnormal conditions and pathologies, and even of simply reversible interference. In six of his fourteen chapters, Haldane discussed the relationship of the blood to oxygen and carbon dioxide, as well as to respiration in general. Yet he spent just as many other chapters on pathogenic circumstances: the want of oxygen, effects of gas secretion in the lungs, dangers associated with high altitude and high air pressure, and abnormal air and its effects. Haldane was particularly concerned with the way organs affected one another and with the coefficient changes resulting across the body from disturbance, especially on the nervous system. This sense of normality and pathology was key.[66]

A physiological reaction was an effort to stabilize and arrange a stimulus; yet by the same token a reaction was both localized in terms of its origins and altogether global. The entire system was at stake. Neither the study of the normal nor that of the abnormal was sufficient in and of itself: individuality between different beings and even across time was important. "The frequency of breathing varies considerably among normal individuals, or in the same individual at different times; and it is easy to vary the frequency while leaving the depth of breathing to regulate itself in a natural manner."[67] Haldane's version of the new physiology, his holistic response to the new biochemistry, was thus grounded in research on the maintenance of stability between different parts that could as easily retain this stability as they could ruin it. His description emphasized the organism's internal coordination of its give-and-take with its environment rather than the give-and-take itself.

What mattered was the capacity of organs to reestablish constancy with their immediate surroundings within the organism—that is, with the blood and other organs. Rather than writing of systems, as many of his colleagues did, Haldane wrote of *centers* and *reactions*. The choice of terms allowed for individual organs and organ systems to be identified as centers that were in constant interrelation. To paraphrase Canguilhem, for there to be an internal environment, there must be internal centers. Haldane had found what he considered the right unit of reference beneath the organism: the center in its relation to multiple other centers.[68] Given the epistemic impossibility of involving all centers in all possible interrelations when examining each center, wholeness was postulated as manifesting in reactions and their coordination.

Reactions constituted the second key concept: because every organ and group of organs reacted to stimuli and amendments in its immediate environment, as well as in the broader internal environment, reactions were constantly mutually dependent: "When we remove any part of the organism from its physiological connection with its environment including the other parts, we at the same time necessarily alter its reactions and the stability of its living structure."[69] A minute change in water content excited, for example, the kidneys, the blood, and other centers; the kidneys in turn affected other organs and centers, as did the blood, and so on. This was not a matter of a buffer for a given stability. Rather, minor stimuli set in motion a long series of different effects that in turn had their own effects: a "vast number of conditions of structure and environment" influenced every reaction, and regulatory and corrective efforts were fundamentally involved in this constant rebalancing.[70] Thus understood, reactions—not properties or processes or even regulatory activities—constituted *the* dynamic force field of organismic activity and unity. Any hermeneutics of the body, for Haldane, required the recognition that these were not derivative; nor were they exactly produced. They were themselves the organism's integrative forces, at play thanks to the interrelations of organs and the chain of reactions at all levels at once.

Reactions also guaranteed what Haldane called the "organic identity" of each individual being.[71] The maintenance and constant reestablishment of this organic identity was at the core of individuality and was the central operation of interiority; Haldane thought it had to be explicitly declared the subject of physiology and even biology more broadly ("The science which traces this organic identity is biology").[72] For this organic identity, it was not so much the life of the organism that was individual; rather, individuality was subtended first of all by the life of each center, because that life was always interdependent with the lives of other centers, and because from that center individuality dove deep to the life of each cell within the center (cells being analogously interdependent as well). The life of such centers also expanded outward toward family and society.[73] At each level, regulation made possible strategies through which the units that claimed internal unity or identity might remain consistent, functional, and differentiated among themselves, thereby facilitating individuality in the existence and constant reforging of each internal environment.

For this argument to hold, Haldane obliged himself to a dual critique of contemporary paradigms in physiology, first, of the nervous system's dominance, and second, of hormonal regulation. Haldane thought the nervous system was unduly credited with control over respiration and circulation. His phrasing here is key to his own holism: "Another tendency has been to regard the nervous system as the primary autonomous regulator of breathing and circulation. The evidence brought forward above has shown, however, that the regulative influence of the nervous system is not autonomous, but dependent on conditions of environment determined mainly by varying tissue activity."[74] Gone was the lure of hierarchy: the body's operations were not forced around by one of its systems, regardless of its cooperative and conflicting impulses. Like the other systems, the nervous system lacked independence and self-rule. Similarly, Haldane found his Cambridge and University College London colleagues to be overly invested in rendering hormones into the catalyst of internal regulation. The hormonal model remained overly mechanical, he insisted, in that it relied on the implicit assumption that hormones alone regulated. But the complexity of responses once again belied the model: "The truth is that every substance which enters into the life processes of any part of an organism is as much a hormone as any other such substance."[75] What was worse, the hormonal model implied that the different organic centers were independent and that hormones and the nervous system were all that united them.

In this regard, the individual organism functioned passively, on the basis of responses and reactions, yet it was active in the sense that it had to constantly manage itself internally, endosomatically, and it did so not through overarching systems, but by way of its centers and reactions. No willful or voluntary activity was required or mattered. Haldane's key phrase became "we are led to the conception of a living organism as the seat of a vast system of mutually dependent reversible chemical reactions."[76] Reusing the adjective *vast*, Haldane facilitated the position on holism that we noted earlier: it was not a classical, self-sufficient holism, but one based on an extraordinary level of complexity that must be understood as a matter of integration, and that allowed for differentiation between individual beings.

Wholeness was thus marked by degrees of variance compounded by continuous reactions and counterreactions occurring in this spiderweb of an internal environment. This constantly evolving interiority, alongside the continuous temporal recalibration of the organism, was constitutive of individuality. It also allowed for an individuality that could be understood only at the social level.[77] Haldane rendered beautifully his sense of the singularity of a self that was premised on reactions and organic identity:

> When I write with the pen the movements of my muscles are determined by the actual presence to me of innumerable past, present, and anticipated future events in both my own individual history and that of mankind. The past events are not simply past and done with, like events interpreted physically or biologically, but they, and not their mere effects, are still present and active. What I have experienced before . . . is still taking on fresh meanings in my mind and directly determining my action now. The same is true of all I have absorbed of the common spiritual heritage and anticipations for the future of my country or of mankind.[78]

"They, and not their mere effects": the traces of past reactions and movements persist into the present, recomposing the individual's everyday activity.

What we might call a physiological phenomenology relied as much on the reimposition of selfhood and experience as it did on an inversion of Ernst Haeckel's phylogenetic recapitulation. The Haeckelian implication of a biologically persistent intergenerational memory was replaced or rearranged here by way of an automated physical memory premised on the ever presence of past movements and reactions—a memory operating unnoticed, which carries with it the internal environment to external activities. This memory was specific to each individual and each moment, Haeckel argued, as it made use of the individual's personal history at the same time as it did the entire history of humanity.

Given his emphatic antimechanism, Haldane's beautiful conclusion—part phenomenology, part recapitulation, part organic identity—was also tricky. Convinced that he was right, he was untroubled about dismissing others from a discipline he clearly regarded as his own province. Lawrence Henderson wouldn't do; nor would Barcroft—then the Fullerian Professor of Physiology at the Royal Institution and himself a researcher on respiration who had self-experimented with asphyxiating gases during the war. At times, as Haldane wrote to Henderson, not even William Bayliss was a good enough physiologist: "I of course don't regard a book like Bayliss's 'General Physiology' as physiology at all, though it contains much useful material for physiology."[79] Biochemists like A. V. Hill and other contemporaries returned the favor, deriding him as a dinosaur: "The poor old chap is getting a little cranky. If this communication to the *Proceedings of the Physiological Society* [the 1929 critique of Henderson] had come from anyone else it would have been rejected out of hand by the editor. . . . But Haldane has a special license to do anything he likes, including never giving notice of a dinner which he attends."[80] Apropos the same event at the Physiological Society, John Fulton asserted: "When all was said and done the chief upshot of Haldane's argument was that [Henderson's] book was not physiology. Someone asked what did it matter anyway, so long as it was the truth. This made Haldane rather angry, and there the discussion ended. I wisely held my tongue, though itching to make a speech. I have nothing but admiration for Haldane's insistence upon approaching the organism as a whole, but his broad philosophical outlook does not prevent from failure to recognize the magnitude of Henderson's great achievement."[81] This allowed Henderson to confirm that their disagreement concerned what was and was not physiology, and to perform graciousness that he had "the greatest admiration for his work and a feeling of profound respect for the man. My Silliman Lectures . . . take his as a starting point."[82]

This allows us to distinguish Haldane from his peers and confirms his self-image as a champion of biological wholeness and individuality. Still, it begs the question anew: Was this self reimposed onto organic identity in a dualistic fashion that suggested the presence of something beyond organic identity—a force specific to social differentiation? Could these presumed biological wholes be deemed properly individual? Did Haldane's claim to be committed to divergence between different beings amount to an assertion of individuality proper?

For now, we leave these questions to hang.

Cannon: Fluid Matrix, Margin of Safety, Higher Mental Functioning

Cannon published an article on homeostasis for the first time just months after finding himself caught in the middle of the 1929 debate about Lawrence Henderson's *Blood*.[83] It was perhaps the most influential article of his career as a physiologist, and it would form the basis for *The Wisdom of the Body*. The article largely consisted of a long list of forms of internal regulation and an explanation of their functions, operations, and properties—a classification, as he put it, of homeostatic conditions. And it was the work of a master diplomat—"*Blessed be the peacemakers!*" wrote Yandell Henderson in response to Cannon's effort to mediate in the affair.[84] Rather than pick sides, Cannon had cited Haldane, Henderson, and Henderson approvingly. Rather than choose among regulatory devices, as Haldane and Lawrence Henderson could be said to have done, he piled them atop one another. Warm to ideas from both thinkers, Cannon deployed his own epistemological tactic, which amounted to outdoing both by assigning them particular places in the functioning of organic and systemic forms of regulation. Together with Haldane, and citing Yandell Henderson, however, he made clear that physicochemical equilibrium did not explain the organism's integration:[85]

> The term *equilibrium* might be used to designate these constant conditions. That term, however, has come to have exact meaning as applied to relatively simple physicochemical states in closed systems where known forces are balanced. In an exhaustive monograph L. J. Henderson has recently treated the blood from this point of view, i.e., he has defined, in relation to circumstances which affect the blood, the nice arrangements within the blood itself, which operate to keep its respiratory functions stable. Besides these arrangements, however, is the integrated cooperation of a wide range of organs—brain and nerves, heart, lungs, kidneys, spleen—which are promptly brought into action when conditions arise which might alter the blood in its respiratory services. The present discussion is concerned with the physiological rather than the physical arrangements for attaining constancy. The coordinated physiological reactions which maintain most of the steady states in the body are so complex, and are so peculiar to the living organism, that it has been suggested (Cannon, 1926) that a specific designation for these states be employed—*homeostasis*.[86]

In other words, if the arrangements in Henderson's blood were "nice," integrated organic cooperation was "so complex" that only Cannon's own work sufficed for designating them. To strengthen his point and organize the past around his own present work, Cannon struck at an intellectual history of the concept of the internal environment since Bernard, and of the purpose of physiology, that was far more detailed than either Henderson's or Barcroft's.[87] Cannon also offered a long, careful definition of *homeostasis* and related terms, notably *fluid matrix*. The historicist effort and the definition mattered as much as his classification of organic and biochemical regulation: asserting his own version of bodily organization, Cannon effectively dismissed existing models proximate to homeostasis as obsolete.

"Organization for Physiological Homeostasis" makes for a curious manifesto. Like Cannon's popularized argument on homeostasis in *The Wisdom of the Body*, it was first outlined in an undated draft that Cannon saved alongside his copy of Ernest Starling's 1923 Harveian Oration "The Wisdom of the Body"—the lecture from which he took the title for his own 1932 book. This early draft, "Some Tentative Postulates regarding the Physiological Regulation of Normal States" introduces his effort to distinguish organic and physiological arguments from equilibrium and to find a name for them.[88] Cannon asks: "What to call these steady states—physiological norms, normal levels, the 'set,' the constants—in the fluid matrix?"

The conceptual and stylistic decisions made in these interventions are important indicators of Cannon's approach to interiority and individuality. His choice of *fluid matrix* over *liquid matrix* and other terms in the draft represented his attempt to square the circle: it avoids terms like regulation or organization. Cannon, who was meticulous about his neologisms, surely appreciated that he was deriving *matrix* from its two main references at the time—the Latin for "womb" and matrix theory in mathematics. *Matrix* adjourns the debates over *balance* and *equilibrium* by virtue of being too complex to be covered by them, yet also adjoins these alongside Haldane's system of reactions and centers, indicating a stable—bordering on static—webbing of the different units and elements of the organism. Put differently, it allowed Cannon to use elements from both Haldane and Lawrence Henderson, identifying both materials and processes as the sites of regulation.[89] It prioritized the fluids of the body (blood, lymph, water) and their contents and properties, including their role as chemical messengers: it carries the sense that organisms "live in the fluids which bathe their cells" which he was pulling from Bernard.[90] In that manner, it avoids what Haldane had focused on, namely the sense of reactions as units of force that might be criticized as too formal or too abstracted from a bodily economy.

Similarly, *margin of safety* closely approximates Lawrence Henderson's *buffer*. Refusing to simply adopt Henderson's readily available formulation, Cannon found an alternative in S. J. Meltzer's 1907 lecture to the Harvey Society in New York: *factor of safety*.[91] That Henderson had also used *factor of safety* barely a year after Meltzer did not seem worthy of reference.[92] *Margin of safety* carries economistic implications and aims to hold the accusation of Romantic holism at bay. It has very different connotations from *buffer*, which points to room, leeway, a certain openness. Instead, *margin* sets a horizon or limit, and *safety* improves as one moves inward from the margin that constitutes its vanishing point. *Margin of safety* signals elements of the fluid matrix being besieged from all sides; the organism is conceived as a beleaguered womb.

In that womb's threatened openness resided the triumvirate of interiority, integration, individuality. As we saw in the last chapter, Cannon used a vivid description of the skin to separate the organism from the outside and to syncopate the external threat by replicating it internally. As a dead container of life that nevertheless partook in life, the skin individualized the threat by demonstrating that what threatened the individual was less a piercing or penetration from the exterior than the recoding of that external stimulus or penetration from the inside. For Cannon the internal environment was profoundly affected by the external one. By translating the external threat in responding to it, Cannon's fluid matrix redeployed this external threat within itself.

In a manner absent from the deindividuating emotions and fight-or-flight response of *Bodily Changes*, the homeostatic organism of *The Wisdom of the Body* was profoundly individualizing: the response to the confronted milieu was partly based on already-known physiological structures. But by relying on myriad homeostatic regulators, not to mention on processes of inundation and storage of nutrients and elements whose details and effectiveness in each case was not known, the fluid matrix was no longer simply replicable across organisms; it was in each case different. The impossibility of prediction had become an epistemological tool for understanding the nature and functioning of the individual, given the sheer complexity of the fluid matrix. Emotions, regulation, and shock helped explain pathologically the interiority that was otherwise too complex, too specified to study.

Individuality was now built in—in interiority and integration. But did Cannon really point to individuals? In his political texts, for example a 1933 article "Biocracy," where he declared himself simultaneously in favor of social homeostasis and opposed to a biological regulation of society that might endanger freedom, he gave a precise sense of the meaning of *individual*:

> Because of [the] assurance of constancy in the fluid matrix, we are freed from paying routine attention to the management of the details of our bare existence. These details are attended to by regulatory devices. Without these devices we should be in constant danger of catastrophe unless we were always on the alert to correct voluntarily what, in the ordinary course of events, is corrected automatically. With these devices, however, the fundamental bodily processes are kept steady. We, as individuals, are thus made free, free to enter into agreeable relations with our fellows, free to enjoy beautiful things, to explore and understand the wonders of the world about us, to develop new ideas and interests, and to work and play untrammeled by anxieties concerning our bodily affairs.[93]

Thanks to Cannon's penchant for redeploying Bernard, this identification of individuality with freedom allowed him to use each concept to the benefit of the other: individual meant free, and vice versa.[94] As he acknowledged in a 1931 missive to Barcroft, Cannon was following his lead in identifying the higher functioning of the central nervous system as the space opened or freed by the stability of the internal environment.[95] The positive end result of emphasizing the malleability and fragility of the organism—and hence its individuality—was the groundwork for a metaphysics of higher mental functioning.

What is lost in this theory of the individual? Cannon's last letter to Barcroft on higher mental functioning offers a clue. If higher mental functioning is largely separate from the physiological individuality Cannon worked so hard to establish by way of pathology, are we speaking of one individuality or two? Is the thinking individual the same as the pathologically conceived physiological individual? Are their individualities piled atop one another? Is the thinking that is exhibited thanks to the freedom produced by the individual, fragile body in any sense conditioned in its contents by this freedom? These questions are not vain ones, especially given Cannon's claims in the previous passage about freedom.[96] If the fragility of the individual body were indeed the premise of all free social behavior, this individual would seem to bridge, conceptually as well as epistemologically, a fundamental soma-psyche dualism. If interiority and integration were taken to be largely self-sufficient, operating thanks to a multitude of homeostatic devices separated from the outside world, beneath consciousness and the will, then the individuality they developed could not simply be extrapolated onto the world of conscious activity and social life. The same system that built individuality undermined its very meaning in society—from which the concept was derived in the first place.

Cannon thought so too: his individual was an individual only when threatened, only when too closed off to function properly in society. This was the individual menaced by shock, but also the individual menaced by voodoo.

Dale: Histamine, Capillaries, Autopharmacology

We turn now to a fourth figure in interwar physiology, the one who most emphatically pursued the effects of the internal recalibration of the organism resulting from external stimuli. Henry Dale is most famous in the history of biology and medicine for studies he carried out with Otto Loewi on acetylcholine, which led to his being awarded a Nobel Prize in 1936. Dale's research presents us with a conceptual opening that at first glance follows a line of questioning initiated by Bayliss and Starling regarding the human organism's pathological response to injury or other crises. The similarity in scientific posture should be no surprise: besides working with them all, Dale set up the same citation system for identifying precursors—beginning with Bernard—when illustrating the physiological value of his work. Because of his work in chemistry and pharmacology, moreover, Dale straddled the divide between Haldane and Lawrence Henderson,[97] linking chemical, biological, and systemic forms of integration and disintegration. His studies of histamine and anaphylaxis connect him to the history of allergy, immunity, and autoimmunity. But they are also relevant for our main inquiry into the conceptual and experimental design of interiority and subjecthood, as Dale was careful to premise his laboratory work on the relationship of the organism to its external and internal environments as well as to itself as a unit. Describing the priority of his work to be autopharmacology, Dale turned the internal environment into something more than a regulated or self-regulating unit: pharmacology of the self by itself meant this unit was an agent observing itself and the changes within, and responding, even leading itself.

After a brief time studying with Paul Ehrlich in Germany in 1903,[98] Dale began his professional career under the mentorship of Starling and Bayliss and conducted his early research as an experimental scientist at the Institute of Physiology at University College London. Facing "bleak" academic prospects, he moved from the university laboratory to a research post in the Wellcome Physiological Research Laboratories of Burroughs Wellcome & Company in 1904.[99] The move provided

Dale with a new perspective by bringing endocrinology and physiology into close alignment with pharmacology. Henry Wellcome steered Dale's efforts toward the pharmacotherapeutic applications of the substance ergot of rye, which was at the time well known for its use in the treatment of postpartum hemorrhage. He became absorbed with the resulting mechanism of histamine because it gave "evidence for its contribution to local and general reactions, by which the organism as a whole and its separate tissues respond to various chemical, immunological, or physical assaults upon the integrity of their living cells."[100]

By the beginning of the war, Dale was well known for his success in isolating naturally occurring histamine, as this is produced by decarbonoxylation of the diamino acid in ergot. In 1910, together with his colleague George Barger,[101] Dale first announced the discovery of naturally occurring histamine, narrowly beating Dankwart Ackermann in Germany who isolated histamine from putrefying pancreas. Three years later, he demonstrated that histamine's effects on the anaphylactic contraction of plain muscle could be caused by prior cellular sensitization to an antigen (one of his "most fortunate accidents").[102] Much of the historical focus has been placed on Dale's research with Loewi on acetylcholine, the chemical phase of transmission in nerve impulses.[103] But it is his work on histamine that formed his unswerving commitment to the problem of bodily regulation between the normal and the pathological, the natural and the terrible.

Why did histamine matter—and why does it matter for our story here? Dale's argument went like this: the stimulation of histamine in cells and tissue was a reaction—a "bringing forth"—of a substance already residing in tissue. The substance was highly variable in terms of its distribution throughout the body. Once it was animated, the severity of its response to the stimulus affecting the body varied greatly between individuals despite the similarity of stimuli. Through the physiological function of histamine leading to symptoms of anaphylaxis, Dale would gradually establish a picture of the chemically regulated body, where substances lay in wait in cells and muscle tissue until they were required. Then they would be liberated by the actions of the environment—not the general environment, but specific actions, irritants, or vectors in that environment. This liberation either would result in a normal protective response or would contribute to catastrophic harm to the whole organism.

Dale later linked his understanding of chemophysiological phenomena that resulted in high rates of battlefield and postbattle mortality to a range of hormonal and autopharmacological actions of the organism;

he developed a model for explaining the general effects of shock by its likeness to histamine shock and anaphylaxis.[104] The observations Dale and his colleagues had made before the war on the place of histamine response during circulatory collapse and respiratory failure provided him with a ready-made framework in which to interpret other efforts, including those by former mentors such as Bayliss.[105]

Laboratory efforts here trumped clinical considerations, and Dale accentuated the distance between the two: "As a mere laboratory worker I know nothing at first hand of shock as a practical clinical problem."[106] Dale distinguished the laboratory and clinic as "the artificial and practical," noting that, for the biochemist, the use of animal experimentation would often lead to an attitude that *tierchemie ist schmierchemie* (animal chemistry is sloppy chemistry).[107] His work thus speaks to the complicated, speculative labor of research in the face of byzantine injuries witnessed on the battlefield. Although he traveled along similar paths of inquiry as his colleagues in the Royal Army Medical Corps, his focus led him to claims of toxemic causation that diverged from the observations of his peers even as his conception of the body was radicalizing theirs.[108] The main operative principle for much of Dale's wartime and postwar work, namely the isomorphism between histamine shock and wound shock, extended only so far. As he himself recognized, he too had contributed to the overwidening of the category of shock that thinned the possibility of describing the phenomenon meaningfully.[109] His claims were questioned at length, but Dale's missteps, as much as his achievements, confirm his contribution to a set of perspectives that forced the reconsideration of the gross distinctions between interior and exterior milieus.

: : :

The key development concerned the effects of histamine, especially in relation to anaphylaxis. Adolf Windaus and W. Vogt had prepared histamine three years before Dale and Laidlaw isolated it, but Windaus and Vogt had "no suspicion" of its physiological properties; they simply carried out a chemical synthesis and prepared it by decarboxylation.[110] Dale and Laidlaw's interest lay in its properties, as they clarified long-held debates on allergy and anaphylaxis. The substance was not new; understanding its action was.[111] Because oxytocin and histamine are part of the natural constitution of the body, experimental procedures were required to demonstrate their effects in pathological reactions.[112] In his 1911 research on anaphylactic reaction with Barger, Dale was able to establish the first chemical isolation of histamine from animal tissue. In these early samples, histamine seemed to be both a vasoconstrictor and a vasodilator, but because Dale and Laidlaw had found what they were looking for, they proceeded no further.[113] Dale would later comment on his findings: "From the point of view of present knowledge, it seems difficult to understand why we did not recognize its potential significance more clearly, or do more to put the finding beyond criticism."[114] In his mind, like those of his critics, the findings were muddled and "still affected by memories" of being in the laboratory of Starling and Bayliss, where Bayliss had been busy investigating "depressor substances" from intestinal extracts containing secretin. Now it seemed to Dale that this depressor action was the same as that of histamine.

In the 1910s, Dale's effort with histamine and histamine shock made further strides and played directly into the questions of allergy and anaphylaxis. Knowledge about histamine contributed to an understanding of the body's excessive response to external stimuli. The discovery of allergy is conventionally identified with Charles Richet and Paul Portier's experiments on mechanisms of hypersensitivity reaction in dogs: when injected with toxin produced from sea anemones, the dogs suffered respiratory distress and sometimes even death.[115] As Mark Jackson suggests, allergy has always figured in a limited way in the history of immunology.[116] Still, in the years leading to Dale and Laidlaw's publications on histamine, there were significant disputes about the mechanism of anaphylaxis.[117] Their key observation was that histamine was present throughout the body, across its different tissues. It was activated by local injury or exposure.

When in 1913 Dale identified histamine as the cause of contractions in cat uterine muscles in vitro, and when during the war he tested histamine in vivo to monitor blood pressure, hypersensitivity, and shock, his work represented a huge step forward in understanding how the tissue-fixed nature of "anaphylactic antibodies" worked in relation to local injury, and not simply by a systemic, intercellular reaction to the antigen.[118] What he demonstrated was that allergic reactions were an interaction between an antigen and an antibody already fixed to cellular tissue, which, in response to the antigen, stimulated the release of histamine.[119] While Dale thus supported the cellular theory of anaphylaxis, he also showed the complex interweaving of local and whole-body activity. This was one area in which he differed very obviously from both Haldane and Lawrence Henderson. The sudden introduction of the foreign protein was not a natural event for the organism, and the localized anaphylactic reaction was ideally protective, defending the organism from invasion at the expense of the tissue immediately affected. Anaphylaxis was coextensive with and not opposed to immunity:[120] "Anaphylaxis, as we see it in the laboratory, is not the opposite of immunity, it is the physiological response of an animal in a certain phase of immunity, to the artificial test which we impose."[121] As a result, while anaphylaxis was still referred to as a syndrome, "it is not the syndrome which is anaphylaxis; anaphylaxis is the production of a common syndrome by a special kind of immunological reactions,"[122] no different from "anaphylactoid reactions," "nitroid crises," or "hémoclasiques crises."[123]

The internal and outer environments of the body had to be rethought to incorporate the idea of an interior milieu of tissue that, once animated, changed the observable chemical characteristics of that tissue. This approach marked a fundamental conceptual turn away from thinking of external stimuli and injury as unidirectional and consistent. Histamine and anaphylaxis did more than approximate a general understanding of shock syndromes. Indeed, they matter because at their core they represented a vital change in how local and generalized responses to injury or irritation were considered within the whole of the organism in its environment. Dale was not simply suggesting (as Cannon would) that a deregulation of the organism's fluid matrix forced the organism to counteract and outdo the intruder in a manner that worsened this instability. The stimulation of histamine in cells and tissue was a reaction—a bringing forth—of a substance already residing in tissue, a reaction that was highly variable between individuals in terms of its distribution throughout the body and the severity of response given similar stimuli.

Through the physiological function of histamine and the symptoms of anaphylaxis, Dale established a picture of the chemically regulated body where substances lay in wait in muscle tissue until suddenly they became necessary, then are liberated by the external environment. Specific actions, irritants, or vectors from the outside triggered the normal protective response, but for reasons unknown, the response enabled the organism to cause catastrophic harm to itself. In other words, there was a *within within*: underneath the first interiority of tissue, naturally occurring substances flowed, hidden from view and unsensed, until they suddenly became animated and dominant in the balance or matrix of force.[124] The pathology that ensued after injury was one in which the body assaulted itself and threatened its own demise because some hidden internal limit had been crossed by the external stimulus. Every defensive response was de facto overwhelming. And this was not an autoimmune reaction, but an autopharmacological one, a self-produced medication or drug with a dose-response or upper limit that was, crucially, different for each individual.

The body of research that grew from Dale's efforts made plain the unresolved contest between the variability of the individual and the human organism as a general member of the collective. Richet asserted in his 1913 Nobel Prize lecture: "Anaphylaxis is thus necessary to the species, often to the detriment of the individual. The individual may perish, but this does not matter. The species must at all times retain its organic integrity."[125] For Dale, the role of the individual was much more fraught. "In the anaphylactic shock we may contemplate the sacrifice of the individual to the purity of the type," and yet what guided his efforts was concern regarding natural variability within individuals and between them and their environments.[126] To understand this concern, we turn to the three elements that defined Dale's understanding of interiority, namely capillaries, smooth tissue, and the occurrence of shock. In avoidance of terms relating to allergy, autoimmunity, and anaphylaxis, these helped organize what he would call autopharmacology.

Despite the successes of early histamine research, doubt surrounded its explanatory value, especially once Dale began to use the histamine model to explain other disorders, notably wound shock as of 1916.[127] Even before the question of shock was raised, there seemed to be something else at work with the basic functional elements of the dilation and constriction of blood vessels, the former of which was associated with the action of histamine. For histamine's effects to be comprehensible, researchers needed to confront the possibility that the functioning of the circulatory system as understood around 1900 was incomplete to the point of misguided. In particular, the how and why of a rise in blood pressure at the local and organismic levels remained only partly comprehensible. Its mechanism in relation to the constriction and dilation of blood vessels as these were prompted by the introduction of chemicals foreign to the body was a site for a wide array of research.

For example, in a series of important experiments in the 1890s, Rudolf Kobert injected sodium salt of sphacelinic acid intravenously into rabbits and observed a rise in blood pressure.[128] About the same time, Bayliss stimulated the sympathetic nerves and looked at inhibitory elements in vasodilator nerves.[129] Working in Bayliss's lab, Dale carried out research on vasoconstriction of extracts of the pituitary glands and observed the rise of blood pressure due to arterial constriction.[130] In 1913, injections of crystalline ergamine phosphate from the chemical labs of Burroughs Wellcome & Company produced an almost immediate cardiac effect;[131] Dale and Laidlaw's introduction of histamine to the circulating blood produced a notable rise in pulmonary pressure, resulting in intense and sudden vasodilator action.[132]

What was the meaning of *vasodilator action, blood pressure rise,* and *histamine dilation*—terms that seem technical enough to stymie historical research? As histamine was added, vessels dilated, and therefore blood pressure dropped. This was mostly a local effect, but it could as well be a broader or organism-wide one. We might ask the question differently, the way Dale asked it: what kind of structure and function did the blood require in order for histamine to be both present in every cell of the body and capable of acting sometimes locally and sometimes globally, with potentially catastrophic results? The question of vasodilation provided a site for studying the work of histamine, and here the research on wound shock brought forth an important surprise.[133]

Researchers were troubled by the question of where blood went within the body during wound shock. Once their experimental results led them to discard theories that veins or arteries were the repositories of the "missing blood," only the capillaries remained. Yet this did not make sense: Cannon, for example, pointed out that the capillaries should not be large enough to hold the blood that had gone missing.[134] The capillaries, after all, were conceived as having a single, straightforward function: disbursing into the organs and muscle the nutrients carried in the blood. Indeed, this was precisely the description Starling had offered in his *Elements of Human Physiology* as recently as 1907: capillaries were there to aid the diffusion of nutrients and chemical compounds to and from particular sites of the body, and in occasions of injury to aid leukocyte defense.[135] Speaking before the Royal Institution in 1925, Dale framed this problem as follows: "Until recently it was supposed that [the capillaries] were adapted to these ends only, and that the mechanical functions of maintaining adequate pressure in the arteries, and of distributing it in accordance with the varying needs of the different organs, came abruptly to an end where the smallest arteries branch out into the complex network of the capillary vessels."[136] Let us leave aside the critique of the mechanistic concept of the body, to which Dale would return time and again.[137] Past teaching limited the capillaries to functions of "filtration, diffusion, and osmosis."[138] But as Dale pointed out, during the war researchers had learned that capillaries could become extraordinarily distended, and they could "hide" the missing blood.[139]

Dale proposed that the contractile function of the capillaries helped them to become extraordinarily distended during times of need or stress.

He argued that Thomas Lewis, Thomas F. Cotton, and John G. Slade in England; Ulrich Ebbecke in Germany; and especially August Krogh in Denmark had also demonstrated that the capillaries, in a manner quite independent from the work of arterial pressure, created resistance to blood flow, responded to "functional activity," contributed fundamentally to the way hormones traversed the body, and indicated that the physiological body was not merely mechanistic but an integrated and integral structure, a "system" whose different parts contributed to and resisted one another.[140] This, Dale felt, had given strong support to his theory of the activity of histamine or a "histamine-like" compound during the otherwise vague condition known as shock.[141] Histamine caused vasodilation and a drop in blood pressure. The dilation of the capillaries contributed to a situation in which, in the absence of nervous or metabolic counteraction, histamine levels could rise at the local level of injury and then very quickly across the organism. This might explain the central symptom, the dramatic drop in blood pressure.[142] After the war Dale continued to link vasodilation with the dangerous drop in blood pressure, and he pursued directly the ground for global effects. In a 1918 article on histamine vasodilation, authored with Alfred N. Richards, he continued to be concerned with the maintenance of vascular and capillary "tone" apart from nervous control, tying histamine activity to shock.[143]

Dale's effort to wrangle the whole of the mechanism and to explain global effects is interesting. As "histamine-like" and "shock-like" became placeholders, the extreme situation of shock for the whole body became a structural requirement of the body's absorption and use of histamine, especially in anaphylactic situations. For example, in 1925 Dale and Laidlaw would note: "When the vasodilator effect is perpetuated by continued slow infusion of the drug the shock-like failure of the circulation soon appears."[144] Action on the heart and the distribution of blood in the peripheral vessels led to the observation of "shock-like" stagnation of blood in microscopic venules. Capillary constriction, the movement and stagnation of blood, and the changing composition of fluids together produced a "vicious circle" that played a part in "histamine shock." Thus Dale cleverly and boldly reversed his earlier claims, now relating wound shock to histamine shock rather than the other way around.[145] Loss of blood volume owing to the escape of plasma and the general dilation of the capillaries appeared to him to be the main factors in the production of shock.[146] They could be a subcategory of histamine or histamine-like action: "traumatic shock," like histamine, isolated and intervened through the body's own chemical response, beginning at the point of injury or irritation, and then potentially throughout the whole organism. Other conditions too—notably ones caused by poisons—might be described in the same terms. In 1920 Dale proposed "shock-poison," where he again ascribed "common action to a large group of poisonous substances of animal or bacterial origin" that would affect the whole organism.[147]

Crucially, capillary distension as a result of histamine made sense with his argument that histamine reactions were not aimed against the body but were part of an immune response that, nevertheless, could not be sustained by the body, especially if it spread. The "relaxation of the capillary vessels" was restorative yet potentially catastrophic.[148] Capillary tone was at once the site for histamine activity and the signal that this activity could spread. Its localized activity, like the extent of vasodilation in the organism, could not be anticipated. Capillary tone indicated that capillaries had an agency of their own, "a power of independent contraction,"[149] which involved, besides the transfer of hormones and other substances, a functional, organism-wide role involving the unification of the body. Capillaries contributed to local activity at the same time as they worked with the whole. Dale argued that bodies were biologically divergent and could not be understood by way of standards and rigid norms; biological comparisons with a standard betrayed the importance of biological variation at the individual level.[150] Medicine and chemical intrusion had to be carefully studied rather than applied across a population on the assumption of the chemical givenness of a body.

It is in this conceptual mixture—of a *within within* and a reconceptualization of the spread from local to global effects, thanks to the structure of circulation—that autopharmacology comes to eloquently summarize the organizing principle of Dale's research.[151] The distinction is essential between autoimmunity and what Dale means by autopharmacology: the natural production and release of chemicals in the body given specific triggers or stimuli in the environment, isolated in substance.[152] As historians, philosophers, and anthropologists have shown in detail, *autoimmunity*, a term that emerged much later in the twentieth century,[153] carries with it a vast assemblage of metaphors about the body—a body against itself, foreignness within, tolerance and intolerance, flexibility. But for Dale in the 1910s and 1920s, nothing about histamine could be imagined as autoimmune in the terms others would later elaborate, even if histamine response involved an attack by the body on itself. The autonomic chemical release in tissues as a result of specific environmental stimuli was a normal function of the human organism that most of the time not only failed to produce crises in the organism, but also was relatively finite and limited to single encounters. Pathology followed suit.[154]

Now the organism possessed the latent power to reenact outside stimuli internally and overenforce its response so it suffered an overdefensive collapse. This constituted a major conceptual turn. The pharmacological and chemical properties of that reaction came exclusively from within, and they could lay the organism to waste.[155] Dale's autopharmacology thus served a specific rationale, one quite foreign to the language of immunity: if different individuals have different thresholds that cannot be predicted, if their capillary capacity cannot be calculated, if the severity of their response to allergens could not be anticipated, then seemingly minor variations were not merely contingent, they were constitutive of fundamental differences between different organisms. Not only could identical stimuli no longer be treated as the same by different organisms, but these organisms operated (quite like Cannon's homeostatic fluid matrix) with distinct capacities to observe these stimuli, to amend and counter them in their responses, to relate particular regions and lesions of the body back to the whole, and to self-observe. *Pharmacology* in autopharmacology suggested not just that the organism was pursuing its own chemical regulation, but that this physiological body, this bodily self, guarded a threshold of resistance, adding extreme volatility to the dose-response relationship that is essential to its functioning. Whatever chemicals did, the body took them up, worked them around, and saw to their effects on its own. Once again, interiority was individuality, though this time, once it was perversely allied with the outside world, it was also precisely what threatened individuality.

Goldstein: From Anxiety to Self-Actualization

We concluded chapter 4 on the profound difference between Head's and Goldstein's ways of according individuality to their patients. For Head, patients were individuals because they were injured and therefore radically unfree, because active compensation for their injuries made them structurally different from normal organisms and from every other injured patient. Each patient was uncategorizable and therefore had to be engaged as truly individual. But for Goldstein, persons were individual because and when they were free, and so long as they were healthy; aphasic patients who found their abstract capacity diminished lost that freedom and found themselves compelled into a struggle with themselves. This conception of individuality organized Goldstein's philosophical thought.

The physiological and philosophical stakes of Goldstein's coupling of individuality and disintegration deserve further elaboration, and it helps to triangulate his position with that of the Russian neuropsychologist Alexander Luria's contemporary exposition of physiology's priorities and of the trickery of bodily individuality. Luria's 1932 book *The Nature of Human Conflicts*, developed over a decade of research at the Institute of Experimental Psychology in Moscow, criticized neurological theory in much the same terms as Head and Goldstein did. Luria turned explicitly to physiological (rather than neurological) writing to provide the scaffolding of his psychology. He began the theoretical introduction of his book by rejecting "the analogy of the machine" in the study of the human body, and with it any approach that would proceed from parts to the whole.[156] Although excitation and inhibition explain some of the functions of individual neurons, an explanation of the nervous system in the body could not proceed by way of accumulation from individual neurons toward the whole or by way of any equilibrium of nervous activity.[157]

Painting the picture of an aphasic patient and recalling (without citation) Goldstein's critique of atomism, Luria required that a highly complex whole be presupposed and asked whether the physiologists' conceptions sufficed. The organism was not a "mosaic" but the complex interweaving of superimposed systems as suggested by the new physiological understanding of a functionally integrated body, a functional correlation of its systems, and a functional inequality of these same systems. So it would be with the brain as well, its organization and disorganization: what establishes the higher psychological processes is the correlation, integration, and inequality of the brain's component structures.

As if this reference to internal systemic regulation were not enough, Luria then cited Head's "brilliant" description of the disorganization of these processes and addressed two implications of the "principle of organization."[158] One was the tendency toward traditional vitalism; Luria seems to have targeted embryology, or von Uexküll's method. (The refusal of vitalism proper was de rigueur both in the Soviet Union and, as we have seen, among integrationists.) The other was the tendency to turn integration into the basis for universal laws: Luria's target in our context would be Lawrence Henderson. Universality in organization slid "too easily" toward

> a certain general law, which appears equally in mechanics, in physics, and in neuropsychiatry and social life; many authors have proceeded along the path of the universality of this principle. The logical consequence of this procedure is the wish to carry the complicated forms of organization of behavior into general laws which have already been observed in physics. . . . The extension of the principle of organization into the domain of a general law inevitably leads us to ignore and misunderstand the details of human neurodynamics and the highest and specific forms of behavior always remain beyond the field of vision of the mechanists.[159]

No clearer repudiation of mechanism, nor of the tendency to look for organization at different levels and then treat those as analogous or isomorphic, could be expected from a neurophysiologist. Luria used Cannon's overcoming of the old division between physiology and psychology, including the James-Lange theory of the emotions, as proof that "separate physiological symptoms do not by any means completely describe affect, and for a satisfactory explanation we must consider the whole function of the organism."[160] Human behavior was specific enough, he argued, for its organization and integration to be impossible to analogize or reduce to general biochemical laws. It required a highly plastic correlation with physiology. Luria justified his attention to pathology by using the same argument: behavior disintegrates when violence is done to the neurophysiological apparatus that undergirds it. Such disintegration could help explain both the structure and functions of the whole because of the effects on seemingly disparate parts. Turning to pathology—to the data that emerge from the situations in which disease dominates the individual subject—Luria carried the entire conceptual apparatus onto a methodological ground founded on the individuality of expression of disorder and the individuality of experiential discovery.

Goldstein, in *The Organism*, approached physiology neither by way of a description of normal functions nor by way of a series of interwoven systems and organs, but instead through the lens of conservative tendencies initiated by the milieu-driven experience of disease. His study of compensation in tonus was exemplary of the role of disease and response in the body; further, he cited Cannon's homeostasis affirmatively and advocated for physiology and neurology to incorporate questions of behavior by engaging with the external milieu.[161] A healthy organism, for him, was characterized by the tendency to face new situations and institute one's own norms. Accordingly, his theory of individuality arose, first, out of his understanding of pathological data; second, from his thinking about anxiety and catastrophic reactions; and third, as a result of his commitment to self-actualization. These directly channeled his interpretation of norms—as produced in efforts at compensation and stabilization—into an understanding of interiority, environment, and integration, which, alone among our thinkers here, he did not credit to Bernard, despite knowing his thought well.[162]

As we have seen, Goldstein was particularly emphatic on the primacy of pathological performances. Useful data arise when we compare the solitary patient in a specific situation with the same patient in a previous, "normal" state.[163] Once ill or disturbed, the organism moves on to a different state, often vastly narrowed vis-à-vis its milieu.[164] It is not in comparative or population-wide examination, but only in self-comparison, that order and disorder hold meaning.[165] (Canguilhem, deeply influenced by Goldstein on this point, wrote: "The borderline between the normal and the pathological is imprecise for several individuals considered simultaneously but it is perfectly precise for one and the same individual considered successively."[166]) This is significant also because order carries a social dimension and thus replaces any notion of merely internal equilibrium. Pathology introduced questions of behavior into the physiology of performances.

With the individual, Goldstein thus offered a unit of analysis that doubled as a critique of generalized criteria.[167] Against the illusions of such criteria, "well-being consists in an individual norm of ordered functioning."[168] The ability to claim health or illness was tied to the promise of autonomy, of a trajectory unaffected by the orbit of threat. Disease and disorder are found in the patient in the sense that the patient's experience of suffering is highly individualized: only out of the patient's perhaps singular welding together of successful or troubled performances can the neuropsychiatrist reconstruct the drift of the patient's wholeness and pursue intervention, treatment, and healing. The organism is thus freed from poorly conceived biological determinism at the same time that it becomes increasingly unwieldy from the perspective of therapeutics. In other words, whereas Goldstein's critique of localizationism and atomism did not reject discernable patterns of general pathological phenomena, particularly in the case of organic damage, it did demand that these patterns be found in the individual. It is only in this individual that they came together into a general system whose parts affect, conflict with, and adapt to one another and to the specific milieu.

This meant several things. First, by arguing for the singularity of

FIG. 6.2. *Woman with left cerebellar lesion* (ca. 1937). Goldstein runs into frame to aid a female patient who, instructed to walk along a path with her eyes closed, staggers, her torso tilting toward the bench. Box 18, Kurt Goldstein Papers, Columbia University Rare Book & Manuscript Library (used with permission).

every case or patient, Goldstein dislodged the identity of the norm with the normal, the average, the "healthy." He directly attacked political and medical positions that would use conventional normality against those deemed nonaverage, nonhealthy, nonnormal. He thus stunted the meaningfulness of classical normativity and the identification of norms with laws, regulative ideals, or statistical averages,[169] in favor of a counteridentification with the individual's order in handling the environment with a degree of freedom. This constituted a departure from liberal conceptions of individuality, and even more from conceptions of norms that allowed for a generalized normalization.

Now, because individualized norms result from responses to disorder and debilitation, the individual is biologically bound to struggle to carve out a space where they can take hold. Illness disrupts the internal environment at multiple points, in a compounding fashion, and in such a way that the organism's behavior—the level at which integration manifests in full—is disarranged into dysfunctional components.[170] Goldstein insists that in response to this individually experienced decomposition that occurs with deep disease and disorder, the threatened organism strives toward adaptation, ideally toward new norms.

The key device for the organism's adaptation was the experience of anxiety, perhaps the most curious concept in *The Organism*. The brain-injured patient who was confronted with overly complicated demands became anxious; lost in this experience of anxiety, the patient buckled, progressively losing the capacity to handle the environment and suffering a catastrophic reaction: "'Catastrophic' reactions . . . are not only 'inadequate' but also disordered, inconstant, inconsistent, and embedded in physical and mental shock. In these situations, the individual feels himself unfree, buffeted, and vacillating. He experiences a shock affecting not only his own person, but the surrounding world as well. He is in that condition that we usually call anxiety. . . . He becomes more or less unresponsive and fails even in those tasks which he could easily meet under other circumstances."[171] Anxiety wove together this body imploding through its reaction, and channeled the scientist's attention toward the biological, and not merely abstract, striving for a return to wholeness and order. In the experience of disintegration, anxiety became a signature of a different individuation—of the patient's difference from every other patient in the moment and of the manner in which the environment rendered his or her pathology overwhelming.

The individual character of the experience of anxiety is particularly evident in Goldstein's construction of the concept. Traditionally understood as a psychological concern, for Goldstein anxiety was by no means a reaction tied to the "abstract" attitude, or a mere emotional outburst; instead, it was exemplary of the "concrete" attitude, an unreflective response in the closure of possibilities. It placed the patient at the limit of the patient's body and performance—that is, beneath any internal self-conception in the experience of disease.

Goldstein first published on anxiety in 1929, shortly after Freud's 1926 *Inhibitions, Symptoms, and Anxiety* and Heidegger's 1927 *Being and Time*.[172] In both the article and its recapitulation in *The Organism*, Goldstein disparaged contemporary psychological theories but concurred with Heidegger and Freud—whose theory of the drives he explicitly disdained—that anxiety was different from fear: it was more general than fear and lacked an object.[173] Goldstein further appreciated Heidegger for the prominence he granted to the concept of anxiety in his reading of Dasein's confrontation with death. But Goldstein then inverted Heidegger's famous argument that anxiety could be identified with subjective authenticity: although the ability to "bear" anxiety indicated the patient's "genuine courage" against the "threatening of existence,"[174] anxiety was for the patient an inexorable sign of pathological destruction.

In an astonishing formulation, Goldstein offered the idea that the organism in a catastrophic situation "'is' or personifies anxiety," as "it has become impossible ... to cope in any way with tasks commensurate to its real nature."[175] Anxiety thus denoted the ultimate "endangering situation," in which the patient became "actually endangered in his existence."[176] In the brain-injured patient, "completely helpless, ... entirely surrendered to the anxiety situation,"[177] this was a constant threat whose recursiveness resulted in a disordering of even compensatory pictures of experience, with the patient experiencing this shattering while being compelled to retain a basic practical or pragmatic unity. To overcome anxiety, Goldstein concluded, was a "characteristic peculiarity of man,"[178] yet to live in anxiety, to embody it, meant to be laced into a pathological experience of individualized disastrous collapse. For the physician it was thus deeply troubling, but anxiety as disintegration also demonstrated the patient's individuality.

Goldstein then proposed to use anxiety to understand the conditions of possibility of its contrary, the organism's tendency toward self-actualization—in other words, the individual's free and creative grasping of possibilities. Goldstein's deconstruction of the spiraling into anxiety inverts here into thinking about the chances—whether normal or reduced—of rendering threatened individuality into a chance at more than survival—at control. If Heidegger had theorized anxiety as capable of enforcing the usually dispersed unity of Dasein—an ontico-ontological unity that reaches beyond the unities of body or consciousness, toward the totality of its existence—Goldstein attributed unity to the normal, healthy organism. His epistemology and philosophical anthropology are fundamentally based on this claim: the individual strove for wholeness.

The living being was not given whole; it could always be broken up, and it was constantly threatened.[179] Still, it always operated as a whole, and when healthy it did not flaunt or even experience its wholeness. Its integrity existed in silence and operated foremost as a methodological, epistemological principle, to be used for comparison by a doctor or a sick person aware of wholeness that had been lost. Wholeness was experienced as such only by an organism that was imperiled, an organism that, in pathological individuation, had lost its original individuality, its autonomy and normativity. Life is at its utmost when it manages to institute its own norms and impose itself on its milieu, thereby also maintaining and guiding its individuality.

Goldstein named this persistence of the individual, this sense of self-sovereignty, this effort to exist and to realize one's possibilities, "self-actualization." Perhaps the most holistic of Goldstein's concepts, self-actualization was structured in such a way as to express the individual's freedom and normativity vis-à-vis her milieu. It was the organism's sole, singular drive: to exist, to realize, and to complete itself. But the synthetic bridging of alternatives that self-actualization offered masked broader intellectual tensions. The concept emerged in *ex negativo* formulations in Goldstein's discussions of disorder and anxiety. Disorder scrambled the tendency toward self-actualization, setting the organism on a conservative path of adjustment, but perhaps allowing new directions for self-actualization to emerge. Anxiety destroyed self-actualization altogether. The concept was further linked to the organism's constancy mechanisms and to self-regulation.[180] Self-actualization was, in that sense, the value in Goldstein that came closest to Bernard's sense that freedom resulted from the internal environment's independence from the external world. It was, however, far more affirmative, reaching so far as to be the purpose of each individual.[181] It motivated the physician in that it was commensurate with the goal of helping a damaged organism, a patient, to transform and to live with comfort and confidence. Hence Goldstein's conclusion: all persons balanced neurophysiological dynamics with complex performances in their worlds, and they became all the more singular—never complete or perfected, but always irreducible to a group, always striving to develop their own norms, always menaced.

Conclusion: The Breakdown of Inside-Outside and the Denouement of Individuality

For physiologists, neurologists, and psychiatrists in the 1920s, individuality was at once an object, an ontological category, and a hermeneutic tool. It was an epistemological and experimental object because the different approaches to integration were explicitly and concretely invested in arguing that living beings were individual wholes integrated, integrating, compensating for, and responding to loss, and that they are threatened as such. In this way its partisans also turned it into an ontological category. Individuality was also central to hermeneutics, in that the principal figures discussed in this chapter (except for Henderson) used it to reshape the body's wholeness, understand and name actors and powers within it, and seek the meaning of collapse. It is this history of compounding interpretations and reinterpretations of experimental results and research objects, in conjunction with shifting epistemological systems, that led to the temporary resolution and refabrication of conceptual problems. Crucially, the alignment of individuality, freedom, and internal environment were forced interpretations of the signs offered by the body to the experimental, therapeutic, and epistemological apparatuses that these thinkers developed. The conceptual configurations that each thinker deployed were in turn at odds with those of the others. More interesting, thinkers used the very concepts they were ostensibly advocating in ways that often caused them to stumble into a dismantling of these same concepts.

Cannon, Goldstein, and Dale, and to a lesser extent Haldane, emphatically staged the individuality of a human body as something that became visible to the researcher thanks specifically to disintegration. The disarrangement of interiority laid the individual bare as an individual. Homeostatic operations, autopharmacological processes, and attempts at self-actualization were shared by every member of the species, and some even by higher animals in general, but only thanks to pathological research had it become possible to identify them within the bodily interior.

In Cannon's early work, emotions were signatures of specific internal disarrangements to achieve reregulation in response to external disarrangements, notably extreme stimuli. The organism did all this beneath the level of consciousness, as one or more systems in the body (the sympathetic nervous system or adrenal secretion) rapidly redirected the rest. Similarly, in his *The Wisdom of the Body* disarrangements like shock showed the way toward understanding the value, procedures, and even strategies of homeostatic regulation. When healthy, however, the body didn't lose its individuality. Regulation was itself a forceful operation that presumed the possibility of instability, proposed competing forces within the organism, and prepared reserves and excess to deal with the constant possibility of radical pathology and collapse. In these efforts, both Dale and Cannon thought, the individual remained just that: individual. Most of the time, however, it faded and hid, in what René Leriche would call in the early 1930s the "silence of the organs." In that silence, individuality did not vanish; it simply could not be easily demonstrated or identified. It remained asleep, just beyond epistemic reach. Goldstein agreed: the healthy individual pursues her self-actualization at the same time as she does not experience her wholeness, neurophysiological individuality, or interiority.

For whom was the organism an individual? Was individuality merely an epistemological category for the researcher, or was it an ontological category, one felt by the patient? The silence on which Cannon and Dale enthroned the normal organism suggested that for the organism individuality was still present but unfelt, silenced perhaps under the weight of the individuality exhibited in higher mental activity and reasoning.

Paradoxically, as Goldstein showed better than anyone, individuality became a category for the pathological researcher at the same time that it was animated and felt by the patient. The same operations that produced this individuality, both for the researcher and for the patient under study, were the ones thanks to which it was dismantled, disordered, and imperiled altogether.

For Dale, Cannon, and Goldstein, the categories of interiority, integration, and individuality were marked by asymmetric counterconcepts, all of which were to be identified with organismic collapse. Not only was integration best studied by way of the occasioning of disorder and disintegration; it was responsible for occasions, like shock, that truly imperiled it. Similarly, what guaranteed and systematized interiority was also what targeted the sufficiency of any traditional internal-external divide. For that reason, the triumvirate of interiority, integrity, and individuality became profoundly unstable, and individuality came unhinged from its traditional sites. Fight or flight, self-actualization, and histamine response were not merely claims on an irreducible individuality of the organism, or terms for the mobilization of internal resources and agencies by an organism; they were occasions on which the individual was being assaulted from the outside but also internally rearranged into carrying out activities (among them self-preservation) in a manner largely, if not entirely, beyond her control. Stimuli that disrupt the external milieu's relationship to the organism generated a profound internal disarrangement. Indeed, when we read the texts closely, this disarrangement involves fundamental and largely unconscious effacements of the inside-outside barrier, even ruptures coming from within.

For Dale, the awakening of histamine for purposes of self-protection—the very apparatus whose thresholds, amount, and severity of response differed in each case—collaborated with the antigen against the organism's own stability. In its animation—indeed, under normal circumstances—the marker of individuality attacked the individual being it was animated to protect. For Goldstein, self-actualization emerged as an idea from the dramatic suffering of individuals who found themselves in anguished and catastrophic situations. The pilot of the organism, self-actualization bridged internal and external environments, guaranteed subjective integration, and served as a measure of individual autonomy until the always-threatened organism was struck anew. One could no longer say "the organism and its milieu," as if this organism were simply a given, enclosed entity: the organism, ostensibly the original individual, was torn apart by internal forms of stability and destabilization—agencies that in health kept it individual. In times of danger individuality, the sovereignty of the subject over itself, came to belong to these systems that both defended and undermined the inside-outside barrier, which had signaled, even kept up, this individuality. Subsequent attempts to reinforce individuality on the social level (as in Cannon's "Biocracy" article and Goldstein's theory of self-actualization) veered between illusion and moralist proclamation.

The category of the individual for these thinkers was itself fragile. It was fragile because it emerged from and meant very different things to each researcher. It was fragile because individual organisms, in their own fragility and faced with the potential of collapse, placed individuality in question. An organism's individuality stood in the crosshairs of its own defensive weapons. But individuality remained fragile for another reason: as we have seen in Cannon's case, individuality did not quite capture psychological or higher mental functioning, but it assumed that this higher functioning was largely determined physiologically and operated in variation from that of other minds, much as the body did. By subsuming physiology into a broader neuropsychology, Goldstein's concept of the individual promised to overcome that. Yet by the same token, Goldstein shackled individuality to anxiety and self-actualization. The great achievement of his thought—imagining an organism that sets its own norms or works even in disease to achieve more restricted norms—also became its weakest point, as the attention Goldstein could pay to individual therapeutics thrust against the possibility of a systematic medical science capable of treating broader groups and populations.

This disentanglement of the individual from a larger group, the recognition in physiology that individuals are integral, threatened, and different from one another, bolstered a defense of the individual against the homogenization that occurred in seeing each one as part of a larger set. As Goldstein put it in his concluding words for *The Organism*:

> Individualization, which always means an emancipation from the superordinated whole, be it species, group, etc., involves a necessary contrast between the individual and his fellow men, and so brings about that *imperfection* which manifests itself in the catastrophic form of all coming to terms of the organism with the world. In that fact is given the transitoriness of all living beings bearing a specific individuality. This may well be the only genuine, real imperfection by the very nature of life, the imperfection which is inherent in life as such. It shows itself in the incompleteness of the individual's participation in that reality to which it belongs according to its nature.[182]

Traditionally Nietzsche, Marx, and Freud are credited with subverting the concept of individuality. We credit that subversion to Cannon, Dale, and Goldstein. As the individual, integrated, internally regulated body became the centerpiece of physiology and medical research, as it came to be celebrated for its individuality and difference, it came unbound from its strict identification with social existence. As it reached a greater distance from biochemistry, physiological individuality became a more precarious paradigm that could be undermined, as indeed it would be with the advent of molecular biology. And as physiology dissected the body for all the elements, chemicals, and performances that guaranteed its individuality, it also made that individualized body solitary and constantly menaced, almost one with the darkness of its own potential catastrophic collapse.

7 Psychoanalysis and Disintegration: W. H. R. Rivers's Endangered Self and Sigmund Freud's Death Drive

When Sigmund Freud announced his concept of the drive in 1905, he defined it as "the psychical representation of a continuously flowing, endosomatic source of stimulus (to be contrasted with simple individualized and externally-produced stimuli)."[1] Despite modifying psychoanalytic theory extensively, and detailing the sources, aims, and objects of the drives, Freud kept this basic definition in all essentials constant. The source of the drive is internal and continuous, not some external trigger. The aim of the drive is regulatory—to annul this organic stimulus, to remove "the state of stimulation at the source of the drive."[2] The main goal is satisfaction, removal of tension, equilibration with the pleasure principle. Until 1919, the pleasure principle ruled the human psyche and the organism as a whole: "The activity of even the most highly developed mental apparatus is subject to the pleasure principle, i.e. is automatically regulated by feelings belonging to the pleasure-unpleasure series," Freud declaimed in 1915.[3] The revisions—to the drives, the pleasure principle, and the reality principle—bridged two tendencies, to see the drives either as motors of human motivation or as the leading techniques of engagement with the world. Freud gradually restricted the original "anaclitic" model of the drives, according to which the libido was propped up on self-preservative drives like hunger or thirst and emerged by way of a loss of their original object (for example, the breast), and replaced it with a "narcissistic" model that saw even the ego as filled with narcissistic libidinal energy.[4] As he did so, he moved all the more emphatically into the regulatory, equilibrating purpose of the drives.

If these purposes can be discerned in Freud's early work, then they are all the more present in the late theory of the drives that contended with wartime violence. In 1919, Freud famously turned to ask, is there a *beyond* to this equilibrating mechanism of the pleasure principle? Does that beyond work perhaps toward a similar equilibrating purpose? We show in this chapter what it means that his answers were affirmative. Psychoanalysts were well aware of the wartime return of the problems of integration and individuality, and in this chapter we propose reading postwar psychoanalysis as a careful response to the contemporary physiological and neuropsychological turns.[5] The emergence of the death drive in Freud's *Beyond the Pleasure Principle* invented a system of integration that brought into psychoanalytic theory the profound traumas and some of the epistemological pressures of World War I. This should come as no surprise, but, like the connection between psychoanalysis and integration, it has been largely ignored.

In the context of theories of physiological integration, the later theory of the drives fits astonishingly well. Like Walter Cannon with wound shock, like Henry Dale with histamine, like Kurt Goldstein with aphasia and anxiety, Freud recalibrated his theory to account for trauma, to reassert the independence of the internal environment guided by the drives, and to ask what happens when a self that is suffering from a war neurosis attempts to handle a new status quo. This chapter shows convergences—even correspondences—of Freudian psychoanalysis and these other medical-physiological philosophies of integration, notably the psychiatry of W. H. R. Rivers. We have already encountered Rivers's importance in the psychiatric treatment of war neuroses. His theoretical work derived in part from Freud's thought and by 1918 had become a theoretical counterpart to Freudian psychoanalysis.

We begin with an outline of Rivers's psychiatric and quasi-psychoanalytic writings on psychic integration, instincts, shell shock, and the unconscious, particularly his "The Repression of War Experience" (1918) and his *Instinct and the Unconscious* (1920), and we compare Rivers's model of the instincts to Freud's theory of the drives.[6] We then pursue Freud's development of *Beyond the Pleasure Principle* in two steps that follow after his own two-stage composition of the text. The first step considers Freud's original displacement of the earlier drive theory, with the introduction of trauma and the death drive. The second engages the strange and belatedly added sixth chapter of the book, notably the phylogenetic and cell-biological revision of the death drive that propped it up on a metonymic linkage of single-cell organisms and "individuals." Finally we reinterpret the regulatory and individuating functions of the death drive.

The dual problem of integration and brittleness helps explain the otherwise perplexing urgency of the death drive in Freud's postwar writing. We argue that his drive for mastery, deriving from the death drive, is the metapsychological analogue of integrative neurophysiological attempts to respond to traumatic aggression by calling up the body's multiple systems in a semicoordinated, holistic response. *Death drive* is a name for the body's response to external shock, an effort to master wounds physical and psychic, to reintegrate the mind after its injury, to control one's own fate rather than simply succumb to violence.

We are far from the first to link Freud to Cannon and company. Looking backward from today to Freud, we can easily reconstitute a whole tradition that has made the point. Recent literature on "the science of stability" has occasionally linked Freud with authors like Cannon.[7] Jean Laplanche, discussing the way in which Freud's 1914 theory of narcissism refracted into a particular conceptual constellation the relationship between the ego and life, argued: "The ego . . . seems to take over the vital order as its own; it takes it over in its essence: constituted as it is on the model of a living being, with its level, its homeostasis, and its constancy principle."[8] Laplanche insisted on the link when looking at the death drive, which he argued "may be attributed to economic concepts derived metaphorico-metonymically from the register of biological homeostasis."[9] Laplanche's use of the term *homeostasis* derived partly from Jacques Lacan, who had declared it essential and explained its lack of explicitness on the grounds of Freud's reliance on Gustav Fechner's "inertia."[10]

Earlier still, the Hungarian psychoanalyst Franz Alexander had declared that "Freud's and Fechner's stability principle, according to which the ego's function is to reduce excitation within the organism, is identical with Cannon's physiological theory of homeostasis."[11] Alexander, considered a founding figure of "psychosomatic medicine" in the 1950s, further identified Freud's psychoanalysis and Cannon's focus on adaptive bodily changes as the "new methodological and conceptual advancements" key for his new science.[12] Ernest Jones, in his biography of Freud, also attempted to swallow up Cannon and cybernetics: in *Beyond the Pleasure Principle*, Freud had identified the pleasure principle with Barbara Low's "nirvana principle," that is, with "reducing to as low a level as possible the tensions induced by either instinctual or external excitation. . . . The common term used nowadays, with a very similar meaning, is that of Cannon's, *homeostasis*. . . . The whole train of thought is a remarkable anticipation of the modern science of cybernetics."[13]

Even in 1916, the psychiatrist William A. White, then superintendent for the Government Hospital for the Insane (St. Elizabeth's) in Washington, DC, attempted at the American Medico-Psychological Association's annual meeting to harmonize Freud with physiological and neurological thought. E. E. Southard responded with outrage at "how many eminent workers have been proved by Dr. White to put grist into the Freudian mill. Sherrington's integrative action; Pawlow's conditioned reflexes; Bayliss and Starling's hormones; Fraser's *Golden Bough*; Bergson's *élan vital*; everybody's pragmatism, and the like are all Freudian. Why not Darwinian or Newtonian physics, Aristotelian entelechy? Why not Faraday and Helmholtz? Perhaps all these worthies were actually engaged in lining a vista leading to Freud. I doubt whether Freud himself would claim so much. I myself believe that to claim so much is nothing but *phagocytosis of theories*."[14] That these systems might be related was not at issue; that White should cannibalize them on behalf of psychoanalysis was.

Freud's own link of biology to psychoanalysis in his discussion of the death drive is perhaps the most famous and enigmatic: "Biology is truly a land of unlimited possibilities. We may expect it to give us the most surprising information and we cannot guess what answers it will return in a few dozen years to the questions we have put to it. They may be of a kind which will blow away the whole of our artificial structure of hypotheses."[15] These references do not make the link a reality. The goal of our own phagocytosis of theories is to demonstrate historically and theoretically the alliance between the late theory of the drives and the integrationist approach that signals an epistemological alignment and even an isomorphism.

There is surprisingly little contextual evidence that Freud knew of Anglo-American research on physiology, the emotions, and drives or instincts (we will use *drive* for Freud and *instinct* for Rivers and other British writers).[16] No Freud correspondence survives with any of the contemporaries we have discussed in this book; nor is there clear evidence that Freud met any of them or used their work. Yet inklings do survive, as does evidence of Freud's evasion of trends endorsed by his contemporaries. Echoing perhaps William McDougall's first reference to "flight or aggression" in his 1908 *Introduction to Social Psychology*, or perhaps Cannon's "fight or flight" after *Bodily Changes* (although that book would have been nearly impossible to procure in the middle of the war, and Freud does not use any of its other terms),[17] Freud spoke of "flight or defense" in his 1916–1917 *Introductory Lectures on Psychoanalysis* and later studiously avoided the term and its staging of a danger situation by focusing on "anxiety" instead.[18] In 1919, Freud unequivocally approved of Jones's citation of Rivers. And perhaps most notably, a year later Rivers sent Freud a copy of *Instinct and the Unconscious*, the thrust of which Freud may have already known even though he received it after he had nearly completed *Beyond the Pleasure Principle*.[19] The structure of Freud's argument betrays some knowledge of the direction of contemporary biological discussion, and of the consequences, especially in Rivers and in response to Rivers, of taking a different path through claims of organismic stability.

The point of the highly detailed close reading of *Beyond the Pleasure Principle* that follows here is to show that as Freud dramatically modified the drive theory, he made his own the problem that had suddenly become paramount across the physiological and related sciences. Integration was affecting psychoanalysis and its claim to scientificity, and Freud worked to outmaneuver that threat. Insofar as our account lacks the conveniences of contextual evidence, it has the benefit of spiraling in on the relationship and meaningfulness of proximate concepts and shared epistemological premises. In Freud's attempt to reconstruct individuality, personal history, and death—in a language that was often sui generis—we see an effort to regain the initiative by designing the death drive so it would respond to a problem raised by physiological advances and their questions of integration and organismic collapse. Through the death drive and its expressions, the organism regained control of the internal scramble that occurred as a result of its wounding and its unconscious pressures.

W. H. R. Rivers's Revision of Psychoanalysis

From the correspondence between Rivers and Jones, only two of Rivers's letters survive. In the first, dated November 21, 1921, Rivers complained about *Beyond the Pleasure Principle*: "As speculation it seems to me very bad—among other things, the reinstatement of lifelessness is not the same as death. I shall be very glad to know if psychoanalysis just connects it to the wish for death, . . . but the speculation in *Jenseits* . . . seems to me to be . . . in the air and illogical at that. Is there any chance of the *Massenpsychologie* being out soon? I should like to take it away with me for my Xmas holiday."[20] It is an odd truncation of the death drive, exemplary of the controversies raised immediately upon the book's publication. Rivers's desire to move past it to Freud's *Group Psychology and the Analysis of the Ego*—a theme for which he prized Freud—is just as evident.[21] It would also be Rivers's last comment on it: only a week or so later, relations between him and the psychoanalysts declined precipitously. Rivers's abrasiveness in this letter has the whiff of a family squabble, but he and Jones moved on to mutual recriminations, ostensibly over whether Rivers was still willing to advocate psychoanalytic treatment or whether Jones was being obnoxious and unprofessional (which he was).[22]

Rivers would die suddenly seven months later while running for office with the Labour Party, and his contribution to psychoanalysis would remain a matter of debate. For Henri Ellenberger, Rivers was essential to the popularization of psychoanalysis in the period immediately after World War I; this is certainly plausible given Rivers's prominence as an intellectual and as president of the Royal Anthropological Institute.[23] Most psychoanalysts and historians discount this, interpreting him as a mere Freud lite who perhaps did more harm than good by bastardizing and desexualizing Freud's work.[24]

The story of Freud and Rivers is tantalizing, not least because of the lack of contact between the two. In *Totem and Taboo* (1913), Freud worked with Rivers's ethnographic research from the Banks Islands; however, he seems to have known of Rivers's work mostly through James Frazer's *Totemism and Exogamy*.[25] As Rivers concluded his contentious *History of Melanesian Society* a year after that, and began work as a Royal Army Medical Corps psychiatrist, he started to read Freud carefully. He recorded some strong objections, effectively asking for independence "not so much based on my clinical experience as on general observation of human behavior, on evidence provided by the experience of my friends, and most of all on the observation of my own mental activity, waking and sleeping."[26] Alarm bells should have gone off for psychoanalysts reading this passage: Rivers was adopting Freud and judging him, propagating analysis without undergoing it, and even pronouncing the sufficiency of his own self-analysis. This slapdash tactic—at least from the perspective of the psychoanalysts—tilled the ground for later confrontations. But at the same time, Rivers unreservedly praised the favorable "alteration in the attitude of psychiatrists towards the views of the psychoanalytic school" and expressed satisfaction that "the partisans of Freud have been led by the experience of war neurosis to see that sex is not the sole factor in the production of psycho-neuroses, but that conflict arising out of the activity of other instincts, and especially that of self-preservation, takes an active if not the leading role."[27] He would repeat these estimations and demands in the years to come.

After the war, Rivers became instrumental to the British Psychoanalytic Society. In his midfifties by then, with a storied résumé of contributions to physiology, psychology, anthropology, neurology, and psychiatry, and according to Myers no longer timid and diffident, he also had no trouble voicing strong intellectual objections.[28] Jones installed "the distinguished anthropologist" as the first president of the British Psychoanalytic Society in order "to heighten its prestige."[29] But Jones, too, by then the policeman of psychoanalysis in England, had no qualms about criticizing Rivers all the while employing him—for example, when Rivers delivered "The Repression of War Experience" as a lecture on December 4, 1917. Jones ironized Rivers's pretense that his terminology and argument differed from Freud's and gave him an impromptu lecture of his own on psychoanalytic theory.[30] By a year after the Armistice, the lack of interaction between Freud and Rivers becomes interesting, especially as Rivers invited Freud (through Jones) to visit and speak at Cambridge in 1920.[31] Jones mentioned Rivers in a couple of letters and cited Rivers's "The Repression of War Experience" in his "War Shock and Freud's Theory of the Neuroses."[32] In subsequently introducing the volume *Psychoanalysis and the War Neuroses*, which included this paper, Freud specifically approved of that citation.[33] Gradually offense became the predominant psychoanalytic reaction to the half-convert's criticisms, as evidenced in Jones's extensive 1920 review of *Instincts and the Unconscious*.[34]

Because of the evident purging of the Jones archive (so that only the letters concluding the Jones-Rivers relationship survive), and because of the later breakup of Freud's and Rivers's archives into incomplete parts, it is difficult to judge the extent of the interaction. As mentioned already, Rivers sent Freud a copy of *Instinct and the Unconscious* in 1920; it likely arrived after Freud's completion of the second and near-complete draft of *Beyond the Pleasure Principle*.[35] What of the period before that? Freud, who so obsessively strategized the spread of his science, cannot have remained indifferent to Rivers's contribution or to the conflicted attraction that led Rivers to renegotiate concepts such as repression, the drives (as "instincts") and censorship.[36] Nor could Freud have remained ignorant of Rivers's many writings and speeches on instincts, the unconscious, and repression, given the promptness with which he read his English and American fellow travelers.[37] Certainly their gist was indirectly conveyed by Jones's own articles, which Jones routinely mailed to Freud. And conversely, the demands Rivers put on analysis by focusing on self-preservation are precisely the ones that Freud sought to obviate in his revision of metapsychology in *Beyond the Pleasure Principle*. To pursue the "encounter," we first turn to Rivers's engagement with and objections to psychoanalytic ideas through his neurological and other work.

Like Freud, Rivers reached the unconscious at least in part through his work in neurology. Freud had begun as a neurologist and had pursued, in his famous 1895 "Project for a Scientific Psychology," the goal of creating a "psychology for neurologists."[38] As he wrote at the time to Wilhelm Fliess, what he sought in that project was to force neurologists to begin anew with a model that would be less anatomical and mechanistic because it would be congruent with his psychology.[39] Rivers's own arrival came late, during the war, and was not due to his work in psychology in the 1890s or in the "disused missionary house" on Murray Island where he, Myers, McDougall and Charles Seligman set up their psychophysical apparatuses for ethnographic observation.[40] The conceptual conduit for Rivers's psychiatry was the year he spent toying with Henry Head's radial nerve.

Rivers and Head derived from their experiment a theory that two fundamental forms of sensibility, protopathic and epicritic, ruled the organism's form and structure. Rivers treated the protopathic sensibility as hitherto unknown: it "would never have been suspected" before Head's experiments.[41] As Head's radial nerve regenerated, sensation was gradually restored to his arm, first in a deep, intense, and undifferentiated mode—that is, as a disproportionately painful and imperfectly localizable jolt.[42] This was protopathic sensibility. Only thanks to gradual relearning did the epicritic return, allowing for a distinction between different kinds of sensations (between hot and cold, for example) and for engagement in a more complex, tender, and nonbinary fashion with the stimuli involved.[43] Rivers and Head derived from this the existence and complexity of an afferent nervous system that recombines sensations through "the combined activity of more than one group" of organs and nerves.[44]

Once Rivers began thinking about psychiatric matters, he hardened the meanings of *protopathic* and *epicritic*.[45] In prioritizing the two systems' distinctness from each other,[46] he claimed that "the two modes of sensibility represent two stages in phylogenetic development," with epicrisis as the more recent evolutionary development.[47] Compared to it, the protopathic system was deeper, more primitive: binary, violent, all or nothing.[48] Rivers expressly identified protopathic sensibility with the unconscious and the epicritic with consciousness and intelligence.[49] His protopathic unconscious was a very different apparatus than Freud's, and at a July 1918 joint session of the British Psychological Society and *Mind* journal, titled "Why Is the 'Unconscious' Unconscious?," Rivers publicly disagreed with Jones (and with Carl Jung's disciple Maurice Nicoll) in explaining the unconscious through the protopathic and epicritic distinction. In this now-unknown lecture, Rivers began by refusing to consider the *why* question in favor of addressing how the unconscious "becomes unconscious and again enters into consciousness."[50]

Rivers relayed the case of a patient suffering from dissociation—a "state of active existence apart from experience readily accessible to consciousness."[51] Because a past experience could interfere with current harmony, it was "by some mechanism . . . shut off or dissociated," rendered unconscious and unavailable. Experiences recalling that original traumatic occasion, given the "vast scale on which injuries of the central nervous system have been produced during the war,"[52] broke free the instinctive, protopathic layer that stored the earlier experience.[53] In the psychophysiological injury and the regeneration that followed, protopathic sensations remained dominant over consciousness; as Rivers later put it, protopathic sensations "persist in a latent form ready to come again into consciousness."[54] Just as severe injury destroyed epicritic sensibility, traumatic experience forced consciousness to obey primitive instinctual behavior. A strict isomorphism could be postulated between physical responses to injury and the processes used by the individual to forget or suppress events that have breached or disordered epicritic sensibility. In all this humans resembled animals, and adult humans regressed to childhood.[55] Rivers then turned to phylogenetic recapitulation to explain the persistence of the unconscious: nerve regeneration was comparable to—even a manifestation of—the early development of the nervous system at both the individual and the evolutionary levels. Its sudden emergence involved processes of both extreme sensitization to external stimuli and loss of intellectual comprehension.[56]

This approach offered to Rivers a fantastic solution for apparent Freudian aporias. If Freud had overemphasized the sexual dimension and misinterpreted self-preservation in terms of hunger and thirst rather than of recoil from danger, Rivers could offer a form of care and a theory of instincts that identified individuals with the traumatic situations they had survived and the subsequent pathology they were enduring. Whereas for Freud the unconscious was the general site of agency, for Rivers the unconscious remained latent in the form given to it by evolution until it was laid bare at times of threat or need. Although it was present throughout our everyday life, its self-protective instinctual function would shoot through on an occasion necessitating fight or flight ("flight-or-defense" in McDougall's original term, which Rivers adopted).[57] The persistence of that trauma required the analyst to help complete the movement from clotted wound to regenerated selfhood. Rather like Cannon, Rivers organized the instincts and the unconscious in terms resembling military reserves—in such a way that they might serve basically as a counterattack and preserve the organism but at the same time overpower consciousness. Particularly forceful stimuli coming from the outside would then be suppressed (often unsuccessfully) into the latent neurobiological unconscious as the organism healed itself.

Rivers further changed the Freudian relationship of consciousness to the unconscious by reformulating repression in such a way as to enable the patient to retain autonomous agency, thanks to which she or he could both repress and (usually) lift the suppressed back into consciousness.[58] In "The Repression of War Experience," on the subject of psychic trauma, Rivers started out with a distinction between *repression* and *suppression*, complaining that Freud's understanding of *repression* left vague a necessary distinction.[59] The distinction, and the definition in which it resulted, amounted to a profound conceptual shift in *Instinct and the Unconscious*, as Rivers's terminology and argumentation destabilized what remained of the theoretical and clinical scaffolding of psychoanalysis.

Rivers held that Freudian repression did not distinguish active, voluntary, conscious repression carried out in a witting, concerted fashion from suppression as a state in which memories are inaccessible to memory.[60] His "repression" is not a pathological process; indeed, it is a "necessary element in all education and social progress."[61] Only when repression fails "to adapt the individual to his environment" does it become harmful. It would be hard to overstate the consequences of this amendment. By casting repression as a voluntary act of "thrusting" undesired elements out of consciousness, of thinking them away to the protopathic reserve, Rivers was affirming the priority of consciousness over the unconscious, identifying the latter not with the source of all mental activity as Freud had, but with a mechanism for the persistence of suppressed memories that remained unknown to the subject.

Rivers posited here that what mattered was less the individual's entire life history, marked by infantile sexuality and the Oedipus complex, than "times of special stress."[62] War experience was of particular value because it compressed "the training in repression normally spread over several years" into "short spaces of time."⁶³ Rivers translated the pleasure principle into a clinically oriented version of consciously experienced pleasure: it was for him the ego's pleasure, rather than that of the entire organism, that mattered to the psychiatrist.⁶⁴ With this in mind, and given his very abbreviated chances at treating soldier-patients who were supposed to be returned quickly to the front, Rivers attempted to isolate "some aspect of the painful experience which would allow the patient to dwell upon it in such a way as to relieve its horrible and terrifying character."⁶⁵ This set the trauma squarely within conscious experience, as it implied that, in principle, the patient could remember the event. It also implied a specific relationship between patient and psychiatrist: the latter's purpose was to enable the patient to face the situation and to cease actively removing it from consciousness.⁶⁶ Whether for therapeutic efficacy or out of the theoretical need for a strong, flexible, adaptive ego, Rivers tended toward a reduction of the transferential situation that so troubled Freud. His patients had to be helped along to carry out the right kind of suppression, and he was there to help them through it.

A further complication related to the place of drives in the Freudian corpus involved the translation of Freud's work into English because the concept of instinct in late-Victorian England forced a set of conceptual shifts that could not be anticipated and afflicted his reception there. As Kathleen Frederickson has recently argued, by 1910 the meaning of *instinct* in Victorian England had shifted away from the mid-nineteenth-century conceptual framework in which it had been contrasted to reason (the latter belonging to advanced, sophisticated Europeans, whereas instinct was to be identified with savages, animality, and sexuality). By the early twentieth century, instinct "offers (in contradistinction to the lack associated with desire) a plenitude," in an assemblage that was far from identical to or coextensive with sexuality.⁶⁷ As important, and against Freud, "the would-be bifurcation between instinct and self-consciousness fail[ed] to hold fast in any sustained way."⁶⁸ Instinct was quite an illusory, ill-placed term for Freud's concepts *Trieb* and *Instinkt*, which ran afoul less of residual Victorian prudishness than of a shift in the web that had fundamentally universalized and largely desexualized instinct.

Rivers's use of *instinct* fits Frederickson's description of its place in late-Victorian thought: it is universal in humans and animals, insects even; biologically innate (whereas intelligence is acquired);⁶⁹ it is neither opposed to sexuality nor submerged in it; and it is what makes rationality possible. Rivers was bound to object to Freud's prioritization of libido as the crucial instinct. He was not simply objecting to Freud's thought; he was refusing the placement of desire at the center of instinct's scope, a categorization that to him seemed forced.

In *Instinct and the Unconscious*, Rivers conjured three classes of instincts: self-preservative instincts (the key ones), instincts dedicated to "the continuance of the race," and social instincts to maintain "the cohesion of a group."[70] The idea that these are different groupings was essential to his argument: not only did this allow him to speak of all sorts of behaviors as instinctual; it allowed him to see some instincts as epicritic and others as protopathic and unconscious. For all his doubts concerning Freud, Rivers based his "social instincts" on Freud's thought and that of his own former student and colleague McDougall, going so far as to replicate Freud's explanation of the origin of the Oedipus complex.[71]

The other two classes of instincts were to be divided between instincts "committed to the continuance of the race," including sexual instincts, and "instincts of self-preservation," notably danger instincts. Self-preservation was the central instinctual response of the organism, affecting the organism in danger.[72] For Rivers, the danger instincts have a specific end: to protect the animal from danger. They were in that sense a particular kind of self-preservation. Self-preservation as a "danger instinct" made little sense in the Freudian schema: it could not work for economic reasons related to the expenditure of energy, nor for dynamic reasons concerning the interplay of the unconscious and consciousness. Rivers proceeded to place an entire homeostatic mechanism within this psychic apparatus—with the result that Cannon's handling of bodily changes was imported into his argument, either intentionally or through the common source to be found in McDougall. In the encounter with great danger, the hidden, latent unconscious broke through to protect the organism or received the brunt of the external assault in such a way that the consciousness might suppress, and later be haunted by, the wound that was inflicted.

The First Draft of Beyond the Pleasure Principle
London, Trauma, Narcissism

After the war was over, intermediaries traveling between London and Vienna, including Otto Rank, conveyed to Freud the greatly improved, albeit revisionist, atmosphere in London. In early 1919 Jones sent Freud his essay "War Shock and Freud's Theory of the Neuroses," which would be included in *Psychoanalysis and the War Neuroses* and constituted an explicit retort to Rivers and J. T. MacCurdy over their revisions of psychoanalysis on the basis of their "touching" accounts of soldiers' struggles. Jones was direct: "The experience of neurotic affections engendered by the war . . . has enabled the critics of psycho-analysis to put forward the view that the factors invoked by Freud in explanation of these affections need not be present, and therefore cannot be regarded as essential, in the way maintained by him, whereas . . . a different set of factors is undeniably present and operative; not only so, but these latter factors are held to be all-sufficing . . . [for] the aetiology of the conditions in question."[73] Jones's text mentioned versions of fight or flight twice, as reactions to "external danger . . . in various activities suited to the occasion—flight, concealment, defense by fighting, or even sometimes by attacking," and destabilized Rivers's terminology of fear by arguing that even unafraid soldiers take "suitable measures of flight, fight, or what not."[74] The upshot was clear: fight or flight and self-preservation do not amount to a theory of psychic functioning.

Freud read Jones's essay, appreciated the threat posed by revisionism, then adopted and critiqued elements of Jones's counterargument in a letter on February 18, 1919, a few weeks before the essay was published in German and just a couple of weeks before he began the first draft of *Beyond the Pleasure Principle*.[75] In this epistolary appraisal, in which he admitted that he had "made no analysis of a case of war-shock," Freud explicitly supported Jones's argument that anxiety prepares the organism to resist being overwhelmed by a fight-or-flight situation:

> Anxiety is a protection against shock. . . . The condition of the traumatic neurosis seems to be that the soul had no time to recur to this protection and is [overrun], taken by the trauma unprepared. Its shield against stimuli is overrun, the principal and primary function of keeping off excessive quantities of stimulus frustrated. Then narcissistic libido is given out in shape of the signs of anxiety. This is the mechanism of every case of primary repression, a traumatic neurosis thus to be found at the bottom of every case of transference neurosis. . . . The War Neurosis is a case of internal narcissistic conflict in the Ego.[76]

It is a telling letter for several reasons. First, it makes clear that Freud endorsed Jones's scruples against non-Freudian theorists of shell shock. Freud's subsequent endorsement of Jones's use of Rivers and MacCurdy in his introduction to *Psychoanalysis and the War Neuroses* indicates that he was by then aware of Rivers's ideas, at least in generic form. Second, Freud's paragraph says a lot about his theory that anxiety prepared the ego for the excessive stimulus of trauma. Whereas fight or flight doesn't reach into the unconscious, an anxious psyche is a prepared one, one in which consciousness does not take priority; the ego is flush with narcissistic libido, and any effort to preserve it is thus based on libido. (Freud then splits the ego into a "habitual" and a "fresh warlike" ego, divided in narcissistic conflict, as though to insert narcissism and libido into the discussion of danger instincts.) But third, the text matters because of what in *Beyond the Pleasure Principle* it does not presage. The letter's fast replay of Freud's theory of the drives and its insistence on the war neuroses as a matter of narcissistic conflict goes only so far. An introductory goal of *Beyond the Pleasure Principle* was to present and subsume this account of the war neuroses into the drive theory, although in the effort to do just that, Freud also depleted and overhauled the theory.

Integration and the Pleasure Principle: The Enigma of the Two Drafts

Freud completed a first draft of *Beyond the Pleasure Principle* in early May 1919.[77] As Ilse Grubrich-Simitis first pointed out, Freud substantially modified a typescript version of the original text to create a second draft, which, according to Ulrike May, he completed in July 1920.[78] May's critical edition of the different versions of the text makes it possible to study the major alterations of *Beyond the Pleasure Principle* from the original 1919 manuscript to its 1920 publication, including the crucial addition, in 1920, of the entire sixth chapter, which by itself amounts to some 40 percent of the text, and also of passages to chapter 1 as well as the book's concluding paragraph.[79] Chapter 6 in particular has long baffled students of psychoanalysis. In chapters 1–5 Freud had constructed a fairly tight argument that upturned the formerly conventional link of the "self-preservative drives" to "life" by treating the self-preservative drives and the new drive toward mastery as components of a drive aimed at death and the restoration of absolute prenatal peace. But then, after sitting on the text for a year, he decided that the limitations worrying him could be resolved by redefining the death drive proper and in reaching outside psychoanalysis, namely to a biology of germ cells. So, in the new chapter, and without rewriting or suppressing previous ones, he moved to "salvage" the self-preservative drives from the death drive.

Some of the rationale behind this may be attributed to a need for legitimation over the thoughts on death. Similarly, eschewing the ostensible conclusiveness of the idea that "the aim of all life is death" and that there is another "group of drives" that lead toward death, Freud employed a two-pronged tactic in the new chapter 6, first to define the death drive, and second to ensure that this definition did not seem unscientific.[80] But in what follows we argue specifically that what was key to the revision was a bolstering of the claim that the aim of the psyche, and hence the organism, was stability, the minimizing of tension. Key to the second draft was a systematic effort to shore up the psychoanalytic conception of constancy and to think through the relation of the death drive to that constancy. Concerns with integration and with a kind of psychic homeostasis premised on the pleasure principle are as central to the deployment of this argument as they are to its tensions and the attempt to resolve them.

Freud opened the book with a strong assertion of psychoanalysis's reliance on the pleasure principle in order to establish the constancy-pursuing form of regulation and integration of mental events. The pleasure principle aimed to reduce tension to a minimum, and Freud noted its general preponderance over the psyche and the course of mental events, including unconscious ones, such that it was the principal apparatus of "automatic regulation": "The mental apparatus endeavors to keep the quantity of excitation present in it as low as possible or at least to keep it constant. This latter hypothesis is only another way of stating the pleasure principle."[81] Freud introduced the self-preservative drives soon thereafter by localizing them in the ego and arguing that they replaced the pleasure principle with the reality principle, thereby forcing a second regulation that involved, beyond internal pressure and desire, the capacity of this ego to render the organism capable of harmony with the surrounding world (what Freud usually called "reality").[82]

From there, the essay advanced in a fairly straightforward test of exceptions to the pleasure principle, which led to the overhaul of the drive theory. War neuroses, children's play, and transference, Freud argued, were situations in which subjects exhibited a compulsion to repeat painful events, situations where subjects' mental actions could not be explained as aiming at a pleasurable effect. Freud then turned toward the problem of trauma and the categorization of the drives as life or death drives.

It would be a great exaggeration to suggest that the completed book amounts to an intellectual bayonet drill with Rivers or anyone else at the end of the blade, but contrasts to a shell-shock-based approach are immediately clear. Freud targeted the major points of any theoretical revision that would prioritize "self-preservative instincts" or consciousness. He repeated that all primary processes come from the unconscious and bind—organize, redirect, control, or give content to—impulses, memories, perceptions, and the processes themselves as they reach consciousness. He further recalled that pleasure for one system (the ego) can be unpleasure for another (the unconscious), and vice versa:[83] but such conflicts between systems should not be taken to mean the abandonment of the pleasure principle.[84] He forcefully reaffirmed the priority of libido over self-preservation and just as pointedly refused the claim that self-preservative drives surface only at exceptional moments. By making clear that repression is unconscious, not under the direct control of consciousness, Freud further rejected anything that might resemble the revisionist "suppression."[85] However, Freud did not explicitly return to the earlier models of the self-preservative drives, namely, hunger and thirst; he simply left the self-preservative drives contentless, as though he were aware of and quietly countering Rivers's danger or hunting instincts.

Instead of allowing any "aggressive" or "danger" drives, as did Rivers, who proliferated them in self-preservation,[86] Freud expressly relocated aggression and sadism within a single drive for mastery, which he staged in cases of war neuroses and in forms of child's play that involve an Oedipal attempt to control abandonment by the mother (the famous *fort-da* game). As he deployed his argument and presented the compulsion to repeat past events as the definitive symptom of a gesture that was not constrained by the pleasure principle, in his most remarkable difference from unorthodox views, Freud overturned the expected wisdom that self-preservative drives were attached to simply advancing the ego or the self through life. On the basis of the conservative nature of the drives, he posited that their attachment lay instead in their guidance of the organism to death, understood as the quiescence that had preceded birth proper.

Trauma, Mastery, Inside-Outside
In this original draft, Freud engaged the problem of psychic integration across the theory of trauma—the "event" in which a breach of consciousness both destroyed existing psychic integration and unleashed internal dynamics for the mastery of unconscious forces unleashed by the event. Whereas Freud claimed that psychoanalysis was largely covered by his explanation of traumatic and war neuroses—as concomitant with the theory of the neuroses—he also found the compulsion to repeat that he saw emerging in play, in the aftermath of trauma, and in the psychoanalytic encounter to be an excess that demanded the fundamental revision of the theory of the drives.

Traumatic events, in other words, were not in and of themselves challenges to the drive theory. Only to the extent that they revealed a situation demanding new mastery that could not be recuperated by the existing drive theory did they open up this ostensible outside to the normal integration of the pleasure principle. To wonder why Freud first recovered trauma and the war neuroses into the existing schema but then perplexingly morphed and expanded them toward what lies beyond the pleasure principle—to see why these couldn't be handled in terms of an economic mechanism or mishaps of repressive or other forces—is to watch him till the conceptual ground, dismiss any pressure to move toward self-preservation, and destabilize the drives to allow for a new organization. The tension of this schema culminates in the subsumption of the self-preservative drives in the new tendency toward death, crowning the first draft of *Beyond the Pleasure Principle*. (Only in the second draft do the new integrative and disintegrative tendencies become clear.)

One further element of the theory of trauma in *Beyond the Pleasure Principle* is particularly significant for Freud's regulatory and integrative efforts, and it exemplifies the tension in the first draft. Here is Freud in his earlier *Introductory Lectures to Psycho-Analysis* (1916–1917), privileging an argument common in wartime, namely that trauma was externally imposed: "An experience which within a short period of time presents the mind with an increase of stimulus too powerful to be dealt with or worked off in the normal way . . . must result in permanent disturbances of the manner in which the energy operates."[87] In his February 18, 1919, letter to Jones, Freud repeated this formulation, claiming that if the psyche's "shield against stimuli is overrun, its principal and primary function of keeping off excessive quantities of stimulus [is] frustrated."[88]

In his introduction to *Psychoanalysis and the War Neuroses*, Freud repeated this point, but he added another crucial one, rendering terribly ambiguous the role of the topographic interior of the psyche:

> In the traumatic and war neuroses the ego of the individual protects itself from a danger that either threatens it from without, or is embodied in a form of the ego itself, in the transference neuroses of peace time the ego regards its own sexual hunger (libido) as a foe, the demands of which appear threatening to it. In both cases the ego fears an injury; in the one case through the sexual hunger (libido) and in the other from outside forces. *One might even say that in the case of the war neuroses the thing feared is after all an inner foe, in distinction from the pure traumatic neuroses and approximating to the transference neuroses.*

The theoretical difficulties which stand in the way of such a unifying conception do not appear to be insurmountable; one can with full right designate the repression which underlies every neurosis, as a reaction to a trauma, as an elementary traumatic neurosis.[89]

In *Beyond the Pleasure Principle*, the claim that "the thing feared is after all an inner foe" became fundamental: trauma was not merely an event that overwhelmed the organism from the outside. As Laplanche argued in a classic analysis of Freud's early theory of trauma, "Everything comes from without in Freudian theory, it might be maintained, but at the same time every effect—in its efficacy—comes from within, from an isolated and encysted interior."[90] Ruth Leys's masterful analysis of the instabilities inherent in Freud's trauma theory in the 1920s similarly accentuates that Freud was very ambivalent about causality in trauma, and that his theory fundamentally destabilized any division of inside from outside[91]—and as a result, any firm subject-object distinction.

In the analysis of trauma in chapter 4 of *Beyond the Pleasure Principle*, Freud was indeed ambivalent, if not on the "necessary" external spark of the stimulus, which broke through the protective shield and into the self, then certainly on the manner in which consciousness was overwhelmed.[92] He had proposed earlier that the external layer of any "living vesicle" was dead, a shield that was part of, but almost removed from, the vesicle itself, so that "the amounts of excitation impinging on it have only a reduced effect."[93] He now told that a breach of the shield set in motion "every defensive measure" and put "the pleasure principle . . . for the moment out of action."[94] Famously, "there is no longer any possibility of preventing the mental apparatus from being flooded with large amounts of stimulus, and another problem arises instead—the problem of mastering the amounts of stimulus which have broken in and of binding them, in the psychical sense, so that they can then be disposed of."[95]

On its own this passage seems to correlate with the idea of an external intrusion of the trauma. However, all is not as clear as that, because in the previous paragraph Freud had just warned the reader of two things. First, no shield protected the ego *from the inside* in the manner that the now-broken shield used to from the outside: excitations "in the deeper layers extend into the system directly," and they were far stronger and unchecked in "intensity" and qualitative "amplitude" than ones that came from without, even if they were commensurate with the "system's method of working."[96] In a manner that recalls the depiction of the internal environment described since Claude Bernard, the inside was far more consequential and significant for the organism's functioning than the outside, from which the organism usually retained relative independence.

Second, and more important, Freud proposed that this same system dealt with unpleasurable excitations from within by projecting them onto the outside to raise the shield against them; the system acted as if excitations came only from the outside.[97] A few pages later, he twice emphasized that any failure to bind internal excitations resulted in a situation "analogous to a traumatic neurosis": until binding was accomplished, no pleasure principle would supersede it.[98] This attention to interiority (the "inner foe" and the permeability of the inside-outside barrier, whether because of projection or the operation of binding) dovetailed with his theory of deferred action, or *Nachträglichkeit*, which postulated that trauma required more than a single event. As is well known, Freud was intentionally unclear as to whether it was the event of psychic injury or its triggered recollection that caused the trauma; plural instances were necessary because it was not a single event but the more-than-one-event structure that rendered the breach traumatic.[99] in other words, in this theater the protagonist was what was already "inside." The unrecollected past participated in the attempted and failed binding of stimuli.

Thus Freud's assertion that the breach of the barrier resulted in a "flooding" of the mental apparatus by "large amounts of stimulus" was anything but straightforward, and his phrasing here matters. It is unclear whether the flood was due to the massive anticathectic defense operation and the internal pressure now released as a result of the external breach, as well as to intrusion itself. (Even the single-cell metaphor he used here recalls more the bursting outward of cell plasma following a breach, rather than a bursting inward.) Freud immediately reverted to a language of invasion and defense, using military metaphors that are obvious throughout, but in arguing that the organism's preparations would avert, scramble, or diminish any breach, he still showed that the heightened cathexis due to anxiety, just like the defense operation, was not the simple consequence of any incoming stimulus, however intense.[100] Even the internal attempts to bind had been released from the operation of the pleasure principle.

Why did this all matter? It dismissed the intrusion model used by theorists of shell shock, including Rivers, in which the goal was for the doctor to help the ego close the wound. Rather, it was not the stimulus that mattered but the internal preparation and response. The massive binding operation that occurred as the organism tried to control the flood of stimuli played its part through its ostensible force and its overwhelming of consciousness. Just as anxiety, as a narcissistic cathexis of the ego, would have protected the psyche (the "system") from being breached, the flood of stimuli was not merely external but also effected from within: it could even be caused solely from within. Freud, who had insisted before that fright in the face of the unexpected object was a necessary component of a traumatic confrontation,[101] now made it clear that fright was not itself the cause.[102] Neither the external breach nor the fright would on its own suffice: it was the lack of preparedness and the subsequent overreaction that made a situation traumatic, and it was the failure *to bind properly* that resulted in trauma.

The very least that can be claimed of this account is that it aligns expertly with the physiological accounts we have discussed throughout this project, for which the external stimulus is far less significant than the organism's attempt to respond to, encode, and control—an effort that occurs in the body's integrative apparatus. In an interesting convergence with Cannon's *Bodily Changes*, Freud abstracts *backward* from the fight-or-flight situation. Cannon sought the biological preconditions that fail to disable the fight-or-flight situation, and Rivers credited traumatic neurosis to a third option: freezing, then failing to take either of the other options. Freud credits the absence of anxiety, the rapid onset of fright, with that traumatic result. The inside then destabilizes the compromised ego and attempts to master the traumatic situation, in the absence or suspension of the pleasure principle. What matters is not the external wound alone, or even the fight-or-flight situation, but the attempt and failure at reintegration. In terms familiar to us through wound shock, Freud proposes the unconscious as an internal psychological environment that operates largely independent of the outside world, as a grouping of forces that expends itself in consciousness and is only intruded on in a forceful manner on traumatic occasions. It involves a shift of the theater of agency that is highly similar to that of the physiologists: for them, most activity is effected internally, even beneath consciousness (by the sympathico-adrenal and other systems, to keep with the Cannon reference), and the trigger of an injury or lesion at once points toward the outside world and turns systems of the whole individual into the motors of internal devastation.

The Self-Preservative Drives and the Death Drive in the History of the Drive Theory

This account of trauma leaves the architecture of the drive theory quite unstable and offers, alongside Freud's attempt to recover the war neuroses within the existing schema, the moment of a decisive "beyond" to the pleasure principle: the organism's attempt to master an intrusion or bind stimuli can be understood neither as part of a self-preservation attempt caused by the ego nor as mere libidinal effort. It belongs instead to a dual moment in which the organism at once contributes to its traumatization and attempts to reintegrate, though for reasons other than pleasure or self-preservation. Binding (succinctly put, the matching of stimuli that enables their control) cannot occur in the name of either of these processes, yet its operation is precisely one of a reregulation as defense, an instance of the drive to mastery practiced within the system and aimed at its control of external impositions.

The second decisive moment of the first draft thus concerns the self-preservative drives, and it is central to the tension in Freud's text between the recuperation of the war neuroses and the deployment of a new conceptual mechanism of disintegration and reregulation. Freud's distinction between self-preservative (ego) drives and the sexual drive began in the *Three Essays on the Theory of Sexuality*. There, he modeled the self-preservative drives not on danger but on hunger and its satisfaction through nutrition: they involved the individual's survival in the satisfaction of her basic needs. By comparison, the sexual drive had a general, diverse function at the heart of virtually all other human intentions and targets, including sublimated ones that did not immediately appear to be influenced by sexuality. Unlike the self-preservative drives, the sexual drive had neither a specific goal nor a singular, well-defined aim, and it emerged out of self-preservative drives through what Freud called anaclisis ("resting upon").[103]

In 1914, emphasizing his distance from Jung and his ilk, Freud wrote "On Narcissism: An Introduction."[104] Freud reconceived the ego as fundamentally narcissistic—as approaching love objects in a manner that could never fully expend or cancel ego-libido, the eroticized conception of oneself. At stake was "an original libidinal cathexis of the ego, from which some is later given off to objects, but which fundamentally persists and is related to the object-cathexes much as the body of an amoeba is related to the pseudopodia which it puts out."[105] This meant that the ego was self-preserving because it was already narcissistically invested with libido. This redistricting of the drive theory rendered obsolete the strict distinction between self-preservative and sexual drives, thus complicating the original anaclitic model.[106]

In his February 18, 1919, letter to Jones, as in the introduction to *Psychoanalysis and the War Neuroses*, which he wrote the following month, Freud refused the self-preservative drives' self-sufficiency in dealing with trauma and specifically handled anxiety, which was aimed at preventing trauma, as the product of narcissism.[107] But in the first draft of *Beyond the Pleasure Principle*, the idea that anxiety helped preserve the organism because it was pervaded with unconscious narcissistic libido—not because it was premised on self-preservation—remained quiet, latent. Freud defined anxiety not in relation to narcissistic libido but merely as a form of preparation that staves off the danger of trauma.[108] This implicitness allowed him to separate anew the self-preservative drives from libido.

: : :

To repeat the cardinal tension of the original text: whereas the challenge of trauma and the not-so-new war neuroses had ostensibly been met, something else remained. Freud posited repetition compulsion as the principal symptom of this other order, this new and larger structure—exhibited in play, in the transference, in posttraumatic dreams. Repetition compulsion showed the psyche trying to master earlier, perhaps traumatic, events that had been bound only partially. It thus confirmed that the work of the drives was fundamentally conservative: by replaying, often ad infinitum, something that had occurred earlier in life, the drives invariably attempted to return to a status quo antecedent to that event. But the pain of the drive to master was inescapable and severe: it could not be covered by the pleasure principle. Repetition compulsion thus betrayed the organism's urge to remove these events from defining it.[109] Like a wound-shocked soldier replaying the event of his injury to imagine the harmony and possibilities before, Freud's traumatized psyche was compelled to restage or repeat earlier events, even though doing so posed no pleasurable gain, feeling, or resolution.[110] Rather than aim at minimal, harmonious tension, as per the pleasure principle, repetition compulsion sought, Freud claimed, zero tension, none whatsoever. Only the removal of all tension would end the painful urge.

Freud further noted that repetition compulsion involved an equivalent of phylogenetic recapitulation, Ernst Haeckel's biogenetic law.[111]

For Haeckel earlier species forms recur in each organism's embryonic life, showing the signatures of evolution out of lower species. Now insofar as repetition compulsion involved a continuity with life before birth, species life, life as potentiality, it also differed from Haeckel's law. For Freud, repetition compulsion marked instead the psyche's opposition to the organism's development, its desire to refuse being forced forward. Adaptation in the midst of a hostile environment compelled the development and transformation of the individual and, through the individual, the species. The death drive, by contrast, carried the law of resistance to this forward compulsion through repetition—or rather through the tension that it forced on the organism by obliging it to bear the pressures of the environment.

This reversion to a status quo ante posited first that death constituted the absolute release from tension, with the end goal of the conservative drives being the return not to a pleasurable state but to one exceeding it toward the absolute rest of inanimacy.[112] At this point in the text, Freud had not yet divided between the drives: he treated all of them as exceeding the pleasure principle and its regulatory, homeostatic structure. In other words, thanks to the restructuring, the pleasure principle was quietly dislodged from its integrative function and eminence—and yet it was granted just that place, except only under "normal" circumstances.

A few words later in the text, Freud recuperated libido as "the true life drives."[113] Using recapitulation again, he posited a single-cell origin of all sexual and reproductive behavior and suggested that the sexual drives "watch over the destinies . . . that survive the whole individual," that they even "provide" these destinies "with a safe shelter while they are defenseless against the stimuli of the external world."[114] These drives, he added, are conservative in that they "bring back earlier states of living substance" (as opposed to prenatal inanimate states) and "jerk back [the psyche] to a certain point to make a fresh start."[115] "A certain point" points to the beginning of a new life, the birth of a new organism, created by sexual union thanks to libido, but also at life at an earlier stage of the libidinal individual when she or he is jerked *back* to the pleasure principle.

And so, libido became attached to a perpetuation of life of the species through reproduction. Yet while making this set of claims freeing libido from death, Freud also ensured that no such escape from the overall rubric of these "other drives" that led to death was forthcoming for the self-preservative drives, even though he repeatedly raised the possibility and, in the second draft, amended the argument at just this point to call a timeout—to "reflect: It cannot be so."[116] Still he concluded: "I know of no certain example from the organic world that would contradict the characterization I have thus proposed."[117] Unlike the life or sexual drives, which obeyed the pleasure principle anew, the self-preservative drives stayed resolutely within the embrace of death. Freud was not immune to the paradox of what we might call his catachrestic use of the term *death*:

> The implications in regard to the great groups of drives which, as we believe, lie behind the phenomena of life in organisms must appear no less bewildering. The hypothesis of self-preservative drives, such as we attribute to all living beings, stands in marked opposition to the idea that the life of the drives as a whole serves to bring about death. The theoretical importance of the drives of self-preservation, of self-assertion and of mastery greatly diminishes, so to speak. They are component drives whose function it is to assure that the organism shall follow its own path to death, and to ward off any possible ways of returning to inorganic existence other than those which are immanent in the organism itself.[118]

Freud's formulation of this passage is peculiar: the second sentence sits awry and in some contrast to the third. How could Freud place a reminder of the "marked opposition" (*merkwürdigem Gegensatz*) of the self-preservative drives to death-oriented *Triebleben*, only to immediately and without note subsume them in the next sentence as component drives (*Partialtriebe*)? It is as if Freud needed to stage the claim of the self-preservative drives a last time before subsuming them, alongside those of self-assertion and mastery (*Macht- und Geltungstriebe*), into the death drive, then retroactively and quietly objecting to the "hypothesis."

In this stark inversion the wide gulf between Freud and Rivers became evident. Freud overturned any Rivers-like hope in self-preservation because the struggle of self-preservative drives against death or injury no longer aimed at the organism healing itself toward a new stability. Instead, it is in the name of death itself, death as chosen internally, that these drives act on the organism's behalf—death not dictated by the environment or by failures of adaptation: "We have no longer to reckon with the organism's puzzling determination . . . to maintain its own existence in the face of every obstacle. What we are left with is the fact that the organism wishes to die only in its own fashion. Thus these guardians of life, too, were originally the myrmidons of death."[119]

Biology, Reintegration, and the Death Drive in the Second Draft
A Second Integration

Without abandoning psychoanalysis to the war neuroses or libido to the "hypothesis of the self-preservative drives," Freud had largely undone the original regulatory system that posited the continued dominance of the pleasure principle. To elucidate this turn, we could quickly look to Jacques Lacan's 1954–1955 seminar in which he offered his listeners a "small key" for interpreting the text, and dramatically mused that "either [*Beyond the Pleasure Principle*] makes not the least bit of sense or it has exactly the sense I say it has." Lacan's "key" was to code "psychic reality" in explicitly homeostatic terms as "an internal organization, which up to a certain point tends to oppose the free and unlimited passage of forces and discharges of energy, such as we may assume to exist, in a purely theoretical way, intercrossing in the inanimate reality. There is a closed precinct, within which a certain equilibrium is maintained, through the action of a mechanism which we now call homeostasis, which absorbs, moderates the irruption of quantities of energy coming from the external world."[120] *This*, Lacan surmised, was the position from which Freud had begun, and lacking the language of homeostasis, it had been merely "Fechnerist." Repetition compulsion disturbed this picture: "This system has something disturbing about it. It is dissymmetrical. It doesn't quite fit. Something in it eludes the system of equations and the evidence borrowed from the forms of thought of the register of energetics as they were introduced in the middle of the nineteenth century." And shortly thereafter: "The principle of homeostasis obliges Freud to inscribe all his deductions in terms of investment, charge, discharge, energy relations between different systems. However, he realizes that something doesn't work in all this."[121]

We should like here to go a step past Lacan and show that, after rendering the old stability mechanism "dissymmetrical," Freud asserted an other homeostasis through the biologization of his metapsychology and the refinement of the death drive. This effort provides the central gain of the second draft: a new and more complex set of homeostatic processes, attached to a different and trickier mechanism of disintegration and individuation.

In May 1920, Freud returned to *Beyond the Pleasure Principle* after a year of intermittent engagement, during which he began the study that would lead to *Group Psychology and the Analysis of the Ego* and that distinguished the social life of the self from the drives.[122] Rivers's *Instinct and the Unconscious* was not out yet, but Rivers had delivered the lectures that were at its heart the previous summer, and the climate in Britain remained tricky, especially for the orthodox psychoanalysts. Jones visited Freud just as the latter was beginning the revisions, and the London situation cannot have been a matter of indifference.[123] In the two months that followed, Freud added the introductory citations of Fechner that buttressed his opening claim that the psychic apparatus tends toward constancy—that is, the original logic that the theory of repetition compulsion and the drive toward mastery undermined. In the chapter he now added, Freud managed to overturn much of the thinking and many of the more imposing conclusions of chapter 5, notably the relation of self-preservative drives to the death drive. Employing the term *death drive* with greater conviction and working to sharpen it, Freud also shifted to discuss biological matters in greater detail.

He opened the new chapter by repeating and phrasing anew the distinction between sexual and ego drives as one between life and death drives. The logic shifted slightly as it became more explicit: while the self-preservative drives place the organism on a path toward an inanimate state (death), the sexual ones "clearly aim . . . at the coalescence of two germ-cells differentiated in a particular way"[124]—to "prolong the cell's life and lend it the appearance of immortality,"[125] "perpetually attempting and achieving a renewal of life."[126] A considerable part of chapter 6 established Freud's biological approach to unicellular organisms, which he had begun in chapter 4 of the first draft. Freud argued now that cell division in protozoa did not amount to eternal life by endless duplication, and hence to an exception to the death of every living being. Cell division thus did not flatly contradict the death drive or the strict division of the life and death drives that Freud sought.[127]

This biological acceptance of the death drive allowed Freud to reintroduce narcissism into the relationship between self-preservative and sexual drives.[128] Having aligned the death drive with the self-preservative drives in the first draft, Freud plied them away from the rigors of death. Narcissism extinguished the original anaclitic foundation of the sexual drives and provided an alternative: if narcissistic libido served as the dynamic foundation of self-preservative drives, then the self-preservative drives could not simply function on the side of the death drive.[129] If, however, the death drive did not work with self-preservation, then of what did it consist?

Haeckelian Biogenesis, Weismannian Embryology, and Death

Freud's "biology," that "land of infinite possibilities," deserves closer attention. Given that much has been written about the relation of Freud's metapsychology to biology, we restrict our discussion to two points concerning his prioritization of the single-cell organism, which Freud repeatedly named "the individual."[130] First, Freud's use of phylogenetic recapitulation, which has often been regarded as compromising the claim of psychoanalysis to scientificity,[131] and second, the place of death in post-Weismannian genetics, which underwrites Freud's argument on the destiny of the biological individual and relates closely to contemporary physiological concerns.

Until this point, Freud had related single-cell organisms to individuals by way of two metonymic relations.[132] First, the cell's membrane stood in for the "crust" of the psyche, the ego—a shield toward the outside but not the inside.[133] (Trauma is usually understood by reference to a wound on a physiological body, but Freud aimed for the cell instead.) Second, the single cell was metonymically related to the whole individual human insofar as each was permeated by the two types of drives, life and death. These could be more easily recognized or modeled in the single cell, so Freud used germ cells for heuristic and recapitulative purposes to reengage the drives.[134]

Physiologists had a peculiar ambivalence regarding the use of evolutionary and recapitulation theories. From Charles Sherrington's *Integrative Action of the Nervous System* onward, the use of evolution as an explanatory device for bodily integration declined. Structure and disorganization took over from genesis. Still, phylogenesis remained a necessary pillar of physiology; phylogenetic recapitulation remained current in physiological and psychiatric texts. Part of the reason was a species-based thinking—as Ernest Starling put it, "Throughout the animal kingdom we find the welfare of the individual subordinated to that of the species."[135] Rivers explained the place and recurrence of instinct in human nature by referring to instinct as the clearest psychological carrier of evolutionary origins and transformations; failing to recognize this evolutionary dimension amounted to failure to handle psychology.[136] Even more explicitly, Cannon noted that homeostatic regulation confirmed "the idea that the history of the individual summarizes the history of the race, or that ontogeny recapitulates phylogeny."[137]

Phylogenesis has been leveled as an accusation against Freud; yet as we have seen, physiologists considered it a useful tool for understanding how integration developed anew in each individual. The prioritization of the individual—the individual's successes and dysfunction—also explains the shift in emphasis away from a racial politics exhibited by, for example, George Crile in his recapitulationist physiology of occupied Belgium. Where Haeckel's biogenetic law—that ontogeny recapitulates phylogeny—had been useful primarily for varieties of Romantic or racial politics, with the "new physiology" it became relevant to an individualism reliant on the individual in peril. Freud's case was highly similar: he avidly endorsed phylogenetic recapitulation, but he used it as a signature of the evolution that each living being resisted when it exhibited pathological symptoms of the drives guiding it. Haeckel's biogenetic law became instead a dual appeal: to the evolutionary basis of the drives, which were indefinitely replayed and reborn in each individual, and to the particularity of the individual who resisted adaptation and sought, through the conservative and recapitulative drives, to recover and return to a lost past.

Freud also approximated the physiologists when he guided the drives beyond the individual, casting them as forms of compulsion that traversed that individual as they traveled across the species. In *Elements of Human Physiology*, Starling had also followed August Weismann and declared that "the production of new individuals [is the] the crowning point of an animal's existence." As far as the parent organism was concerned: "The rest of the cells of the body afterwards die, having served their function when they have reared the new family of individuals. Thus from a broad standpoint all the complicated processes that we study in physiology, all the toil and turmoil of human existence, are nothing but 'the by-play of ovum-bearing organisms.' The biological destiny of man is accomplished with the production and rearing of a new individual."[138] Starling further distinguished cells aimed at the persistence of individual life from germ cells aimed at the perpetuation of the species:

> As we go higher in the scale we meet with more highly differentiated organisms, consisting of cell colonies, each member of which has its own appointed task to fulfill; and here we find that the office of reproduction also is confined to one cell or group of cells. The immortality of the amoeba has been transmitted to this group of cells. From this point onwards, in the scale of animal life, we may regard the reproductive cells or germ-plasma as being continuous through successive generations. With the production of each new generation the germ-plasma divides into two parts: one part, the somatic half, forming what is generally understood as the individual and being differentiated into various forms of cells, to perform the multifarious functions of reaction associated with life; and the other half, persistent in its primitive form as the reproductive part of the individual, ready

when the time comes to divide again and give birth to a new generation.[139]

With "biological destiny"—a human being's "achievement" being the production of a new individual—each individual was defined as much by death as by the complexity of these cells and their appointed tasks.

Starling was not exceptional: textbooks discussed at length the different ways cell death and organismic death could occur as a result of a modification in the environment, yet most of them left some space for a kind of death that was internally caused, without injury, disease, or external intervention. This natural or "physiological" death—immanent death, not death caused by decay or external intervention—even where it was postulated only *ex negativo*, served physiologists well.[140] Freud, in *Beyond the Pleasure Principle*, recalled August Weismann's original construction of this argument and commented:

> [Weismann] considers that unicellular organisms are potentially immortal, and that death only makes its appearance with the multicellular metazoa. It is true that this death of the higher organisms is a natural one, a death from internal causes; but it is not founded on any primal characteristic of living substance and cannot be regarded as an absolute necessity with its basis in the very nature of life. Death is rather a matter of expediency, a manifestation of adaptation to the external conditions of life; for, when once the cells of the body have been divided into soma and germ plasm, an unlimited duration of individual life would become a quite pointless luxury. When this differentiation had been made in the multicellular organisms, death became possible and expedient. Since then, the soma of the higher organisms has died at fixed periods for internal reasons, while the protista have remained immortal.[141]

The affinity with Starling's account is striking. Starling, too, tracked back and forth between talking about each individual cell and the whole body—even "the conscious being"—just like Freud did.[142] The two claims were isomorphic where matters of life and death were concerned.

Physiologists did not continue down this path for long. In Starling's *Principles of Human Physiology*, as in William Bayliss's 1915 *Principles of General Physiology*, passages like this were largely left out, perhaps as a result of the influence of Mendelian genetics, which displaced

Weismann-like concerns for a new technical language.[143] It is likely that Starling's choices in his earlier work were explicitly Weismannian for the same reasons as Freud's in 1920: they articulated the common vision of continuity across time and offered a "scientific" explanation of death in relation to life. Freud's own achievement was to codify the drives in a way that facilitated their relation to his understanding of sexuality, all the while providing for them a formal analogy and a properly *longue durée* logic, from the first cell to modern humans. Having no interest in Mendelian genetics, he could discuss germ cells and embryology, as well as the biology of protists, with specific reference to authors whose engagements did consider the problem of death or decay. Once again like Starling, Freud had come to see the death as a biological necessity given the uselessness of "an unlimited duration of individual life."[144]

Emphasizing the proximity of Freud's biology to the contemporary handling of death in physiology resolves several puzzles at once. It undermines the sense that what he called his "speculation" was in contradiction with the physiological understanding of individuality and life. On the contrary, it was largely in accord with presuppositions concerning the individual's place in the species, just as metaphor of the living vesicle in the discussion of trauma approximated the treatment by Cannon and others of the organism's totalizing response to injury. The biologistic revision also allowed Freud to distinguish the supraindividual species character of the drives—specifically the death drive—from their functioning in the individual human being. In the same gesture, Freud organized the revised death drive of the second draft as the mechanism central to the individual's self-integration and self-regulation. Integrated individuality, which Rivers had identified in the self-preservative instinct and its conscious reconstruction of autonomy, could now appear in inverted form in the death drive, in the individual's self-subjection to repetition compulsion.

The Death Drive in Freud's Chapter 6

It is time to begin putting the puzzle together. At the beginning of Freud's chapter 6, the life-death opposition between the libido and self-preservative drives seemed to be clearly established: although the self-preservative drives place the organism on a path toward the inanimate state that is death, the sexual ones "clearly aim . . . at the coalescence of two germ-cells differentiated in a particular way" to "prolong the cell's life and lend it the appearance of immortality."[145] It is at this point that Freud turned to embryology, to find in it ciphers that would show the death drive to be as fundamental as processes of cell division and conjugation, which he identified with a primitive form of the life drive. Eventually proclaiming the death drive to be biologically acceptable—that is, to work with his metonymy between whole individual and single cell—Freud returned to the relationship between self-preservative and sexual drives but found it now mediated by narcissism.[146]

This completely disorganized the schema offered by the end of chapter 5. Calling germ cells complete narcissists, Freud forced narcissism back into the drive theory.[147] Narcissism finally extinguished the original anaclitic foundation of the sexual drives and provided an alternative foundation whereby narcissistic libido could be considered a partial foundation of self-preservative drives.[148] It thus forced Freud to reconsider his alliance of the self-preservative drives with the death drive, all the while maintaining that the fundamental dualism of his thought precluded any monistic reduction and preserved the death drive.[149]

The second draft's explicit pivot to narcissism did not undo Freud's entire conceptual construct. Had narcissism been brought into the first draft of *Beyond the Pleasure Principle*, imported from the 1919 introduction and Freud's letter to Jones, and had it been cast as the reason underlying anxiety's defense of the ego, or the bindings effected by the "massive anticathexis," then Freud's appeal to the suspension of the pleasure principle during trauma would make no sense. Narcissism would become just another name for self-preservation. By precluding that solution, by maintaining trauma and the war neuroses in his earlier schema, Freud could create the category of the death drives and index a "beyond" the pleasure principle and recuperate the war neuroses. Now, once Freud pulled narcissism back into the argument, he very quickly reordered the drives to reimport the self-preservative drives into the category of life drives.[150] Did this not empty the death drive of content? No, Freud responded, recalling the drive for mastery and foregrounding sadism.[151] Sadism had originally appeared as a libidinal pathology; Freud "forced [it] away" from libido and allied it with repetition compulsion.[152] He instead interpreted sadism as a displaced instance of the drive toward mastery that attempted to master and bind past, unbound experiences. It was "displaced" in that it targeted another and not oneself.[153] Trauma, play, sadism, and resistance in the transferential encounter all involved repetition compulsion in a manner that gave form to the death drive as their underlying force.

The death drive could now be said to have a well-defined place and organization in psychic life, as the individual's attempt to handle, encode, and bind events the individual had suffered, memories that had been restored, and intentions that had been inflicted and controlled. In a revision of chapter 5, Freud buttressed his note that the organism wishes to die only in its own fashion: "Hence arises the paradoxical situation that the living organism struggles most energetically against events (dangers, in fact) which might help it to attain its life's aim rapidly—by a kind of short-circuit."[154] Instead of allowing the environment to dominate, the living organism followed the death drive.

Promising "one or more optima,"[155] Freud constructed a quite peculiar formulation:

> How is it that the coalescence of two only slightly different cells can bring about this renewal of life? The experiment which replaces the conjugation of protozoa by the application of chemical or even of mechanical stimuli enables us to give what is no doubt a conclusive reply to this question. The result is brought about by the influx of fresh amounts of stimulus. This tallies well with the hypothesis that the life process of the individual leads for internal reasons to an abolition of chemical tensions, that is to say, to death, whereas union with the living substance of a different individual increases those tensions, introducing what may be described as fresh "vital differences" which must then be lived off. As regards this dissimilarity there must of course be one or more optima. The dominating tendency of mental life, and perhaps of nervous life in general, is the effort to reduce, to keep constant or to remove internal tension due to stimuli (the "Nirvana principle"), . . . a tendency which finds expression in the pleasure principle; and our recognition of that fact is one of our strongest reasons for believing in the existence of death drives.[156]

Not only was death the aim of individual life but libido, serving the species, unexpectedly *increased* tension. Freud proposed here that conjugation, which by metalepsis he identified with sexual union between human beings, involved the raising of tension, so that fresh stimulus both demanded and could not guarantee a reduction of tension. Libido cannot promise any resolution in the pleasure principle. In this passage, the death drive promised exactly that: it fell to the operations of the death drive to control or systematically reduce all internal tension that was not easily released—to even hope to abolish it. To the extent that this was still a matter of pleasure, the pleasure principle was "one of the strongest reasons for believing in the death drives," and Freud restated that "the life process *of the individual* leads for internal reasons to an abolition of chemical tensions, that is to say, to death."[157] By contrast, at this point the life drives were to be identified with "fresh 'vital differences'" (*neue Vitaldifferenzen*) and the increase in tension.

This was an astonishing turnabout from the first draft of the book and, in fact, from the ostensible identification of libido with the pleasure principle. Shackling the death drive to the pleasure principle and the reduction of tension led Freud to expend several pages in doubt of this "third step in the drive theory."[158] Nonetheless, he returned to it in his concluding paragraph to the book—another passage substantially reworked from its original draft—when he claimed that the pleasure principle worked "to serve the death drives," as opposed to the life drives, which in a later revision he would even call "the breakers of the peace."[159] The irenic, supposedly pure prehistory of the individual stood in contrast with the way libido sustained life in its originality and created new life through reproduction and the achievement of pleasure.

Pace Lacan and Laplanche, for whom the death drive undid an undoing of homeostasis, Freud had not abandoned homeostasis at all but rather complicated it considerably.[160] By the conclusion of the second draft, the language of individuality was resolutely attached to the death drive—the psyche's attempt to master trauma and endosomatic forces and to die in its own controlled manner. Freud advocated the death drive *as an individuating power*, both through his metonymic deployment of biology and through his retrieval from the death drive of the internal rationale that aids control and mastery in the individual. Systemic equilibration occurs not because in the death drive the individual seeks to be maintained in "self-preservation," but because the death drive obstructs the forward movement of the species, seeks an optimum of pleasure tied not to reproduction but to stark harmony, and seeks to (mostly unsuccessfully) master traumatic, incomprehensible events and render the individual whole again.

The "Myrmidons of Death," or the Individuating Meaning of the Death Drive

As opposed to the life drives, which, by the conclusion of the second draft, had come to govern the perseverance of the species and the capacity to go beyond or survive the individual,[161] the death drive met demands specific to "the individual" or "the organism as a whole." Already in the first draft there were inklings of a contrast between this whole individual and the life drives, and these went well beyond the poetic reference to each individual's desire to control his or her own death:

> [The self-preservative drives] are component drives whose function it is to assure that the organism shall follow its own path to death, and to ward off any possible ways of returning to inorganic existence other than those which are immanent in the organism itself.[162]

> The pleasure principle long persists, however, as the method of working employed by the sexual drives, which are so hard to "educate," and, starting from those drives, or in the ego itself, it often succeeds in overcoming the reality principle, *to the detriment of the organism as a whole*.[163]

From the moment when Freud began engaging biology in chapter 6, he systematically aligned terms such as *the whole organism, the living organism*, and *the individual* with the death drive, except when he purposely distinguished two individuals to discuss each one entering into conjugation or sexual union. In the latter case, he explicitly and repeatedly contradistinguished germ cells from the individual soma. Thus, the metonymy of the individual cell identified libido with germ cells, leaving each individual body behind on its own.[164] By the same token, the individual or soma related to the libido and the species only asymptotically. To make the former case, and to link the death drive with individuation, Freud looked to render the claim that "the aim of all life is death" more precise: "the life process *of the individual* leads for internal reasons to an abolition of chemical tensions, that is to say, to death."[165] Never did he link the life drives to the individual:[166] even if libido granted (or aimed to grant) pleasure to the individual, it always slid toward the species. It situated this individual within the lineage belonging to the species, and even if there were no such thing as a species-specific drive, the libido bore no individuating characteristics. Even if the self-preservative drives (hunger, thirst) resembled markers of individuality infused with narcissistic libido, they too could not, so to speak, be trusted.

This point means little unless we read it together with Freud's argument that the death drive made itself known through its symptoms. These symptoms were quite specific to a life's history—to the reception or suffering of external stimuli as these had broken through differently for each individual, and to the establishment of each individual *as* an individual, as a being who lived a particular life that occasionally disintegrated while always reintegrating and thus trying to control itself. The component drives of the death drive—namely, mastery (in both drafts), sadism and aggression (in the second draft), and self-preservative drives (until narcissism stole these away)—may have pointed beyond individual life (in the sense of pointing before birth), but they emerged only in the effort to handle and master past events as these had been specifically experienced by the individual psyche, events that had to do with traumatic breaches, memories, and responses, and that in Freud's economic terms had everything to do with failures and pressures of binding. In this regard, the death drive and individuality coappeared in a movement that for the individual concerned herself alone. Repetition compulsion—the key symptom of the death drive—marked the psyche's opposition to the organism's continued phylogenetic development and externally imposed adaptation.[167] Its main and continuous process, the internal effort at mastery of earlier, only partly bound, and perhaps traumatic events, pointed both backward toward prenatal peace and forward toward mastery: "Each fresh repetition seems to strengthen the mastery they are in search of."[168]

Was this a new homeostasis? Did the argument that the pleasure principle served the death drive, when combined with the attempt to regulate, order, and control traumatic intrusions and recollections, suffice to indicate a thrust toward individuation in the death drive? Did the perennial dream of controlling oneself and imposing oneself on the world by pulling back a libido that would always seek to move forward and offer new lives help close the circle and assert the impulse toward death as the signature of the self? Was it plausible that individuation and perpetual reintegration could be separated from self-preservation? Did the collapse of this integration resemble homeostatic theories through the argument that trauma fields the mental apparatus as one trying and often failing—hence often *worsening*—a breach from the external world? Or would the invocation of Fechner and the principle of constancy in chapter 1 still be the only form of homeostasis for Freud?

To us, that early reference to the principle of constancy had come to take on a very different character: after chapter 6, it had become a reference to a past theory that had once been relevant to psychoanalysis but now had to be pinpointed in the introduction for that version of the pleasure principle to be overcome. Indeed, the death drive had become homeostasis without the name. Its expressions and symptoms integrated and reintegrated the patient and individual in such a manner as to hold her together, in the same gesture with which they forced her to move toward her past and at the same time to foresee her end. Love pushed on, always beyond the individual or the self, whereas the violent work of systemic equilibration and individual selfhood—mastery, imposition, and violence—fell to the death drive, with its mastery of tension, trauma, the ego, and the other.

The nearly contradictory imperatives that the concept of death took on satisfied the surprising synthesis it obliged: first, the return to prenatal harmony, with its zero energy, a situation that would precede the coming of this life, and second, the compulsive replay of partially bound earlier life events in the futile hope that one would finally be done with them. Death also forced a third imperative, if only thanks to Freud's catachrestic lexical maneuver: this was the movement toward the actual end of life, the path to this end, the perennial attempt by the psyche—indeed, the unconscious—to control how life moved against external impositions and evolutionary movements. Suspended from these three temporal regimes was the life of the individual as individual: always seeking an impossible harmony in the complete absence of environments internal and external, always seeking to divert its ultimate and everpresent death in such a way as to master it, always failing to do so and trying again. Remaining always other to itself—in that Freud explicitly denied the death drive the name of *life*, granting that moniker only to libido—the death-bound life of the individual scratched and pulled at the chance to remain integrated and masterful, never finding more than new threats and intrusions to belie this chance.

Conclusion: The Effect of World War I on Psychoanalytic Theory

Compared to the first epistolary conversation with Jones and even the original 1919 draft, the completed *Beyond the Pleasure Principle* displays greater awareness of the threats posed by the English psychiatrists and the war neuroses, as well as awareness of the promises of emphasizing individuality and integration in a structure organized by trauma and disintegration. We might thus conclude our speculation concerning Rivers and Freud by stressing the aggressive manner in which Freud refused and rendered impossible every single theoretical contribution that Rivers made. The future of psychoanalysis hung in the balance of these different individuating integrations.

The self-preservative drives could be redirected to the ego, as per Rivers, but to Freud this meant limiting their mechanism severely. The patient could not simply "remember" in the transferential encounter, as Rivers hoped; instead, Freud insisted, the patient repeated compulsively and deceived the analyst. Trauma involved less a breach than an upsurge of the unconscious to both displace and control conscious matters. Integration occurred not thanks to an autonomous consciousness, but thanks to a highly complex unconscious drive that imposed individual death at every turn. Rivers's system of instincts could not have made sense to Freud. It disobeyed as much the pleasure principle as the place of the unconscious, the limited place and claim of the self-preservative drives, and the basic elements of impetus, aim, object, source, and pressure involved in each drive. But by the same token the new schema of the drives that the war neuroses and their treatment forced on Freud could not have looked like anything but "bad speculation . . . in the air and illogical at that" to someone like Rivers.

The same was the case for Goldstein, who, in his focus on self-actualization, the "single" drive of the organism, could not bring himself to read a "death drive" as anything but an absurdity, and who saw libido as obviously excessive.[169] A blend of clinical, biological, and metapsychological reworking had led Freud to a concept of death that managed the same concerns as his contemporaries, that produced a new concept of selfhood and individuality, and that nevertheless could not be immediately recuperated—despite its remarkable similarities—into their theorizations of the body. And yet Freud wrote as though he knew he had to master a proto-homeostatic theory and to individualize the human being that faces death, even at the cost of his own concept of the libido and its alignment with each human being. Binding and unbinding, the work of death in the drive theory, became the work of psychoanalysis in the new physiology.

Part Three

8

The Political Economy in Bodily Metaphor and the Anthropologies of Integrated Communication

The protagonists of this book endorsed political positions ranging from liberal to socialist. Ernest Starling, a self-made man and scientific humanist, was disgusted by the English class system and spent time struggling against conservatism in the medical establishment as much as in the army during World War I.[1] Horrified by the air quality he studied in mine shafts, John S. Haldane advocated for the Miners' Union from 1907 until his death. Kurt Goldstein was incarcerated and tortured by the Nazis in 1933 more for his political engagement on behalf of the Union of Socialist Physicians than for being a Jew.[2] Five months before his death in June 1922, W. H. R. Rivers sought political office in the House of Commons with the Labour Party. In a curious effort to shore up his candidacy, he gave public lectures on the role of psychology in politics replete with analyses of the group mind, education, and mental hygiene, and he argued that socialism is the politics appropriate to human nature because preagricultural societies were more naturally socialist than individualist.[3]

Walter Cannon spent the later 1920s and 1930s admiring the efficiency of the Soviet one-up on the European welfare state. As Peter Kuznick has shown, after hosting Ivan Pavlov and helping to shield him from the Soviet government, which Pavlov detested in its early years, Cannon warmed to Bolshevik support of medical research.[4] In 1935 he visited the Soviet Union for the International Congress of Physiology—an impressive propaganda affair showcasing Moscow's commitment to socialized medicine—and he returned impressed, comparing that commitment favorably to the American disinterest in physiology. While his conservative colleague Lawrence J. Henderson "strongly disliked the policies of the New Deal," Cannon celebrated the welfare state's capacity to ensure social stability.[5]

Shared political affiliations were the result of several commonalities: in academic and intellectual milieus, in medical ideology and scientific humanism, in attentiveness to which parties and political theories might support improved medical care. The war played a role as well, especially as they encountered it as doctors in the field, the clinic, or the laboratory. The tidier pictures of the organism they had developed over the prewar decade collided head-on with the experience of watching so many men suffer, fall apart, and die. This experience conditioned efforts to project outward from science toward politics. In a 1922 tribute to British physiology, physiologist Walter Fletcher concluded that it had no limits. He called for the application of physiological knowledge and aims in the public domain. However, those who heard this call did not simply translate physiological fantasies into political aims. The relationship between biology and politics was navigated more across the metaphoric and metonymical qualities that bound together the discourses of biology and politics than across biographical, directly political, or otherwise causal lines.

In what follows we are not interested in the political preferences or biographical details of these authors. We describe instead the world of political and social thought in which physiological integration emerged and made sense; the claims against which it stood out; the politico-economic realms it affected; and the course of its influence on theories of communication, language, and symbolization in anthropology, cybernetics, and philosophy. Thus we look first at the rise of the political, legal, and internationalist language of the integration-crisis dyad in the 1920s. As integration became a key category in international law, economics, and theories of authority, so did its counterconcepts: crisis, collapse, disintegration. Biological thought aided and framed much of this discussion by offering many of the principal metaphors necessary to its language and conviction, updating the parallel of the body physical to the body politic, offering an easily comprehensible and citable model of cooperation and crisis, and insisting on the priority of the totality of a system under threat, not just of its points of injury.

Second, we trace Cannon's political and anthropological thought in the 1930s, both because of its postulate of a social homeostasis and because of its part in the development of Norbert Wiener's cybernetics and Claude Lévi-Strauss's structural anthropology. Wiener and Lévi-Strauss retained versions of Cannon's social thought that made possible the transposition of his conception of the body onto theories of communication and social control.

Third, with Marcel Mauss as our guide, we look at the emergence of a concept of the symbolic in philosophy and anthropology, which was explicitly based on the neurological theories of Goldstein and Henry Head, and which became definitive of the philosophical and anthropological works of Ernst Cassirer, Mauss, and Lévi-Strauss.

These revisions spiraled outward from the transformed—even deconstructed—concept of the individual, and they breached and carefully redirected social and political thought—theories of norms, normality, and normativity; of symbolic exchange and linguistic coding; even of society and social structure.

Integration, Politics, and the Interwar Period

When the "war to end all wars" broke out, Starling, a fervent Germanophile, reported "an overwhelming feeling of treachery" and declared that he would never speak German again. He took up French, which, his friend and mentee Charles Lovatt Evans remarked, he "masticated teutonically."[6] Five years later Starling would need his German again: in 1919 he twice visited Germany to prepare a report for British Parliament.[7] At the time he was particularly interested in dietetics.[8] Using interviews and statistics, he assembled a history of German production, imports, and rationing, and traced the steep decline in caloric consumption since 1914, which was still perilously low. He offered a dietitian's explanation of the economic consequences of the peace and even predictions regarding German politico-economic sympathies. At the heart of this argument was the disruption of German economic organization by the war. With metaphors of integration and complexity prominent in his arsenal, he declared, "The more highly evolved the economic organization of a State, the more sensitive it is to injury by interference with its normal routine."[9]

The quality of the soil had declined as a result of the failure to import or produce appropriate manure—and therefore to maintain and reinforce the disrupted agricultural system. Interference and crisis in the distribution of food had rendered it unusually inequitable. Germany's flat-rate rations had not paid sufficient care to the nonhomogeneity of individuals, and this had led to a near-fatal corrosion of the economic and social fabric: "The German Authorities felt themselves obliged to attempt to ration strictly the whole food of the population. No such system as this, involving all the main articles of food, can succeed unless it is combined with mass feeding. Although we may speak of an average man and assign to him a certain ration, the amount of food required by an individual depends on his stature, age, sex, occupation and environment. No two individuals are alike. A ration sufficient for one individual will be too much for another, and too little for a third."[10] We begin to see Starling's priorities in the way he held up a mirror for the national body to see itself in individually malnourished bodies. Individuals suffered because of their difference—even more so if they sought to support kin who, because of divergences from the norm, needed more. The country in turn suffered from the imbalance as the population became far less suited to fighting off disease or maintaining its socioeconomic life.

The results were disastrous. While outlining pathologies of malnutrition—especially for women—Starling added vivid images that buttressed the sense that a continuing sociopolitical disruption was reflected in the physical constitution of individuals and the nation: "In a great city such as Berlin, . . . two-thirds of the population are living on a low level of vitality; they are much wasted, and when stripped, they are seen to have no fat, the neck being hollow and the ribs distinct. They move slowly, are dull and apathetic."[11] Malnutrition had a wilting effect: individuals became standardized, indistinct, then faded into themselves. The national body increasingly became an undifferentiated mass that led the country into a vicious cycle of the collapse of socioeconomic complexity. "The subversion of authority in Germany, the increased moral laxity and diminished confidence in the country which have followed defeat and revolution, will render it in the coming years still more difficult to control the producers." Hence Starling's policy proposal: adequate provisioning of food would restore the German economy and would in turn produce good democratic patriots who were averse to excessive nationalism. Starling inferred the political situation explicitly out of the argument on food:

> Among the lower and middle classes the chief defect noted is the general apathy, listlessness and hopelessness. We found no spirit of resentment among the workmen we spoke to, but simply a condition of dull depression and lassitude. Nor was there any sense of shame at defeat or feeling for national honour.

> The men wanted more food, especially fat, and all other moments seemed of little or no importance. In them, as in many of the upper classes, the feeling of German nationality, which has been created and fostered with such great care during the last 50 years by the rulers of Germany, seems to have entirely disappeared. . . . [The leading] men too seem to have lost their nationality. From many sides we hear that they are not placing their capital in their business, but are sending it out of the country, and the workmen complain that this capital, which represents war savings and which has not been taxed, is fruits of their labour, and should be divided up among the workers. This change in mental attitude cannot be ascribed wholly to food conditions, though in many of these men there has been a considerable loss of weight and they show distinct signs of undernutrition. . . . The impression we have derived is that the nation of Germany is broken, both in body and spirit.[12]

Starling so wished to credit diet with Germany's ills that he presented the insufficiency of food as shaping a listless, irritable population and all its politics. From his Edwardian integrationist perch, there was neither surprise nor contradiction in this claim: rebellion, which he thought unlikely, would simply accelerate the vicious cycle of internal collapse. The reorganization of an advanced economy, the return of political sentiment, the stitching together of a broken nation—all relied on the rapid improvement of caloric intake.

Starling's warning did not go unnoticed. In *The Economic Consequences of the Peace*, John Maynard Keynes quoted Starling's food politics in support of his famous claim that the Paris Peace Conference had all but destroyed the chance of European economic rehabilitation.[13] Keynes had little time for corporeal details but ample reason to enjoy Starling's disdain for reparations. Keynes's language put organizing stabilization at the heart of his argument:

> Very few of us realize with conviction the intensely unusual, unstable, complicated, unreliable, temporary nature of the economic organization by which Western Europe has lived for the last half century. We assume some of the most peculiar and temporary of our late advantages as natural, permanent, and to be depended on, and we lay our plans accordingly. On this sandy and false foundation we scheme for social improvement and dress our political platforms, pursue our animosities and

particular ambitions, and feel ourselves with enough margin in hand to foster, not assuage, civil conflict in the European family. Moved by insane delusion and reckless self-regard, the German people overturned the foundations on which we all lived and built. But the spokesmen of the French and British peoples have run the risk of completing the ruin, which Germany began, by a Peace which, if it is carried into effect, must impair yet further, when it might have restored, the delicate, complicated organization, already shaken and broken by war, through which alone the European peoples can employ themselves and live.[14]

This language was not alien to economics or specific to medicine alone—Starling's scientific work played a part in his concurrent deployment of crisis and "delicate, complicated" organization, but Keynes did not wait for the physiology of fluids to teach him the point. Alongside metaphorical registers of building and planning, the register of socioeconomic life relied on the contrast between restoring integration and enabling further impairment, whose direct application was distinctly medical. The two texts are nothing if not analogous: Keynes stages the life of German and other European peoples as the result of economic organization, emphasizes the derivativeness of political hopes and goals, and regrets the capacity of economic disorganization to complete the ruin initiated by crises and the system's wartime disarrangement.

For jurists across Europe since the late nineteenth century, international organization had been a serious concern.[15] In the later 1910s and 1920s, the rhetoric of integration—specifically formed around an integration-crisis duet—allowed for the transposition of body politics onto economic and political practices. That transposition was intentional, frequent, and consequential. Economics and international law offer perhaps the most relevant parallels for the framework of a brittle, integrated body: economic integration adopted the integrated body as its prime metaphorical register.

Before we proceed to the World War I and post–World War I, however, a short pause. Of the competing biological narratives that found an echo outside the realm of biology in the interwar period—narratives that worked at times like communicating vessels—integrationism was by no means the most popular. As a biopolitical logic, the systematic use of social metaphors of biological and medical integration was new, and the kind of structures it proposed were different from other contemporary forms of politicized biology. To clarify the value of this choice of the integration-crisis duet and its philosophical reference points, three other sources of metaphors deserve an extended parenthesis.

The first of these is, of course, evolutionism, which provided crucial political devices, particularly regarding national or racial cohesiveness and hierarchy. Evolution played diverse roles in Anglo-Saxon, French, and German politics and political theories from liberalism to Victorian imperialism, from social hygiene to crowd psychology—and of course to National Socialism. To operate on an explicitly evolutionary or phylogenetic narrative was to speak of the historical depth of a race or nation and to emphatically argue that individuals bear the race within themselves and belong to it. Evolution, perhaps the dominant holistic narrative after 1900, carried the political weight of racial science and the specter of racial, social, and bodily degeneration.

A second influential imaginary derived from bacteriology. As Donna Haraway explains, "Expansionist Western medical discourse in colonizing contexts has been obsessed with . . . the hostile penetration of the healthy body. . . . The colonized was perceived as the invader."[16] Even though wound shock discussions coded the soldier not as an aggressor but as a penetrated defender who was collapsing internally,[17] the imaginary of self-sustained collapse is very unlike the frame a foreign organism penetrating and destroying the body from the inside. Similarly, venereal disease, especially syphilis, offered a cornucopia of gendered, racial, and polemical imagery; as Klaus Theweleit argued in his account of German Freikorps soldiers' memoirs, it provided the sexualized, politicized sense of bodily intrusion leading to systemic physical and social degeneration and justifying forms of sexual violence.[18] Despite Paul Ehrlich's 1915 discovery of Salvarsan, "the first specific and radically effective cure for syphilis,"[19] and despite tremendous rearguard battles to devise systems of control, treatment, and quarantine, syphilis became an obsession among officers, less because it revealed poor social or moral hygiene in soldiers than because its spread eroded order on the simplest behaviorist, gendered, and sexualized grounds.[20]

Cell theory furnished a third, organicist paradigm for the analogic extrapolation of supraindividual subjects out of bodily biology. In the late nineteenth century, organicist sociologists on both sides of the Maginot Line had directly brought Virchowian cell theory into sociology. In Germany this was Albert Schäffle's *Structure and Life of the Social Body* (1875); in the Russian Empire, Paul von Lilienfeld's *Social Psychophysics* (1877); and in France, Émile Durkheim's early writings.[21] For them, the analysis of society was to be pursued precisely on the basis of this tie of the organic to the social. The ensuing picture usually abetted a quasi-Hobbesian, quasi-Virchowian depiction of the national or political body as constituted by citizen-cells—as a whole that is greater than the sum of its parts. Such arguments often relied on *Völkerpsychologie*, race psychology, or evolutionism, without being reducible to them.[22] Persons were a nation's cells; their origins made up its corporeal pedigree.

Marcel Proust, in 1922, described the countries roused to war eight years earlier as massive agglomerations of tiny creatures, which appeared

> in the same scale as the body of a tall man would appear to infusoria, more than ten thousand of which are required to fill one cubic millimeter. Thus for some time the great figure of France, filled to its perimeter with millions of little irregular polygons, and the figure, filled with even more polygons, of Germany, had both been engaged in such a quarrel. Seen from this point of view, the body Germany and the body France, and the other Allied and enemy bodies, were behaving to a certain extent like individuals. But the blows they were exchanging were governed by the multifarious rules of boxing, . . . and since, even if one thought of them as individuals, they were also giant agglomerations of individuals, the quarrel took on immense and magnificent forms.[23]

Yet where Proust and organicists envisioned "colossal masses," physiologists generally spoke of mechanisms of self-regulation, not cells. Where Proust undermined internal differentiation and structure in order to reformulate the Romantic fantasy of the grand body of the nation towering over the many individual bodies it comprised, Starling, Goldstein, and Cannon were interested in internal forces and counterforces, a sort of Newton's third law in political physiology that could serve as a dialectics of history.

These different applications of biology to social theory were often interwoven. Phylogenetic recapitulation was particularly easy to adapt and blend: we have noted how in physiology, it persisted as a necessary developmental device, whereas for conservative radicals it held racial-political bearing.[24] In the crisis of authority of the 1920s, group psychology theories first proposed by Gustave Le Bon in 1895 were recovered by William McDougall, Sigmund Freud, and others, who retained Le Bon's reliance on recapitulation theory, as well as his argument on the collapse of the crowd in moments of panic—a collapse that furthered the disorganization of society that the "era of crowds" had "heralded."[25] Hans Driesch's "entelechies" adduced the teleological holism he postulated for embryonic development as much from evolutionary thinking as from cell theory.[26] Some such approaches submerged holism within Nazism. Driesch called entelechy the *Führer* of the organism.[27] Others sought an individualist, often socialist politics. Biologistic similes, particularly regarding the body and the history of the nation in analogy with the healthy "normal" biological body, diverged in intent and consequence. Rather than appeal to an imaginary of degeneration or to biological history, the integrationists feared an internal, speedy, compounding collapse, but they attributed it specifically to a system at once insufficiently organized and far too organized to accurately counter the disruption.

: : :

It was in the shade of these other literary and political commitments that physiological integrationism spread its politico-economic wings. In its appeal, physiology walked alongside legal and economic theory, as well as political and theological accounts of a threatened social equilibrium, including Karl Barth's theology of crisis and Jacques Maritain's pursuit of integral Christian humanity against its secular fragmentation. The human sciences followed: linguistic movements of a quasi-structuralist variety, and the new anthropology cast by Franz Boas's antievolutionism and Bronisław Malinowski's functional understanding of society approximated the claims and purposes of the largely closed system imagined by physiologists.[28] As we shall see, the physiological language was at least as frequently and forcefully adapted as any other account of crisis. The historian of concepts Reinhart Koselleck recalled that *crisis* originated in both ancient and modern medical terminology before acquiring theologico-political, anthropological, and economic implications in the early nineteenth century.[29] In the 1920s, the jurists and economists to whom we now turn looked less to the ancient world or a Romantic nineteenth century for organic analogies on crisis, and they did not need to do so: their contemporary physiologists gave them all they needed.

The Great War was precisely the kind of crisis that required of them a theory regarding equilibrium and regulation, and this provided room for biological and biologistic language on integration and crisis. Almost from its outset, the logic pitting integration against crisis mobilized first in national affairs, then in economic and legal matters.[30] In his 1915 book *Europäische Wiederherstellung*, the Austrian pacifist Alfred Fried argued that the war had demonstrated "that the world has become an economic unity, which, if disturbed at one point, suffers in all parts of the organism, because political interrelationships are already so highly developed that—at least in Europe—'localized wars' are no longer possible."[31] Fried structured his internationalism around this idea of a threatened organism.[32] The English political theorist Leonard Hobhouse similarly set aside his earlier evolutionist metaphors to concentrate on an isomorphism between social and bodily pathology for his argument on individuality and statehood.[33] Hobhouse began his critique of G. W. F. Hegel, *The Metaphysical Theory of the State* (1917–1918) as follows:

> People naturally begin to think about social questions when they find that there is something going wrong in social life. Just as in the physical body it is the ailment that interests us, while the healthy processes go on without our being aware of them, so a society in which everything is working smoothly and in accordance with the accepted opinion of what is right and proper raises no question for its own members. We are first conscious of digestion when we are aware of indigestion, and we begin to think about law and government when we feel law to be oppressive or see that government is making mistakes. Thus the starting-point of social inquiry is the point at which we are moved by a wrong which we desire to set right, or, perhaps at a slightly higher remove, by a lack which we wish to make good. But from this starting-point reflection advances to a fuller and more general conception of society.[34]

Disease—digestive ailments in particular—served as an analogue for social oppression; both required regulation.

The jurist Harold Laski, in his 1917 critique of theories that postulated the indivisibility of sovereignty, furthered the contrast between organization and crisis in the body of the sovereign. He regarded singularity as an illusion created by crisis: "a time of crisis unifies everywhere what before bore the appearance of severalty."[35] Two years later, in his *Authority in the Modern State*, Laski would use more explicit bodily analogies when he offhandedly remarked about social diversity that "within every organization, as within each individual, there must be a continuous struggle between life and tradition. . . . To ascribe the whole life of a society entirely to one element or the other is . . . to mistake life for anatomy or physiology. The body cannot function without its background; and a skeleton is still dead matter even if it have living form."[36] The simile was complete: life is not a scientist's fossilized form but a totality that is coextensive between society and the individual life surrounding this skeletal form. Fabian socialism, committed to improving this life at the low cost of such skeletons, offered another clear point of reference, as its association of terms like *organism* and *regulation* carried much the same sense as that proposed by the physiologists. Leonard Woolf's *International Government* (1916), for example, used language that could have easily come from the physiology of hormonal regulation. The previous century had "led to the spontaneous creation and evolution of a large number of new organizations, international organs and organisms, the functions of which are either to regulate through agreements the relations of States or administrations, or individuals or groups of individuals, belonging to several States; or, looked at from another point of view, to promote international interests and harmonize national interests."[37]

Writing from a very different angle in 1916, the Italian sociologist Vilfredo Pareto was even more explicit about the punctured-system motif and its biological relevance. As opposed to statisticians, and social scientists who "meander" about for causes, "the equilibrium of a social system is like the equilibrium of a living organism, and of the latter it was noticed in very early times that an equilibrium that has been accidentally and not seriously disturbed is soon restored. In those days the phenomenon was, as usual, given a metaphysical colouring by reference to a certain *vis medicatrix naturae*."[38] Pareto's reference to the physical body was not just heuristic, like Hobhouse's. It involved an explicit transposition—a physiological ground capable of providing a philosophical grounding for a social system.

By 1919, the integratedness of the postwar international economy, law, and sovereignty were imagined with the language of integration and crisis and their bodily metaphors. The end of the war yielded a substantial harvest of claims on the social potential of physiology, whose coupling of integration and crisis offered by far the most direct basis for the spread of metaphor, and many thinkers were rhetorically and intellectually prepared to adopt and transpose it. Jurists and politicians not only were well aware of the discourses surrounding bodily regulation (particularly hormones and blood equilibria) and physical disintegration; they knew well which implications they were spared when they appealed to these terms. In the early postwar period, internationalists sought in the Paris Peace Conference and the League of Nations to imbue a new postwar order with a spirit and force of coordination capable of welding together different national and international units to avert anarchy and wartime violence. They spoke of the international order in terms of integrated systems whose parts (national bodies, mostly) were functionally interconnected, often by other bodies, organs, or organisms. In his *Europe in Convalescence* (1922), the political scientist Alfred Zimmern wrote of Europe and the international order as a patient suffering through a stunted healing process. Although he often mixed metaphors, these corporeal analogies were precise. Zimmern literalized states into wounded, exhausted, neurasthenic bodies: "The problem for France was to heal the wounds of body and mind so as to be free to pick up the threads of her old life"; Russia was "disintegrating."[39] He promptly scrutinized the new international *bodies*, *organs*, and *organisms* that were working toward a "cure": international structures embodied particular ideas, ideals, and methods, and Zimmern consistently asked whether their contribution to *integration* was proving adequate. This kind of phrasing identified national and international patienthood with the international crisis and convalescence with (re)integration.

Note, for example, the distinction between *organ*, *body*, and *organism* proposed in Zimmern's critique of the League of Nations, a distinction that doubled as a superimposition of each onto the others: "As a standing organ of policy, as a perfected and organized world-concert, it has proved impracticable. On the other hand, as an administrative body for problems of world-wide concern specifically remitted to it, as a forum for mobilizing and giving expression to opinion on world-problems, and . . . as an organization for the preservation of the world's peace and the adjustment of controversies by mediation rather than by aggression, its work is of great and increasing value."[40] These terms and concepts were injected with the value of naturalistic operators justifying internationalist goals, especially in their capacity to normalize and represent these goals and to vividly and automatically portray a catastrophic alternative.[41] For example, the functionalist school in British political thought foregrounded the language of integration and disintegration and occasionally pulled in biology when tracing back the late imperial and international orders. In David Mitrany's theory, order between the different parts needed to be negotiated and organized by bodies and forms that guaranteed technical and material stability across the system. As of the mid-1920s, Mitrany began to offer a well-known account of the paradoxes involved in the interconnectedness of peace with international sanctions and the threat of anarchy.[42] With

The Progress of International Government (1933), he began introducing his theory of a functional unfolding of international government. He called for "a functional integration of technical services upon the largest possible international scale," commenting: "The epidemic character of the unparalleled economic crisis of the last few years has itself been an ominous demonstration of the world's economic interdependence."[43] Once again, the metaphor was all too easy, the words convenient. What matters is that it was *these* words, *this* metaphor, and not others among the then-available ones, not *syphilis* and *degeneration*, not *facial injury*, not *simple injury*, not *respiratory difficulties*. Sociologists like Marcel Mauss in France and G. E. G. Catlin in Britain were still more explicit in their approach to an integrated medical body and politics, and the link of holistic medical care to political convalescence became uncontroversial among scholars.[44]

Subsequent attempts at international cooperation, including the 1928 Kellogg-Briand Pact (General Treaty for Renunciation of War), contributed to the internationalists' depiction of a fragile order in need of bolstering. Important among the internationalists was Henri Bergson, who presided over the League of Nations' Committee on Intellectual Cooperation before eventually despairing of its abilities. In a well-known passage in which he abdicated the hope of interrupting an annihilatory "war-instinct," Bergson lauded the founding of the League while doubting both optimists and pessimists who "agree . . . in considering the case of two peoples on the verge of war as similar to that of two individuals with a quarrel. . . . Even if the League of Nations had at its disposal a seemingly adequate armed force, . . . it would come up against the deep-rooted war-instinct underlying civilization."[45] What matters for us is Bergson's language: beginning with the comparison of individual to social bodies, he described war (an "instinct" akin to Freud's) as having a "disintegrating effect" that opposed every effort toward a well-regulated international agreement that would benefit humanity as a single whole.[46] Where individual bodies might avoid their quarrel by submitting to a court, recalcitrant national ones would not. Bergson's language balanced organicism and integration, and his deliberate structuring of "disintegration" and "war instinct" into the system was paramount.

Bergson's thought also held some sway over Jan Smuts, the premier of South Africa, who used Bergson's concept of élan vital in his 1926 cosmological theory in *Holism and Evolution*.[47] Smuts saw biological holism as the basis for theories of social behavior: evolution and holism complemented each other, generating wholes from the lowest molecules to the entire universe. Quoting Haldane to render this argument relevant to organic regulation, Smuts proclaimed a biologically derived cosmology across universal, historical, and even political lines.[48] His holism bound together a theory that contrasted well-developed, highly complex integrations to inferior nonmodern, non-Western ones, which he presumed to be behind on the evolutionary scale, and which sought further, higher wholes that would guarantee further development. *Holism and Evolution* explains some of the incongruity that Mark Mazower and others have emphasized—namely how someone who contributed so profoundly to racial hierarchy in South Africa could be an enthusiastic supporter of League of Nations internationalism.[49] Even though in Smuts's case we move away from integration-crisis duet and toward evolutionism, the end goal of his theory was the tight and complex integrated whole. The 1920s international order relied on parallel registers of bodily and biological metaphors for the meaningfulness of the relations it described.

Why should corporeal integration and disintegration be relevant at the international level? For one, they were constantly visible in postwar society. Across Europe, the injured were a confrontational sight. The sense that everyday life had been fundamentally disrupted and that political and economic troubles were contributing to its deterioration was enhanced by repeated crises throughout the 1920s, especially in Germany and the former Habsburg lands.[50] When Goldstein described his patients as moving "from catastrophe to catastrophe," the temporality he proposed was palpable to Germans, Jews, and other Europeans who were warm to his non-Marxist socialism.[51] He might as well have been using the expression to speak of the country from which he had just escaped. Particularly pregnant and symptomatic in this regard are elements that turn *The Organism* into a metaphor for Weimar culture: disease as a disorder that may be stabilized or may suddenly bring about a catastrophic situation; the individual as a whole, not merely a sum of parts, and as capable of autonomy vis-à-vis the milieu; and the therapeutic focus on facilitating, for the injured self, a new if restricted totality. When Goldstein was writing in exile in 1933–1934, Weimar had undergone its worst "failure of self-actualization."[52]

At the international level, *integration* denoted far greater interdependence than *cooperation* and sounded more palatable than *international regulation*. The term was also more capable of dissimulating power imbalances under pretenses of participation and quasi equality. Internationalism had an easy time carving similes and structures from the sciences of the body and mapping them onto advocacies of integration in international economy and politics, with heightened urgency after the 1929 crisis and the international political crises of the 1930s.[53] The language of integration averted the dilemmas of both nationalists and communists, generally providing the rationale for an organization that allowed the extension of liberal hopes. This is not to claim that integrationism's politics was somehow pure or that it was unburdened by hierarchy, colonialism, or race. Rather, it conveniently dissimulated imbalances of power and economic as well as imperial hierarchies under the rubric of cooperation and coexistence while allowing for heavy-handed activity for purposes of stabilization or control. Integration thus became a particularly potent and safe language to mobilize; it was for this same reason that physiology was so convenient.

In international economic affairs the physiological language was more explicit and far less metaphorical. The Crash of 1929 and the international domino effect of the Depression (to continue with the corporeally derived metaphors) demonstrated the fragility of existing international self-regulation and made integration into a common topic of discussion. Already in 1909 William Beveridge had famously described cyclical fluctuations as "the pulse of the nation."[54] In 1923, Keynes cited the *vis medicatrix* when he disparaged currency devaluation as a default policy for politicians unable to make decisions: devaluation, he wrote, "follows the line of least resistance, and responsibility cannot be brought home to individuals. It is, so to speak, nature's remedy, which comes into silent operation when the body politic has shrunk from curing itself."[55] By 1930, studies of the international economic system focused consistently on integration and disintegration, and the frequency and complexity with which they referred to biological integration was striking. Mitrany may not have intended *functional* in a biomedical sense, but others clearly understood the term in the valence offered by *functional disorder*—a disease without a localizable source that required organism-wide treatment.

The highly influential German statistician Ernst Wagemann, head of the Weimar Statistical Office and the Institute for Business-Cycle Research, explicitly lifted the term *functional* from "medical terminology" for his 1930 theory of business cycles titled *Economic Rhythm*, which he proposed as a study of "economic symptomatology."[56] This symptomatology was of a decidedly "biological-organic character" and replicated exactly the argument on internal integration's priority over external stimuli—even stimuli as substantial as wars: "The independent organic character of economic fluctuations is . . . particularly noticeable in the manner in which they receive outside influences; they stand up to such influences and transform them, so that the latter's effect is merely that of stimuli."[57] Wagemann raised the integrated and coordinated response to stimuli to the status of one of the two guiding principles of his work. The other guiding principle was that the economic system needed to be understood as biological-organic system.[58] The decline in imports at times of crisis was "a functional disturbance in an otherwise healthy economic body," and the search for causes during economic shocks was like the work of a doctor deriving information about pathogenic agents from symptoms.[59] Wagemann praised the German Business Research Institute for "representing the medical, or better, the organic-biological point of view. It regards the national economic system as an organism the course of whose inner life can never be adequately interpreted by concentrating attention on a single point of its external surface. . . . [In] much the same way as a doctor uses the numerous methods of diagnosis . . . the [Institute's economic] diagnosis is based on a complete study of individual signs and symptoms, and it serves as a basis for forecasting the future course of the malady."[60]

Another influential economist, Wilhelm Röpke, later one of the major figures of ordoliberalism, had few qualms about using biological metaphors to blame liberalism for causing the "progressive disintegration and atomization of the body politic."[61] He called for a synthetic, holistic methodology, opposed to specialization in the social sciences, that "belonged to that stage of the development of modern society in which the sound constitution of the whole body could be safely assumed, so that treatment could be confined to specific parts."[62] Time and again, Röpke used explicitly physiological references to make his point. For example, on hormones and vitamins: "The actual world trade . . . works like a hormone or a vitamin, the biological value of which is quite out of proportion to the active quantity."[63] On integration, and body temperature as a defense mechanism: "social over-integration . . . occurs like the fever of the human body in a quite normal and beneficent manner whenever an emergency . . . requires all the defensive forces of society."[64] And on anemia: "Every modern war . . . demonstrates the extraordinarily swift powers of regeneration inherent in our economic system, provided that its inner driving energy and ultimate material forces are left intact and that the actual loss in man-power does not amount to a sort of anaemia in the body of the nation."[65] Fundamentally, for Röpke, "the market economy is living on certain psycho-moral reserves, which are taken for granted when everything is going well, and only reveal their supreme importance when they are giving out. From this we see at once that the ultimate conditions for the working of the economic process lie outside the strictly economic sphere. . . . Modern medicine has learned that major troubles are usually bound up with the ultimate somatic and mental conditions of our existence, conditions which must be changed if there is to be any definite cure."[66]

With the market crash of 1929, this became an American story, too, especially as a rhetoric joining integration and crisis helped create a "culture of crisis." As Cynthia Russett first showed in her *The Concept of Equilibrium in American Social Thought* (1966), the system-crisis binary and the political and economic desire "to maintain equilibrium" became especially pressing.[67] Herbert Hoover's administration accentuated a language of regulation, cooperation, and integration in part to avoid interfering in the economy.[68] In the shadow of the Depression, the Harvard Business School facilitated a number of interdisciplinary exchanges and circles, seeking to develop models of business stability, equilibrium maintenance, general systems, and industrial, labor, and social administration. Physiology, most significantly that of Lawrence Henderson, answered the call.[69]

Henderson played a key part in the story of equilibrium in American social sciences. He had already proposed in 1917 that, from molecules to society, the entire universe is constituted of integrated systems.[70] Expanding his own work on chemical and biochemical equilibration toward social equilibration, Henderson struck up a close friendship with Alfred North Whitehead, who had been advocating his own form of holism. Henderson read Pareto shortly before publishing his study of the physiology of blood in 1928, and he translated and included in his text the passage from Pareto's *Trattato di sociologia generale* that we cited earlier.[71] In subsequent years he not only lectured on sociology (starting with Pareto and his own theory on the link between biochemical and social equilibria) but also famously organized a "Pareto circle," enthusiastically introducing his writing to select faculty in economics, sociology, and industrial administration.

The Pareto circle included influential contemporaries from Elton Mayo to Talcott Parsons. Henderson's extensive collaborations with Mayo at the Harvard Fatigue Laboratory and beyond became a guide for the Harvard Business School's research and experimental practices, and exercised broad influence on the university as a whole.[72] Meanwhile, during the 1930s Cannon expanded his interest in what he called social homeostasis to treat political and economic matters. In a 1933 article for the *Technology Review* that resembled a sermon, Cannon began, "Civilization is in a hell of a mess," then mobilized his homeostatic argument, including his usual rhetoric of corporeal instability, to propose research "of the utmost social importance" toward "the discovery of sensitive indicators in the commercial stream."[73] Cannon feared that biocracy might be just another technocracy that would erode democracy.[74] He nevertheless glimpsed the ideal model of socioeconomic organization in biological regulation:

> To assure in the social organism the same degree of stability that has been obtained in the animal organism, our bodies suggest such control of the social fluid matrix that its constancy would be maintained, a condition which involves certainty that the moving stream will deliver continuously the necessities of existence. Food, clothing, shelter, the means of warmth, and assistance in case of injury or disease, are, of course, among these necessities. . . . In the light of biological experience, therefore, social stabilization should be sought, not in a fixed and rigid social system, but in such adaptable industrial and commercial functions as assure continuous supplies of elementary human needs.[75]

It is difficult not to read this position, in its dialectical opposition to biocracy and its advocacy of a welfarist "social organism," as a critique of Henderson's much more assertive (and conservative) claim on biology in social matters, and as an attempt to promote the body as a model but not as a politics.

Biological integration allowed for metaphors, analogies, and superimpositions. It peopled a complex enough imagination for different Harvard elites to "interstitially" (in Joel Isaac's phrase) fill social theory with different dreams of biological and organic equilibria and characterizations. By the mid-1930s, the crisis had shaped the integration-crisis duet into a central component of social thought: biological alternatives were all differently available depending on where one looked. In 1936 John Dewey lectured at the Harvard Tercentenary on "Authority and Social Change." He was preaching to the faithful when he declared himself willing to find an "organic unity" for the social organization of individuality and freedom: "The scene which the world exhibits to the observer at the present time is so obviously one of general instability, insecurity, and increasing conflict—both between nations and within them—that I cannot conceive that any one will deny the desirability of effecting and enstating some organic union of freedom and authority. Enormous doubt will well exist, however, as to the possibility of establishing any social system in which the union is practically embodied.

This question, it will be justly urged, is the issue that emerges."[76] By the time that Goldstein gave the William James Lectures in 1939, speaking against "isolation" and in favor of a holism based on the "abstract attitude," his sociological conclusions must have sounded somewhat banal.[77]

Physiological integration also became strikingly useful to sociologists, who in the 1920s had bristled at the use of a phylogenetic or cell-theory-based language, not least because of its racial implications. Overviews of the field, such as Pitirim Sorokin's *Contemporary Sociological Theories* (1928) and George Lundberg's *Foundations of Sociology* (1939), were particularly disparaging, but *not* toward the similes of bodily integration.[78] In the later 1930s the "organic theory of the state" began to make a comeback, for example as the Frankfurt School sociologist Franz Neumann critically examined what he designated as the Nazi state's organicism in his 1942 *Behemoth*.[79] Meanwhile, Parsons was working out the influences of the Pareto circle, while Lundberg examined "Human Social Problems as a Type of Disequilibrium in a Biological Organism," following Cannon in transposing homeostasis to social problems and arguing that biological and social integration were no longer merely analogically related, but were the same concept. "There is no longer any argument as to whether, under these circumstances, the use of the word 'equilibrium' in biology, economics, or sociology is 'figurative,' and is illegitimately based on analogy."[80] Instead, equilibrium was a "concept which denotes a common and significant characteristic of behavior of special cases of widely disparate types of phenomena." This is not to say that the organism-society analogy was uncontroversial—on the contrary, it was a matter of contempt-ridden debates[81]—but that sociologists central to the development of the discipline found in the adoption of integration and crisis metaphors a major site for the understanding of social systems, just as internationalist jurists had done twenty years earlier.

By the end of World War II, it was possible to celebrate biological integration and homeostasis as nothing less than the essential components of human nature. Lewis Mumford did exactly that at the beginning of his *The Condition of Man* (1944), when he praised Haldane, Cannon, and Claude Bernard for having "established the importance of man's internal environment":

> Its delicately maintained stability is the condition of his being set free to think and feel and exercise his senses without keeping too sharp an eye upon the bare necessities of survival.

> Disturbances in this internal environment affect the psyche long before they cast a burden on man's other organs; and there is much reason to think that the opposite process also takes place: a succession of investigators from Janet and Freud onward have established the fact that psychic disharmony may disrupt the equilibrium of the whole physical organism and even cause drastic disturbances of function in one or another organ. Man's sanity and health consist, in fact, in his maintaining this double balance: an even internal environment that frees the mind for independent explorations and a balance of mind that enables the body to function as an effective whole, despite continued changes of circumstance, changes of occupation, and changes of physiological equilibrium due to growth itself.[82]

Georges Canguilhem had claimed much the same a year earlier, when, in *The Normal and the Pathological,* he declared that the integration of ever more complex, always stable internal environments was the very goal of evolution and the starting point for any comparison of biology with the historical character of human societies.[83] And Georges Bataille, writing in the late 1940s of war as "a Catastrophic Expenditure of Excess Energy," and lauding the grand—even grandiose—expenditure of wealth and energy as creative, celebrated regional and global integration beyond the state level as a field in which excess could be better equilibrated.[84] He added:

> History ceaselessly records the cessation, then the resumption of growth. There are states of equilibrium where the increased sumptuary life and the reduced bellicose activity give the excess its most humane outlet. But this state itself dissolves society little by little, and returns it to disequilibrium. Some new movement then appears as the only bearable solution. Under these conditions of malaise, a society engages as soon as it can in an undertaking capable of increasing its forces. It is then ready to recast its moral laws; it uses the surplus for new ends, which suddenly exclude the other outlets. . . . Industrial economy is involved in a disorderly agitation: It appears condemned to grow, and already it lacks the possibility of growing.[85]

Disequilibration and *catastrophe* formed the language of thinking the limits of human society and humanity itself.

Cybernetics and Voodoo: Cannon's Metonymic Bodies and Their Afterlives

Cannon's homeostasis resolved the problem of biological and social integration in a way that took into account his worry about biological technocracy. Social homeostasis came on the heels of a theory of internal systems; it purported to apply to politics and economics at once, yet it did not literalize the connection of biology to society—it treated the two as only analogous. It offered neither sociology proper nor a reductionism, and it could be lifted out of biology with little organicist or evolutionist residue. By claiming that the biological organism was of a different order from physicochemical operations, and by hinting that the case was the same with the social body, Cannon could argue that the complexity of homeostasis provided a structure for social concerns especially about those in need, but he did not endorse a particular politics or a full biologization of the political.[86] For these reasons, homeostasis also influenced theories of communication and individual-society relations ranging from anthropology to cybernetics.[87]

Cannon pivoted to social homeostasis in *The Wisdom of the Body* much in the same way that he had looked at war in the conclusion of his 1915 *Bodily Changes in Pain, Hunger, Fear, and Rage*. There, similes and metonymies of war, peril, and fight had colored his description of the organism as an internally determined, often overflowing form. In *Bodily Changes* he sought alternatives to warfare, given the ineradicability of the "fighting emotions" or "fighting instincts"—notably anger "and its counterpart, fear"—in human life:

> The business of killing and of avoiding death has been one of the primary interests of living beings throughout their long history on the earth. It is in the highest degree natural that feelings of hostility often burn with fierce intensity, and then, with astonishing suddenness, that all the powers of the body are called into action—for the strength of the feelings and the quickness of the response measure the chances of survival in a struggle where the issue may be life or death.... They are the emotions and instincts that sometimes seize upon individuals in groups and spread like wildfire into larger and larger aggregations of men, until entire populations are shouting and clamoring for war. To whatever extent military plans are successful in devising a vast machine for attack or defense, the energies that make the

machine go are found, in the last analysis, in human beings who, when the time for action comes, are animated by these surging elemental tendencies which assume control of their conduct and send them madly into conflict.[88]

If such emotions or instincts were responsible for the periodicity of warfare,[89] Cannon surmised, then the need for alternatives was all the more urgent, and physiological research could and should lead directly to the advocacy of such alternatives.

Cannon had a ready-made option—namely William James's "moral alternatives to war," which he cited approvingly and dismissed.[90] Physiological rather than moral alternatives covered his purpose, and in a fundamentally monistic gesture, Cannon chained society, like the mind, to the physiological regulation of the emotions. Power, war, fear, and their social consequences flowed in society out of bodily changes and feelings. The limits and structure of such a gesture would obsess him in his later years. *The Wisdom of the Body*, which he began writing for a series of lectures at the Sorbonne in the months after the 1929 article "Organization for Physiological Homeostasis," just as the Great Depression hit, carried all the force of the physiology of *Bodily Changes* into the anxious integration-crisis duet. It opened a period during which, for all the customary apologies Cannon was expressing for delving as a biologist into political matters, his output would reach the zenith of its political and philosophical influence. Three texts in particular focused on the relations of body and society: the last chapter of *The Wisdom of the Body*, on social homeostasis; Cannon's 1940 presidential address before the American Association for the Advancement of Science; and "Voodoo Death," a 1942 essay on physiology and witchcraft.

"Perhaps," Cannon wrote in the introduction to *The Wisdom of the Body*, "every complex organization must have more or less effective self-righting adjustments in order to prevent a check on its functions or a rapid disintegration of its parts when it is subjected to stress."[91] He opened the political discussion by recalling that "the integrity of the organism as a whole rests on the integrity of its individual elements, and the elements, in turn, are impotent and useless save as parts of the organized whole."[92] This gave him space for a physiological account of human history: Cannon distinguished between hunter-gatherers, who for him were analogous to heterotherms too dependent on their environment, and agricultural societies, "large aggregations" that have "the opportunity of developing an internal organization which can offer mutual aid and the advantage, to many, of special individual ingenuity and skill."⁹³ The latter facilitate individuation among their members because their internal organization makes the innate regulation and control of those same members indispensable. Cannon thus maneuvered toward explaining social homeostasis in broad terms by imagining society as a form that carries "some indications of crude automatic stabilizing processes." Modern politics entered the argument at this point:

> A certain degree of constancy in a complex system is itself evidence that agencies are acting or are ready to act to maintain that constancy. . . . When a system remains steady it does so because any tendency towards change is met by increasing effectiveness of the factor or factors which resist the change. . . . These statements are to some degree true for society even in its present unstabilized condition. A display of conservatism excites a radical revolt and that in turn is followed by a return to conservatism. Loose government and its consequences bring the reformers into power, but their tight reins soon provoke restiveness and the desire for release.⁹⁴

The claim is quite astonishing: political causes, choices, and effects amount to recalibrations of action and reaction, pressure and counterpressure within the structure of stability and instability. The metapolitical quality of this claim—the moral despair over the post–1918 world, the indifference over which political side would have the upper hand—was grounded here in a pendular back-and-forth caused by the presence of agencies or forces that managed the "self-righting" adjustments.⁹⁵ These agencies were not external to society, nor were they somehow a force hidden deep within society—an Adam Smith–like invisible hand. Rather, they were the result of predictable ways in which uncertainties were handled, a stabilization that generated new uncertainties, and so on.

After 1932, Cannon largely desisted from translating his physiology into a politics, except in his 1934 article "Biocracy," which grudgingly recast this argument as an economic and technocratic one, and in his strong support for the Soviet Union during the International Congress of Physiology in Moscow and Leningrad in 1935.⁹⁶ In the later 1930s, and especially after the arrest of Otto Loewi during the Anschluss, he became an antifascist activist, thanks to the frequent prodding of Franz Boas and the American Committee for Democracy and Intellectual Freedom.

The farthest reaching of Cannon's political texts was his 1940 presidential address to the American Association for the Advancement of Science. Styled as an expansion of the last chapter of *The Wisdom of the Body*, the address allowed him to deemphasize biologism and be more explicit on the search for stability, control, and order in society:

> States of stability and instability in the nation, in industries, and in homes are closely linked. When a break occurs in the even course of activities of larger aggregations, having widespread influence, the harmful consequences are likely to reach down to the ultimate units in the social structure, human beings.... Recognition of that fact is the stimulus for the efforts to devise an organization of society which will furnish a greater assurance than hitherto that mankind shall not suffer distress and privations which are of human origin and which might be avoided by applying human intelligence.[97]

Cannon routinely emphasized occasions when the destruction of such stability took place, referring to "harrowing uncertainties," "business catastrophes," "death or painful and debilitating wounds" on great populations, and—perhaps most significant—the "stressful conditions" that lead "masses of men" to war.[98] The subject of Cannon's theory of social organization had become unusually fluid: under the banner of a metonymic transposition of "the body physical" onto "the body politic," a series of other tropes sutured his earlier parliamentary analogy in *The Wisdom of the Body* to models of production, economics, a nation, or the distribution of goods.

In explicit "comparison of the body politic and the body physiologic," Cannon articulated "correspondences" or "analogues":

> What corresponds in a nation to the internal environment of the body? The closest analogue appears to be the whole intricate system of production and distribution of merchandise. It would include the means of industry and commerce—the agencies of manufacture and agriculture, the rivers, the roads and railroads, trains and trucks, wholesale and retail purveyors—all the factors, human and mechanical, which produce and distribute goods in the vast and ramifying circulatory system which serves for economic exchange. Into this moving stream products of farms and factories, of mines and forests, are placed, at their source, for carriage to other localities. In the operations of this stream one is allowed to take goods out of it only if one puts

back into it goods of equivalent worth. . . . Thus money and credit become a part of the internal environment of society.[99]

The passage is rich in linguistic play. Cannon twice refers "internal environment" to both physical body and society. He identifies society with its economic system and describes this by way of a series of rhetorical figures that bridge the social and physiological registers or, more accurately, that dissimulate both the differences and the operation thanks to which he moves from the "body physiologic" to the "body politic." On top of that first operation—which was, after all, not uncommon—Cannon offers a metonymy of blood circulation—"the vast and ramifying circulatory system which serves for economic exchange"—and instead of describing transfer, transmission, or transportation, he apostrophizes these operations with another metonymy, the "moving stream." This renders the interceding list of elements of industry and commerce into a relatively closed unit, for which circulation becomes the appropriate trope. The complex synecdoche that closes the passage, "the internal environment of society," sneaks in the idea that society itself has an internal environment, closed from the world, and acting in supraindividual, functional terms so that the entire social body is tied to economic components and practices.

Cannon's desire for social and economic policies that would regulate divergent or centrifugal tendencies was further in evidence when he repeatedly temporalized social homeostasis as an emerging principle for scientifically minded governance. In passage after passage, he linked irregularity and instability with the rational imperative of more emphatic control over stabilizing mechanisms:[100] "In the body politic the control of processes still has far to go. Instead of a change of rate a frequent alternative is a total stoppage of the process; men are thrown out of their jobs and drift in idleness. How industry and commerce can be adjusted to inconstant or temporal demands has not yet been discovered. Various schemes for coordinating production and consumption and for assuring regularity and continuity of employment prove that thoughtful men are deeply concerned with the fear, the despondency, and the hardship which result from uncertain economic existence."[101] This argument was intended as a critique of the New Deal's failures. Lest the approach proposed by social homeostasis appear totalizing or politically undemocratic, Cannon specifically appended a claim that "more control is tolerable if it results in greater human freedom."[102] This had been his concluding claim in "Biocracy," which waxed grateful that democrats had the votes.[103] Cannon's theorization of the economy similarly focused on preparedness against the capacity of crises, like that of 1929, to generate "self-aggravating damage to social stability."[104] The concept of a margin of safety took priority: dictators prey on insecurity,[105] which is neither desirable in society (Cannon expresses a classic Hobbesian fear of disorder) nor an accurate description of the brain's governance over the body.

Cannon found a strong ally for this ideology of homeostasis in Goldstein, who, in his 1939 William James Lectures at Harvard, pronounced on times of crisis by translating his philosophical concepts of catastrophic reaction and anxiety into societal-level causes of tyranny. Crisis generated "catastrophic reactions and anxiety":

> If it is extensive enough, the confusion which arises from [a crisis] can be used by a minority to claim spuriously, in the name of society, the right to take over all power in order to protect society. If such a group is victorious, tyranny develops. Thus society comes to be divided into groups that rule and groups that are ruled in opposition to their own wishes, a state of affairs which is incompatible with the essential character of human nature because it contradicts freedom. The members of such a society are rather like the patient who lacks the ability to abstract and who thus becomes the victim of an abnormal response to outside stimuli. Like the sick person, they suffer from the limitation of their freedom and from anxiety, with its consequences (the shrinkage of the environment, dependence upon and submission to other people). Such disturbances affect both the individual and society.[106]

Cannon and Goldstein pronounced the avoidance of crisis to be perhaps the essential aim in treating a patient as much as an effective course of progress toward greater freedom. In Cannon, greater efficiency in homeostatic coordination was essential to achieving not social efficiency but the economic security necessary for social complexity, and hence freedom. Freedom also operates in society thanks to the quiet acquisition and automation of stability and security. Further metonymies guaranteed the identity of individual and political freedom.[107]

This appeal to a largely liberal concept of freedom became all the more pressing for Cannon as war loomed in 1940: "In a world where predatory nations, powerfully armed, are ready to attack, the ideal of security often is not adequately respected. Internal forces are not trained for action."[108] Just as he had done in his physiological work, Cannon prioritized defensive operations in tandem with both contemporary accounts of Nazi aggression and World War I–era American and French propaganda regarding the defensive engagement of the Entente powers and the United States.[109] Mobilized, the body politic was "unified, integrated, and for one purpose: self-preservation."[110] Human freedom, he concluded, was a matter not of efficiency but of a security that allowed for leeway, for a "margin of safety."

Freedom, margin of safety, and social homeostasis were not theories steeped in political hope: there was little liberal promise left in Cannon, and it was anything but Mill's individual who is "sovereign over his mind and body." This became clearest in Cannon's late and spectacular essay "Voodoo Death," which continues to be cited as an origin of psychosomatic medicine.[111] Published in the *American Anthropologist* in 1942, it fundamentally altered anthropological conceptions of the use and effects of magic, and in particular the ways magic could physically affect the human body in the absence of apparent inflicted injury. Rivers had noted magic's capacity to cure or inflict harm, and he had urged anthropologists to abandon the positivist postulate, according to which magic had value merely as a matter of superstition or placebo, but he had proposed no theory as to its actual biological functioning. Similarly, Malinowski had worked on witchcraft in *Argonauts of the Western Pacific* (1922), but he had not articulated a reasoning behind its efficacy.[112]

"Voodoo Death" was the product of long and meticulous effort that began in 1934. It was an auspicious moment for voodoo: use of the word had nearly doubled in frequency over the previous decade, and in early May two films were released that featured scenes with it: *Blue Steel* (starring John Wayne) and *Drums o' Voodoo*, the first film based on a play by a black playwright (*Louisiana*, by J. Augustus Smith).[113] In late May, Cannon, ostensibly intrigued by a passage in John Houston Craige's 1933 memoir of Haiti, *Black Bagdad*, which described the inexplicable death of a man who lacked any visible wound, began to send letters to physicians, anthropologists, and government officials—some thirty letters by October. Keeping a record of replies and follow-ups, Cannon tracked down the autopsy surgeon in the death Craige described, while amassing books and references on magic in Africa, Australia, Polynesia, and the Caribbean, and among Native American populations. In every letter he asked for reliable ("rationalist") eyewitness experiences among travelers, missionaries, and ethnographers.[114] Rationalism was not in short supply, indeed to Cannon's taste the contributors had too "little interest . . . in matters ethnological."[115] In his letters Cannon consistently referred back to wound shock, once citing two wartime cases in which no wound had been identified at all; he specifically identified shock from the casting of a spell as the likely reason behind the deaths.

The correspondence brought in strong results in terms of both cases and atmosphere. Virtually everyone confirmed the prevalence of indigenous beliefs in sorcery's effects, even though several correspondents were skeptical of Cannon's ideas.[116] Several reported secondhand evidence or potentially explainable cases of death following witchcraft.[117] By July Cannon had a report from a medical doctor in Hawaii, regarding a man who fell ill after assaulting a sorcerer: "The 'illness' lasted for weeks—months perhaps. There was only a progressive asthenia without focal symptoms or signs of any sort. The man simply *knew* that he was going to die, and this delusion was not amenable to reason. . . . A rather casual autopsy which sought evidences of foul play only failed to reveal any etiology for the death."[118] By November Cannon had received case reports from missionaries in Australia regarding subjects who had fallen ill after becoming convinced that they were cursed,[119] as well as confirmations from ethnographer J. B. Cleland in South Australia. Thereafter he began to receive direct evidence in the form of medical reports of witchcraft victims dying without a medical explanation, and more cases poured in the following spring from Harvard anthropologists in Monrovia, Nairobi, Hawaii, and Fiji.[120] Cannon became concerned with showing that no poisons but only magic could be responsible.[121] At the advice of his correspondents, he went back to Herbert Basedow's book *The Australian Aboriginal* and culled several examples.[122] Correspondence and ethnographic material continued to flow in.

Before completing the article in 1941, Cannon reconfirmed a number of citations and returned to the Haitian connection by way of W. Lloyd Warner's sociology of magic *Black Civilization*, which had just appeared.[123] Cases from many of his principal correspondents were cited in the article's opening pages, alongside Basedow's vivid account of a bone-pointing death. Cannon summarized the different accounts, referencing superstition as a contextual matter. He explained that, associated with indigenous groups' efforts to "render themselves resistant to the mysterious and malicious influences which can vitiate their lives,"

> is the fixed assurance that because of certain conditions, such as being subject to bone-pointing or other magic, or failing to ob-

serve sacred tribal regulations, death is sure to supervene. This is a belief so firmly held by all members of the tribe that the individual not only has that conviction himself but is obsessed by the knowledge that all his fellows likewise hold it. Thereby he becomes a pariah, wholly deprived of the confidence and social support of the tribe. In his isolation the malicious spirits which he believes are all about him and capable of irresistibly and calamitously maltreating him, exert supremely their evil power. Amid this mysterious murk of grim and ominous fatality what has been called "the gravest known extremity of fear," that of an immediate threat of death, fills the terrified victim with powerless misery. In his terror he refuses both food and drink, a fact which many observers have noted and which . . . is highly significant for a possible understanding of the slow onset of weakness. The victim "pines away"; his strength runs out like water; . . . and in the course of a day or two he succumbs.[124]

Cannon then offered a theory of sorcery grounded in his work on shock and the emotions. Injured soldiers suffering from wound shock, like decorticated cats in whom "sham rage" had been induced, would wither away in a matter of hours as a result of the radical drop of blood pressure and the systemic failure, including organ failure, it caused.[125] This was in all essentials the same process, Cannon asserted, as in bone pointing: wound shock could kill men with even "trivial" or "negligible" injuries.[126]

What is more interesting for our purposes here is the description of individual and social existence that Cannon indirectly proposed. In situations characterized by a social margin of safety, individuals, including patients, have the strong support of their community and remain largely free. This was not the case for the accursed or the injured with no social margin of safety: there the body could, without the least physical force being imposed from the outside, be brought to the point of death as the threads linking physical to social existence wizened. When the foundations of well-regulated social life shattered, the capacity of the organism to resist ostracism and profound emotional disarrangement could quickly prove minimal. It is no surprise that "Voodoo Death" played into the hands of thinkers such as structuralist anthropologist Lévi-Strauss, whose concern with words' ability to affect and determine social existence benefited immensely from Cannon's demonstration of their physiological power.[127]

: : :

Another theory concerned with the complete control of communication and the ensuing stranglehold on the free individual subject was Norbert Wiener's cybernetics. Wiener presented *Cybernetics: Control and Communication in the Animal and the Machine* (1948) as an inquiry into the "no-man's land between existing fields" and totalized its scope into an effort to cover the region ranging from mathematics to physiology through the control and design of communication in terms of "the region as a whole."[128] Offering a science for the regulation and dissemination of information that would "usurp a specifically human function" for applications in engineering, computing, and modeling, Wiener turned to systems of integration to propose programs that would address nonlinear problems.[129]

To understand and engage the complexity of particular integrated systems, Wiener thought it necessary to use neurophysiological models, not only geometric or statistical ones.[130] The integrative structure and action of the organism, and especially of the nervous system, was essential, even paradigmatic, for the argument on communication. Wiener wrote of the "performance of the nervous system as an integrated whole," by reference to a pathological disturbance (the purpose tremor) proposed to him by Arturo Rosenblueth, who, as Wiener admiringly noted, had worked closely with Cannon.[131] Wiener embraced Cannon's pathological-to-normal rationale, which saw disintegration lurking everywhere as a threat to highly integrated and organized systems.

Looking at the purpose tremor as an example of the integration of a complex and imprecise system, Wiener appealed to Cannon's homeostasis for the intellectual basis for understanding the regulation of information passing through a system via feedback mechanisms:

> A great group of cases in which some sort of feedback is not only exemplified in physiological phenomena but is absolutely essential for the continuation of life is found in what is known as homeostasis. The conditions under which life, especially healthy life, can continue in the higher animals are quite narrow . . . our inner economy must contain an assembly of thermostats, automatic hydrogen-ion-concentration controls, governors, and the like, which would be adequate for a great chemical plant. These are what we know collectively as our homeostatic mechanism.[132]

A "thorough, detailed discussion of homeostatic processes" seemed beyond the scope of the book. While agreeing that homeostasis was "the touchstone of our personal identity," Wiener shifted the language and value of Cannon's concept.[133] Cannon had emphasized the margin of safety, storage and abundance, and overflow, but Wiener looked at assemblages and control mechanisms—"an assembly of thermostats, automatic hydrogen-ion-concentration controls, governors, and the like." The purpose tremor patient, alongside a second, imagined patient suffering from ataxia, became central to the book, staging Wiener's articulation of one of his major concepts: negative feedback as it occurs in voluntary and involuntary, including inhuman, activity.[134] Negative feedback was a control mechanism akin to Cannon's homeostatic regulation via compensation. In describing the human body as an "integrated whole" or as a "complex additive system" that attempted to balance out oscillation, Wiener transferred onto information and communication the intellectual premises of the discourse on hormones and homeostasis.[135]

In discussing the sociopolitical connotations of feedback and homeostasis in the final chapter of his book, Wiener showed both his debt to Cannon and his distance from him. Wiener started out his treatment of society with the concept of homeostasis and circled around Cannon's understanding of a social body that is at least analogous to a physical body. This opening move seemed to closely parallel Cannon's assertion that societies often seem to operate like individual bodies when it comes to integration between their parts.[136] Yet Wiener offered a different rationale from Cannon's regarding homeostatic activity in society. Cannon had postulated that as social complexity rises, so does homeostasis: "archaic" groupings lack control over their external environment, which makes them less internally stable than complex societies. Wiener was concerned with the management of information, not control of the environment. Thus, small indigenous communities exhibited a "very considerable measure of homeostasis": "Strange and even repugnant as the customs of many barbarians may seem to us, they generally have a very definite homeostatic value."[137] By comparison, modern society had difficulty controlling information and action at a distance, a problem compounded by the capitalist constriction of the means of communication, thanks to the power dynamics that supported "the very limited class of wealthy men." Thus, "that system which more than all others should contribute to social homeostasis is thrown directly into the hands of those most concerned in the game of power and money, which we have already seen to be one of the chief anti-homeostatic elements in the community."[138]

In other words, despite using organismic integration as his premise, Wiener was not simply extrapolating from human neurophysiology. He jettisoned Cannon's metonymic transposition of the human, homeostatic body onto the social or political body, retaining instead the term *homeostasis* (and some of its implications) for a sociopolitical model based on regulated communication. Social homeostasis was threatened all the more in societies with an imbalanced or privatized control of power: "It is only in the large community, where the Lords of Things as They Are protect themselves from hunger by wealth, from public opinion by privacy and anonymity, from private criticism by the laws of libel and the possession of the means of communication, that ruthlessness can reach its most sublime levels. Of all of these anti-homeostatic factors in society, the control of the means of communication is the most effective and most important."[139] This passage is particularly significant in the context of Cannon's influence because it indicates that far more than Cannon, who rolled his eyes at the supposedly democratic promise of Hendersonian biocracy, Wiener was positively anguished by the transposition of homeostasis onto society. It was imperative to study, for it explained a more rational mode of social organization.[140] It allowed one to unmask "statisticians, sociologists, and economists available to sell their services" as "these merchants of lies, these exploiters of gullibility," and to describe society's failure to satisfy its humanist promises as the essential problem of its self-organization.[141] But the result was a complex of doubt and political anguish, not a cure. "Who is to assure us that ruthless power will not find its way back into the hands of those most avid for it?" he asked of the promise of cybernetic moderation.[142]

The rational, homeostatic, cybernetic approach might be politically troubling, but the alternative was downright vicious. Those seeking a measure of control over their circumstances were trapped, imprisoned in an emotional dilemma ranging between fear and wishful thinking, mice "faced with the problem of belling the cat." The individual at whom the sorcerer points the bone seemed the more appropriate reference point in a society obsessed with control: in need of properly distributed information, the individual was all the more emphatically deprived of the security and freedom it offered.

The Careers of the (Aphasiological) Symbolic: Cassirer, Mauss, Lévi-Strauss

Head and Goldstein were influential beyond their immediate discipline, especially through their parallel use and institution of concepts of the symbolic (Head) and of the abstract or categorical faculty (Goldstein) that, to them, were at work in the healthy mind. These concepts would be picked up by Cassirer (Goldstein's cousin) for his *Philosophy of Symbolic Forms*, and—more consequentially—by the anthropologists Mauss and Lévi-Strauss. Indeed, to the extent that Lévi-Strauss's understanding of the symbolic derived from Mauss, it bore all over its logic the marks of Goldstein and Head's work on the functioning and disturbance of the symbolic in normal and aphasic minds.

Goldstein deployed the philosophical implications of his work explicitly in *The Organism* (1934) and *Human Nature in the Light of Psychopathology* (1940), but he had started down this path already with his publications on aphasia in the late 1910s. This work would be particularly salient for Cassirer's third volume of the *Philosophy of Symbolic Forms*, notably his signal concept of symbolic pregnance.[143] The two carried on a substantial correspondence, and Cassirer visited the Frankfurt Neurological Institute, where he conversed with Goldstein and Adhémar Gelb, as well as with their patients, including Johann Schneider, whom he declared impressive in "the clarity and sharpness of his thinking, the aptness and formal soundness of his inferences."[144] This encounter would become fundamental for the *Phenomenology of Knowledge* (1929), the third volume of the *Philosophy of Symbolic Forms*. Cassirer used the pathological data and interpretation that Goldstein provided him, as well as the work of other researchers recommended by Goldstein, notably Head, Liepmann, and von Woerkom, to explain how a coherent experience of space and reality comes about.[145]

Cassirer's use of Goldstein's work allowed him to treat the relationship between consciousness and the world as resting on *symbolic pregnance*. Reframing his earlier work, which had focused on knowledge within the frame of scientific rigor, Cassirer now asked what kind of knowledge scientific inquiry offers in order to explain humans' ways of formulating their experience of the world.[146] Cassirer was acutely aware that philosophy (including Immanuel Kant's) and psychology (including eighteenth-century sensationalism and nineteenth-century associationism) had been plagued by how a phenomenon became content for consciousness—how it "speaks to consciousness and tells it something."[147] Placing aphasiology under the banner of the "pathology of symbolic consciousness," Cassirer structured the interaction between consciousness and the world as an experience that is rendered coherent by "symbolic pregnance"—that is, "the way in which a perception as a sensory experience contains at the same time a certain non-intuitive meaning which it immediately and concretely represents."[148] Cassirer resolved the problem of the interwovenness of consciousness and reality by presenting symbolic pregnance as what unites in advance the object to its representation in consciousness, and what is lacking in pathological cases like Schneider's where the "close-knit unity" achieved thanks to symbolic pregnance "loosens" or "disintegrates."[149] For such patients, perception fails to produce coherence and representation collapses.[150] Cassirer quoted Goldstein: perceived stimuli "arouse a chaotic impression and not, as in the normal individual, the impression of a specifically formed whole."[151] Whereas normally perceptual data are held together in meaning like words in a sentence, pathological consciousness cannot render the world as one.

Cassirer wound the pathological tightly around the normal.[152] What granted coherence to the various experiences of reality and the stimuli perceived was neither an association nor a mental syllogism, but the very function that Goldstein's patients lacked: the immediate extrapolation from a partial reception of objects toward a coherent experience of the world. Knowledge makes an inference toward the object from a definite characteristic, which, in symbolic perception, results in "unity of view."[153] In other words, Cassirer adopted Goldstein's understanding of the difference between merely concrete experience and the properly abstract attitude that can manage such a symbolic synthesis.[154] *Symbolic synthesis* is Cassirer's term, pointedly echoing Head, whose research he saw confirming his own sense of the prerequisites for perception and communication.[155]

Why is this adoption of the new aphasiology significant? For Cassirer its importance lay in its separation of "merely vital" functions from "spiritual" ones—and hence the confirmation of the priority of symbolic forms in intellectual-spiritual life and experience. Cassirer and Goldstein finally concurred in asserting that human freedom resided in the abstract attitude, and hence in the symbolic and cultural organization of the world perceived. The interweaving of symbolic forms experienced through symbolic pregnance allowed a free, healthy individual to go beyond merely vital functions, to contend with the world, and to represent it.[156]

: : :

The role of Head's aphasia research was doubly significant for Mauss. In his post–World War I efforts to devise an ethnological canon and method that would complete and surpass Durkheim's sociology, Mauss relied elaborately on the work of Head and Rivers on symbolic expression to devise a theory of symbolic assemblages and the symbolic life of the mind. This theory would reach from the most basic ways in which human beings semiautomatically handle their bodies to the elaborate constructions that generate a totality of human experience—what Mauss called the total man.

Mauss had a long-standing relationship with Rivers, and after World War I he began following Head's work closely. He knew of Rivers's research in the Torres Strait at least as early as 1903, when he cited Rivers's approach to collecting in a well-known study, coauthored with Durkheim, on what they called primitive forms of classification,[157] and perhaps already in 1898 when he first traveled to Oxford while Rivers was away from Cambridge on the expedition. Mauss reviewed the publications resulting from the Torres Strait expedition mostly positively.[158] When in 1912 the two met in London and became friends, he showered Rivers with praise for his study on the Toda people.[159] Through Rivers, Mauss met Head "in the admirable gardens of New College at Oxford in 1920," and he reported in 1924 that, regarding the symbolic function of the mind, "I even had the happiness of being in perfect agreement with Head and our dear Rivers during one of these scientific conversations that are the purest joys of our lives as scientists."[160] Mauss's necrology for Rivers was similarly exuberant—especially on the nerve regeneration experiment with Head, and Rivers's efforts to bridge psychiatric approaches—even though he expressed strong reservations over Rivers's wartime conversion to the diffusionist school in anthropology.[161] For Mauss, Rivers was the all-around intellectual, a man who worked in neurology, psychology, and psychiatry; who had invented the symbolic alongside Head; and who at the same time had refounded sociology and comparative religion.[162]

But it was Head who gave Mauss the crucial conceptual gift by conceiving the symbolic as a field coextensive with the social and the physical.[163] It indicated the existence of a consistent, ever-present chain uniting social cohesion and bodily function. It did not so much confirm the Durkheimian attention to symbols as restructure the relationship between the individual body and society. In his famous 1921 paper "The Obligatory Expression of Sentiments (in Australian Oral Funerary Rites)," Mauss offered an interpretation of crying and other lamentations as phenomena that were not "exclusively psychological or physiological, . . . but social phenomena marked by the sign of non-spontaneity and by the most perfect obligation."[164] Because "the entire group understands" cries, chants, and other such obligatory performances of sentiment, they were "signs, expressions comprehended, in short, a language."[165] Immediately following this passage, Mauss added: "It is essentially a symbolic [*une symbolique*]. Here we agree with the very beautiful and very curious theories proposed by Head, Mourgue, and the best informed psychologists on the naturally symbolic function of the mind."[166]

Aiming to unite psychology and physiology on a shared terrain with sociology, Mauss found precisely the opening he sought in Head's concept of the symbolic, not least because this allowed for physiological and psychological phenomena to carry social meaning, be encompassed by social phenomena, and become what Durkheimians like Mauss considered "social facts." Not content with using the notion of the symbolic to criticize most psychologists' general closed-mindedness toward sociology, Mauss repeated this exact discussion, complete with references to Head and Raoul Mourgue, in an intervention at the Société française de philosophie.[167] Philosophers and Durkheimians had already spent nearly two decades sniping at one another, scorning the seriousness and scientificity of one another's endeavors. Now Mauss dismissed as amateurish the philosopher Louis Weber's effort (and the pretense of Kantian philosophers more broadly) to do social theory by using notions like liberty and individuality to postulate how society and politics work.[168] To deny the self-sufficiency of Weber's political ideas, Mauss unsheathed Head's symbolic: "Nowhere is the function of thought more symbolic than in the elaboration" of notions like freedom.[169] The symbolic had become a combative concept, not merely a hermeneutic one: a concept with which to battle other disciplines and their claims on territory that could be deemed social. In this second formulation, it also became a given: imported just a few months earlier from "the most advanced" psychology and physiology, it had come to serve Mauss as synonymous with the cohesiveness of the social field. It drew meaning out of the convergence between the body and the self.

This was a significant development for Mauss and for Durkheimian sociology; before 1918, and like Durkheim, Mauss had spoken of symbolism and collective representations, but only exceptionally and unsystematically of the symbolic. By 1925, Mauss had come to own the concept enough to critically review three books addressing symbolic thought. The first, the repeatedly reissued 1923 *The Meaning of Meaning*, by C. K. Ogden and I. A. Richards, included an essay in which Ogden summarized his own essay "Word Magic" (1933) and cited Head's "famous research on aphasia."[170] Mauss praised that citation but otherwise offered only a backhanded compliment to Ogden; his "useful and spiritual generalizations" were dated. Mauss preferred another essay in the volume, by his friend Malinowski, and noted "the coincidence of his research and his opinions with those of Gardiner the Egyptologist and Dr. Head the neurologist."[171]

The second book for review was the first volume of Cassirer's *Philosophy of Symbolic Forms*, which concerned language and preceded Cassirer's visit to Goldstein's Neurological Institute. Mauss wryly commented, "We are promised an entire philosophy extracted out of symbolic forms," and then he let loose, targeting Cassirer's method for understanding symbolization: "We certainly believe that Mr. C's method is too philosophical, too generalizing, not sufficiently sociological, and also that he remains a prisoner of the classical opposition between symbol and intuition, between culture and life. Culture and symbol are themselves in nature because they are, and are in, society, which is in nature. And there is no intuition that could be detached from them."[172] Mauss credited Durkheim with this last point on intuition and culture, but he had himself decidedly revised the theory of the symbol toward a hinge between the corporeal and the social. He was no longer working, as before the war, with Durkheim's symbol or symbolism; he was working with Head's symbolic. Mauss was articulating this claim with increasing readiness; it was a central part of his methodology and conceptualization concerning the total man.

A year earlier, again while presenting to psychologists on the interdependence of psychology and sociology, Mauss had spoken as though conceding to "my late friend Rivers" the point on symbolic life binding the social to the corporeal. Because of it, "We join on these points with physiology, the phenomena of the life of the body, which between the social and physiology, it seems that the layer of individual consciousness is very thin: laughter, tears, funerary lamentations, ritual ejaculations are as much physiological reactions as obligatory gestures and signs, obligatory or necessary sentiments, either caused by suggestion or employed by collectivities with a specific goal, with a view to the physical and moral discharging of its expectations."[173] Who says antihumanism had to wait for Michel Foucault, Louis Althusser, even Martin Heidegger? To legitimize the wasting away of consciousness through the meeting of psychology and sociology, Mauss again recounted his "happy" 1920 meeting with Head and Rivers, and again turned to the notion of the symbol and the "essentially symbolic activity of the mind": "Here, Head's works find a warm welcome in us."[174] He stressed gratitude because, like Durkheim, he had long taught that "human beings can only commune and communicate by symbols, by common, permanent signs, exterior to the individual mental states that succeed one another. We had come to wonder where these symbols originate."[175] It was a call that psychologists had finally answered. But more depended on this answer than interdisciplinary influence; Mauss paid back the debt, offering up sociological studies of the mythical and moral symbols as "psychological facts."[176] On the shoulders of that counteroffering, he began formulating his theory of the total man.

Four major concepts define Mauss's later career and his anthropological fame: the account of reciprocity and sovereignty in gift giving, famously in the potlatch; the total man as the object of ethnological work; the person as a relative concept in need of genealogical, performative, ethnological, and psychological study; and the techniques of the body as a culturally and technically specific education and use of the human body. With the exception of gift giving, where they remained implicit, these concepts were welded together through the symbolic's mediation of the social and the psychophysiological. Mauss explicitly foregrounded Head's symbolic in each of them.[177]

The total man is always the subject of "our" research, Mauss declared in his 1924 account, even when the ostensible object was a minor fact. For example, "rhythms and symbols put in play not only the aesthetic or imaginative faculties of man but all of his body and all of his soul. In society itself, when we study a special fact, it is always with the psychophysiological *complexus* that we are dealing."[178] Mauss's holism simultaneously undercut and defended the individual in that same presentation.[179] Instead, the total man constituted the only sphere in which the different elements and objects of research could be fused and thought together—from instincts to rhythms to individuality within society.[180]

Bodily techniques, the second major concept involving the symbolic, reemerged (once more hand in hand with Head and Rivers) in Mauss's famous essay "Techniques of the Body." Every kind of culturally specific activity involved a particular mixture of training and efficacy that operated automatically and was involved even in the most basic of human forms, in that they affected the physical and psychological way that individuals operate their bodies and recognize their bodies as operating. Staking out a definition of his titular concept, Mauss wrote:

> Man's first and most natural technical object, and at the same time technical means, is his body. . . . Before techniques using instruments, there is the ensemble of techniques of the body. . . . The constant adaptation to a physical, mechanical, and chemical goal (e.g. when we drink) is pursued in a series of assembled

actions, and assembled for the individual not by himself alone but by all his education, by the whole society to which he belongs, at the place he occupies in it. Moreover, all these techniques [are] easily arranged in a system which is common to us, the notion that is fundamental to psychologists, above all Rivers and Head, of the symbolic life of the mind; the notion we have of the activity of consciousness as being above all a system of symbolic montages or symbolic assemblages.[181]

The body was at once natural and technically reconstructed, in an ensemble of "psycho-sociological" operations, it had within it the room to adapt, through at-first-order invisible educational operations, to the demands and practices of the "whole society to which he belongs." This ensemble of psychosociological operations was in turn theorized and experienced thanks to the symbolic. The "symbolic life of the mind" encoded into a shared system those semi-unconscious everyday practices and techniques that were carried out beneath language or other social structuring mechanisms, rendering meager and basic forms into intentional and clear activities. That such techniques could be unique to each society, yet shared and symbolically valuable, was only one matter. Others included how techniques of the body worked with social authority; how they constructed an "enormous" biological assemblage; how they were not simply comprehensible but constantly and consistently employed, without this comprehension needing to be explicit.

Finally, in the article "A Category of the Human Mind: The Notion of the Person," in which he credited neurology and psychology with having made "immense progress," Mauss once again cited Head (this time, oddly, alongside Théodule Ribot) to show that his own contribution aimed toward the structure of the "I" across societies, specifically toward how this I contributed to the ways in which "meaning is firmed up, functions, decays, deviates, and decomposes, and on the considerable role it plays."[182] Mauss then grounded the historical, ethnographic, and juridical comparison of forms of personhood on the synthesis offered by its psychophysiological arrangement, a synthesis that was fundamentally symbolic in character and for which Mauss could not and would not avoid Head. Mauss exorbitantly declared it plain "that there has never existed a human being who has not been aware, not only of his body, but also at the same time of his individuality, both spiritual and physical." Thanks to the symbolic fusion of body, psyche, and society, the neurology and psychology of the I allowed Mauss to claim that some part of this I was universal: this total individuality, a structure of selfhood, resulted from its integration in the symbolic—internally through psychological and physiological elements, externally through sociological and psychological ones.

: : :

It is difficult to overstate Head's quiet influence on subsequent, notably structural theories of the symbolic. Mauss's "symbolic" was what Lévi-Strauss adopted and adapted in his well-known 1949 introduction to Mauss's work and in subsequent writings, thus making it a key component of structural theory. Lévi-Strauss's allegiance to Ferdinand de Saussure's linguistics served as the foundation, but recently historians have helped clarify how Lévi-Strauss's structuralism relied on and responded to Mauss's approach to the symbolic.[183] If, as we have argued, Mauss shifted from an understanding of symbolism to one of the symbolic after his discovery of Head's work, Lévi-Strauss was not altogether excessive—as some have claimed[184]—to see in Mauss's work a system joining language and experience. Indeed, this system derived in large part from the aphasiological work of the early 1920s that identified a wholeness of human experience through its force against brain injury and disease.

"In Rivers," Lévi-Strauss once declared, "ethnology found its Galileo, in Mauss its Newton."[185] While completing *The Elementary Structures of Kinship*, Lévi-Strauss collated a series of Mauss's most significant essays and introduced them with an eye to presenting Mauss as a protostructuralist and himself as Mauss's heir.[186] Lévi-Strauss opened his introduction as follows:

> The first thing that strikes us about Mauss's thought is what I would like to call its modernity. The essay "L'Idée de mort" takes us to the heart of matters which psychosomatic medicine, as it is called, has only made topical in recent years. Of course, W. B. Cannon's physiological interpretation of the disturbances which he called homeostatic is based on work which dates from the First World War. But only much more recently did the famous biologist include in his theory those peculiar phenomena which seem to put the physiological and the social into unmediated contact. From 1926 onwards, Mauss had been drawing attention to those phenomena, . . . as one of the first people to stress how genuine, . . . how widespread, and above all how

> extraordinarily important they are for the correct interpretation of relations between the individual and the group.[187]

The citation of Cannon was far from incidental: Lévi-Strauss deemed him essential for an understanding of the physiological efficaciousness of language. But to remain with Mauss (and Head) for the time being, Lévi-Strauss cared about a symbolic activity thanks to which physiological effects and social circumstances touched—indeed, about how the self is physically trapped at once into social, symbolic, and bodily matrices. Mauss, Lévi-Strauss later argued, reached the webbing of the social and the physical by way of the symbolic. His understanding of the symbolic was the central concept that Lévi-Strauss followed in his introduction, and it became the main one he inherited. Lévi-Strauss combined Mauss's symbolic with Saussurean and Jakobsonian linguistics to create a structure that would be neither external nor transcendental vis-à-vis those speaking it; nor would it be merely empirical. It would be a totalizing structure, molded into one with the Maussian symbolic and covering in each case the entire social field. Following Mauss, Lévi-Strauss argued that the intersection of the social and the biological localized the interaction between experience and its symbolic organization: "It is natural for society to express itself symbolically in its customs and its institutions; normal modes of individual behavior are, on the contrary, *never symbolic in themselves*: they are the elements out of which a symbolic system, which can only be collective, builds itself."[188]

Would this suffice to suggest that Head's approach lay latent, yet was nevertheless imposing within a conception of the symbolic that bridged social life with normal neurophysiological behavior? Lévi-Strauss never cited Head, but he did include in the Mauss anthology all three of the aforementioned texts citing Head;[189] then he praised Head's twin, Goldstein, in his major 1952 essay "Social Structure." The praise could not have been more emphatic or striking. Goldstein had formulated in the most lucid manner "the rules of the structuralist method":[190]

> On the observational level, the principal—one could even say the only—rule is that all the facts should be observed and described with precision, without allowing any theoretical preconception to alter their nature and relative importance. This rule in turn implies a second: facts should be studied in themselves . . . and also in relation to the whole. This rule and its corollaries have been explicitly formulated by Kurt Goldstein in

psychophysiological studies, and it may be considered valid for any kind of structural analysis. Its immediate consequence is that, far from being contradictory, there is a direct relationship between the detail and concreteness of ethnographical description and the validity and generality of the model which is constructed after it.[191]

With this gesture, Lévi-Strauss joined Goldstein's integrationist method and his therapeutics of the neuropsychiatric patient. Those might concern different registers, as for Lévi-Strauss they were coextensive and coinstituting: they could apply the differential relations between elements of a system directly onto a global social analysis—precisely what he saw Mauss leading to as well.

Finally, no less direct and explicit was Lévi-Strauss's reliance on Cannon, whom Lévi-Strauss cited both alongside Mauss when introducing the latter and separately in his attempt to explain the sociopsychological interactions of the hostile environment at the physiological level. In his 1949 essay "The Sorcerer and his Magic," Lévi-Strauss began with Cannon's "Voodoo Death":

> Since the pioneering work of Cannon, we understand more clearly the psycho-physiological mechanisms underlying the instances reported from many parts of the world of death by exorcism and the casting of spells. An individual who is aware that he is the object of sorcery is thoroughly convinced that he is doomed according to the most solemn traditions of his group. His friends and relatives share this certainty. From then on the community withdraws. Standing aloof from the accursed, it treats him not only as though he were already dead but as though he were a source of danger to the entire group. . . . First brutally torn from all of his family and social ties and excluded from all functions and activities through which he experienced self-awareness, then banished by the same forces from the world of the living, the victim yields to the combined effect of intense terror, the sudden total withdrawal of the multiple reference systems provided by the support of the group, and, finally, to the group's decisive reversal in proclaiming him—once a living man, with rights and obligations—dead and an object of fear, ritual, and taboo. Physical integrity cannot withstand the dissolution of the social personality. How are these complex phe-

nomena expressed on the physiological level? Cannon showed that fear, like rage, is associated with a particularly intense activity of the sympathetic nervous system. This activity is ordinarily useful, involving organic modifications which enable the individual to adapt himself to a new situation. But if the individual cannot avail himself of any instinctive or acquired response to an extraordinary situation, . . . the activity of the sympathetic nervous system becomes intensified and disorganized; it may, sometimes within a few hours, lead to a decrease in the volume of blood and a concomitant drop in blood pressure, which result in irreparable damage to the circulatory organs. The rejection of food and drink, frequent among patients in the throes of intense anxiety, precipitates this process; dehydration acts as a stimulus to the sympathetic nervous system, and the decrease in blood volume is accentuated by the growing permeability of the capillary vessels. These hypotheses were confirmed by the study of several cases of trauma resulting from bombings, battle shock, and even surgical operations; death results, yet the autopsy reveals no lesions.[192]

In sorcery as in "battle shock," the body obeys down to the level of the sympathetic nervous system when commands emerge from a shared signifying system to which all submit.

The importance of this argument is revealed not in the overview of the victim or patient's suffering, induced by magic, but in Lévi-Strauss's extension of this suffering toward both a physiological and a social explanation in which the body figures as part of a semiotic economy. Lévi-Strauss knew that a structural theory of witchcraft that attended to the symbolic dimension of illness and the force of words needed to steer away from the collective representation model and the reliance on psychology in which Mauss's approach was steeped. Cannon had all but abandoned such considerations for an analysis of the homeostatic alternative. This made physiology available for an anthropological terminology that could once again inflict language onto the body, that could work on linguistic and symbolic efficacy while drawing on Cannon's explanation of the neurophysiological underpinnings of its effects.[193] Thus, a shaman who was pathologically involved in his own linguistic abilities and the particularity of his experience "might symbolically induce an abreaction of his own disturbance in each patient," quasi-hypnotically inflicting his own pathology.[194] Lévi-Strauss's economic alternative to Mauss's focus on symbols and magical performance in the latter's "The Physical Effect on the Individual of the Idea of Death Suggested by the Collectivity" goes as follows:

> An equilibrium is reached between what might be called supply and demand on the psychic level—but only on two conditions. First, a structure must be elaborated and continually modified through the interaction of group tradition and individual invention. This structure is a system of oppositions and correlations, integrating all the elements of a total situation, in which sorcerer, patient, and audience, as well as representations and procedures, all play their parts. Furthermore, the public must participate in the abreaction, to a certain extent at least, along with the patient and the sorcerer. It is this vital experience of a universe of symbolic effusions which the patient, because he is ill, and the sorcerer, because he is neurotic—in other words, both having types of experience which cannot otherwise be integrated—allow the public to glimpse as "fireworks" from a safe distance. . . . It is this experience alone, and its relative richness in each case, which makes possible a choice between several systems and elicits adherence to a particular school or practitioner.[195]

"This vital experience," "this experience alone," "its relative richness"—in its singularity, each physical experience joins with the dynamic, "continually modified" total structure in which individuals find themselves. On the hinge between the biological experience and the social situation resides language, which individuals absorb and deploy in an often pathological, neurotic fashion that accords meaning to each particular situation.[196] The shift from Mauss is minor. It retains the force and place of the symbolic at its disposal. Yet it is efficient in that it formalizes the claim, which he had stretched to attribute to Mauss himself, that "modes of individual behavior are *never symbolic in themselves*: they are the elements out of which a symbolic system, which can only be collective, builds itself."[197]

The symbolic here had become a system, a logic of language's efficaciousness on the body in the absence of any organic lesion. With the organic and the natural subverted as unchanging—with the refusal of history and evolution—Levi-Strauss reached the point at which physiological integrationism was taken for granted; the body in its functioning was being dominated and rendered pathological almost wholly from the outside:

> We must see magical behavior as the response to a situation which is revealed to the mind through emotional manifestations, but whose essence is intellectual. For only the history of the symbolic function can allow us to understand the intellectual condition of man, in which the universe is never charged with sufficient meaning and in which the mind always has more meanings available than there are objects to which to relate them. Torn between these two systems of reference—the signifying and the signified—man asks magical thinking to provide him with a new system of reference, within which the thus-far contradictory elements can be integrated.[198]

Like Wiener, whom he also brought into his thought at this point, Lévi-Strauss used physiology in order to move to a system of meaning, information, and communication that structured knowledge, memory, and interaction. Like Wiener he wondered about retaining the notion of the individual, and even more than Wiener he embraced an antihumanism that was as much methodological and epistemological as it was political. But if Wiener held to examples linking pathology to the failure of feedback, and to a metaphorical and metonymical use of homeostasis when he theorized the governance of information and provided analogies from neural systems to individuals to society, Lévi-Strauss was far more explicit on the physiology of structuralism.

The passage that opens "The Sorcerer and His Magic" flaunts a figure at once ejected from and trapped by the social and the symbolic. The human being suspended from the symbolic construction of a reality that reaches deep into the workings of physiology is the individualized human being—a being who is sick and therefore separated, rendered visible within the social, and deprived of the means for the continuation of its individuality. Survival entailed reabsorption into the social—a convincing assumption, redistribution, and stabilization of the symbolic, a symbolic harmony that allowed for individuality, physiological stability, and social reregulation. In other words, Lévi-Strauss's antihumanism required an extremely unstable individual, or rather it asserted individuality alongside instability.

9

Vis medicatrix, or the Fragmentation of Medical Humanism

On October 18, 1923, St. Luke's Day, Ernest Starling delivered the Harveian Oration before the Royal College of Physicians. His career was at its zenith. Starling titled his speech "The Wisdom of the Body" to echo and answer the question posed in Job 38:36, "Who hath put wisdom in the inward parts? or who hath given understanding to the heart?"[1] Starling's answer was clothed in custom: he quoted at length from "our Founder" William Harvey and heaped praise on him for the discovery of circulation and for delivering physiology from darkness. The lecture catered to an audience of fellows intent on self-congratulation and on "reckoning among our saints" even scientists who had no club membership—Robert Boyle, Claude Bernard, Louis Pasteur.[2] It gave credit to contemporaries who provided experimental protocols and helped advance circulation-related research, notably August Krogh in Denmark, Magnus Blix in Sweden, and Walter Cannon in the United States, as well as John S. Haldane, A. V. Hill, Arthur Keith, Ross Harrison, Edward Sharpey-Shafer, and Thomas Lewis among his British colleagues.

Starling clearly meant that no one was more capable of asking Job's question anew, or of displaying for his audience the body's wisdom, than himself. The bulk of Starling's oration was devoted to his achievements in the physiology of these inward parts, especially the hormones and the law of the heart, which he had articulated a few years prior. He identified with Harvey, but he staged the latter's aporias so he could answer on behalf of his own generation and so he could crown his own research. He compressed the entire history from Harvey to his discovery of secretin with William Bayliss, "another chapter in modern physiology that can be said to have grown out of Harvey's discovery of the circulation of the blood."[3] "There is still much to be learned; the ocean of the unknown still stretches far and wide in front of us," Starling said, but there was no longer any doubt that "we know the directions in which we would sail, and every day, by the co-operation of all branches of science, our means of conveyance are becoming more swift and sure. Only labour is required to extend almost without limit our understanding of the human body and our control of its fate."[4] Scientific humanism and positivism meld together in this passage: order was found thanks to the achievements of physiology, and the remaining blind spots would soon fill with light.

Toward the end of his discussion, Starling brought into the picture integration, the decisive advance of his generation. He clarified integration's contribution to adaptation for the purpose of the organism's well-being and noted the nervous system's control over it.[5] Then he turned to that "other chapter of modern physiology" traceable to Harvey:

> In the dedication to his work Harvey compares the heart to the sovereign king, and throughout he continually recurs to what we should now describe as the "integrative function" of this organ. In virtue of the circulation which it maintains, all parts of the body are bathed in a common medium from which each cell can pick up whatever it requires for its needs, while giving off in return the products of its activity. In this way each cell works for all others—the lungs supply every part with oxygen and turn out the carbon dioxide which they produce, the alimentary canal digests and absorbs for all, while the kidneys are the common means of excretion of the soluble waste products of the body. Changes in any one organ may therefore affect the nutrition and function of all other organs, which are thus all members one of another. But, in addition to enabling this community of goods, the circulation affords opportunity for a more private intercourse between two or at any rate a limited number of distant organs.[6]

In this sweeping paragraph Starling forced the heart and hormones to perform his conceptual labor. First, the integrative function was something that Starling could easily link to Harvey while implicitly lionizing his own moment: no one could plausibly attribute integration to the seventeenth-century isolation of circulation, but such reference was essential for the political dimension of biology that Starling would turn to at the end of his lecture. In this manner Starling could locate today's secular "miracles" at the conclusion of a long medical history that doubled as a political history and a history of secularism, in that medicine had always been humanity's physician. It had now reached its positive age, attending to disease as past ages had regarded sin. This goal became politically paramount. He promised, "The mastery of disease and pain . . . will enable us to relieve the burden of mankind."[7]

Second, by treating circulation as an integration-aimed structure, Starling was also giving it a second twentieth-century meaning: it was the enabler of a "community of goods" and the medium for action between different organs. The passage was intentionally economic in character, with cells "working for all others" and organs being "all members of one another." They were measurable through their work and their value as goods in the internal economy of the body. Starling had no difficulty using similes that were oriented toward work and consumption: the heart was like a car, the blood like petrol.[8] He also had no difficulty interpreting a healthy life in terms of productivity: "The sorrow of life is not the eternal sleep that comes to everyone at the end of his allotted span of years, when man rests from his labours. It is the pain, mental and physical, associated with sickness and disability, or the cutting off of a man by disease in the prime of life, when he should have had many years of work before him. To us falls the task of alleviating and preventing this sorrow."[9] The similes were far from gratuitous; integration needed to be thought of as political economy. At the same time, this schema betrayed a new model patient: male, productive, in the prime of life. Where doctors had dealt with people growing and aging, men and women suffering all through life, from often quietly ravaging diseases and not from injuries, now they encountered masses of patients who were different in terms of their individualized reactions but still similar in a sense: male, young, with "many years of work before" them, and who came in quite quickly after injury, requiring rapid care. If a return to "normal" was conceivable in 1923, and a utopia of health, this normal was now attached to the white male youth, who needed to be restored to his status as pillar and productive leader of society.

Third, the heart was no longer "sovereign king" over the rest of the body. The language of community and interdependence of organs ran directly counter to a language of monarchic sovereignty that had been so central to (and vulnerable in) Harvey's own lifetime. Indeed, the point of the law of the heart, like the force of hormones, was that there was no single seat of power, no univectoral direction of force out of the heart, no hegemon and no subjects. The language of economized circulation expressed Starling's politics precisely against an antiquated figure of monarchical physiological rule, and it did so with an eye firmly on questions of destabilization and care. Starling spoke of a "community of goods" as the result of physiological integration, but this was by no means his only reference to "community," apropos of a value he apparently appreciated, that "virility does not mean simply the power of propagation, but connotes the whole part played by a man in his work within the community."[10] Integration and interdependence played a distinctly political, and distinctly masculinist, role: against Harvey, the body politic was a mechanical, integrated, economic body whose parts relied almost entirely on one another.

Starling's political metaphors were as dramatic as they were limited: he had shifted to a language for which integration as the basis of the organism's unity had to be put into political terms, but his hopes were those of a classic medical humanist for whom the curative ideal took priority. He was far from alone. Work, human-motor metaphors, classical economics, and religious miracles formed the basis of this humanism, but Starling felt no reason to step beyond his mandate and inflect all thinking of politics, society, gender, norms, the economy, or the practice of medicine on their basis.

For Cannon, Henry Head, and Kurt Goldstein, this would not be the case, and, as we have seen, their cross-disciplinary impact and political thought would bear extraordinary fruit. At the same time, this influence would lay bare the contradictions involved in the therapeutic ideals they espoused.

These contradictions and their afterlives are the subject of this chapter, and here we turn to the implications of integrationism for medical theories after Starling: theorists of integration posited time and again a highly individualistic treatment of the patient that refused the standardization of welfare and downplayed its effectiveness on the grounds that the individual patient was the only possible object of care. The terms of this care are the concern of the present chapter.

Standardization and the Welfare State

Historians commonly identify the interwar period with the dramatic expansion of the British and French welfare states, including the transformation and standardization of medical and hospital care. Social hygiene proposals, family dependence, the demands of ever-more-complex industry, and the war and its aftermath, including the outbreak of the Spanish flu epidemic, were central to this expansion.[11] So too was the hardening grip of statistical thinking, which, as Theodore Porter has shown, was instrumental for taming various forms of uncertainty.[12] In Germany, Britain, and France, the 1920s was a decade for statistics, standardization, economic intervention, and the systematization of existing welfare elements, including medicine.[13]

The dream of social or socialized medicine was also the key to physicians' swiftly improved opinion of the Soviet Union. The historian Henry E. Sigerist, for example, devoted his 1934 *American Medicine* to examining the disjunction between technical achievement and social efficiency; he asserted that "the technology of medicine has outrun its sociology."[14] Sigerist, who identified the "new physician" as "scientist and social worker,"[15] was most interested in the social part. In two books on medicine in the Soviet Union he celebrated Soviet socialized medicine as the future of social welfare.[16] "To serve society more efficiently and improve the people's health" was his motto, and a society that had shown "that human nature can be changed," especially by virtue of radically improved medical access and care, was the most advanced.[17]

Medical specialization, long central to care and research, was in the midst of becoming a standard part of the landscape of rehabilitation.[18] In this, physiology was just one among many disciplines, and perhaps the one least invested in specialization. In his 1914 *An Introduction to the History of Medicine*, Fielding H. Garrison placed physiology in a labyrinth of specialties, including genetic biology, biometrics, the study of proteins, psychology, psychiatry, parasitology, serology, chemotherapy, and surgery.[19] Improvement in these fields compounded the sense that medicine aimed at both precision and socialization. George Weisz has shown in detail how the new public health movement in the 1920s took on prevention and treatment of tuberculosis, venereal disease, and diseases in children; standardization of surveillance control; and clinical treatment as a single effort.[20] As of the 1910s, the well-controlled experiment meant the regulation of patient behavior as a major component of experimental conditions, the uniformity of treatment no matter the outcome or course, the value of expertise on the side of experimenters, and the blinding of experimenters to the details of data until collection was completed.[21] Harry Marks points out that the "therapeutic revolution" of the late nineteenth century was met in the 1920s with "a rational therapeutics" due largely to the development of purified drug compounds.[22] The rise of the drug industry brought reform to clinic and marketplace.

Yet something is lost in a picture of medicine and social welfare in the 1920s that is presented strictly in terms of standardization, quantification, and specialization: namely, the forceful, even obsessive attention given to individualized therapeutics by physiologists, neurologists, endocrinologists, and psychiatrists. These fields consistently provided critiques of any social medical humanism whose success might come at the expense of individual patients. The construction of the individual patient as a category and the modification of the doctor-patient relationship went hand in hand with the emergence of a neo-Hippocratism, a new *vis medicatrix naturae* (or "healing power of nature"), which put the doctor in aid of reequilibrating the patient. The revolution in physiology was entwined with the neo-Hippocratic recognition that nature contained much of the power of regulation and healing of an injured organism.[23] Similarly, Goldstein's theory that the organism exhibited a single drive, self-actualization, conditioned a therapeutic fantasy that eschewed standardization for absolute attention to *this* patient *here*, yet also found its own ideal neutered by the pragmatic limitations in therapeutic practice. Cannon and Goldstein were only two among several contributors to this new therapeutics and to quasi-transferential protocols of care. Despite the fact that the major physiologists were frequently of a socialist persuasion, a politics of individuality took over.

Thus, without an evidentiary crisis breaking out in medical practice—after World War I practitioners continued to use the same methods and standards of evidence—suddenly a newfound anxiety about the individual led to a crisis of standardization, or of the value recognized in standardization. The figures we are discussing became obsessed with translating their own projects into terms that offered heightened therapeutic value precisely because of their way of handling individuals. Their work constituted at once the grand gesture of the new physiology toward medical humanism and, paradoxically, the fragmentation of that humanism. Where Starling's positivist utopia had made it possible for the social and individual dimensions to thrive together, the two would gradually break apart. Some thirty years later the same authors were being cited time and again in the arguments against institutionalized health care. Ivan Illich, in his 1975 diatribe *Medical Nemesis*, would follow this neo-Hippocratism to its radical conclusion: "The medical establishment has become a major threat to health."[24]

Regulation, Humoralism, and the Vis Medicatrix Naturae: *Homeostasis, Stress, and the Physician*

In 1935 the historian of medicine Arturo Castiglioni noted that around the end of the previous century, "when the return to the clinical conception and to general pathology assumed a real importance, when the study of functions began to prevail over the study of forms, a new tendency towards ideas which had been thought dead began to take shape, while the coming of a new humoralism and of a definitely biological conception appeared on the horizon."[25] *The new humoralism* was a regular nickname for the new physiology of hormones, which, like the four humors of ancient doctrine, were imagined to operate by regulating functional matters to such a degree that organic operations were rendered secondary.

The new humoralism was only part of the story, even for Castiglioni, who mostly credited German surgeons like Ferdinand Sauerbruch, Friedrich Kraus, and Alfred Goldscheider, and Italian clinicians, notably Francesco Galdi and Mario Donati. In 1936, René Leriche, concluding the sixth volume of the *Encyclopédie française*, which centered on the body and medicine and was styled as an account of *The Human Being as a Whole*, similarly appealed for a closer study of the new humoralism that went hand in hand with pathological physiology as an "indispensable discipline of control." This "chemically and physiologically defined" humoralism was less anxious about localized conditions than about the patient's "permanent humoral terrain, . . . the threshold of physiology, a threshold that nothing marks out but that is ceaselessly crossed."[26] To treat the body as humoral was at the very least to accept the spread of nonconventional strands and their fusion with medical practice.[27] Christopher Lawrence and George Weisz have noted that in mainstream medicine, clinical critiques of "rigorous scientific procedures" as reductionist practices also included advocacies of "constitutionalism, psychosomatic medicine, . . . social medicine, Catholic humanism and . . . the convergence of various forms of alternative healing."[28] For Allan Young, the general goal in British psychiatric hospitals during World War I came to be to help patients complete their transformation away from their old broken self to a new one. Psychiatry and neuropsychiatry (like Head's) pursued a similar goal of conformity with nature and its impositions.[29]

Hormonal "humoralism" and neo-Hippocratism went together well as a logic, and for Cannon, who was highly attuned to classical Greek concepts and medical practices, proximity to the ancients was particularly significant. Cannon was explicit, even repetitive, on the healing power of nature. In his first paper discussing homeostasis, in 1929, he asserted that "biologists have long been impressed by the ability of living beings to maintain their own stability. The idea that disease is cured by natural powers, by a *vis medicatrix naturae*, an idea which was held by Hippocrates, implies the existence of agencies ready to operate correctively when the normal state of the organism is upset. More precise modern references to self-regulatory arrangements are found in the writings of prominent physiologists."[30] Cannon's concepts advanced the sense that the very nature of the human body, notably in its self-regulation, tacitly involved a collaboration between physician and environment and demanded the adaptation of medical technique: this was now meant not to interfere, but to facilitate reregulation; not to fix the organism, but to adjust it so internal and external environments could operate in a new harmony. He meant "new" in the sense of postdisease and postintervention. Intervention and medical practice, however deforming they were, were meant to reinstitute peace in a natural and social environment, not directly restore the status quo ante. Asking "if the body can largely care for itself, . . . what is the use of the physician?" Cannon referenced the *Corpus hippocraticum* and replied that in addition to supplying his own cures, the physician could utilize the body's means for its own good: "The physician, then, plays his part in making effective the self-regulating adjustments of the body that have been disordered or that are in need of reinforcement, understanding that, as a rule, nature herself is working with the curative agencies which he applies."[31] As Georges Canguilhem later noted, Cannon's "modern interpretation of the *vis medicatrix*" generated optimism regarding the doctor's cooperation with nature: the guile of attaching the physician to nature staged the doctor-patient relationship as one that indemnified both from the needs of the health care system, recognized the limitations of integrated life, and arrogated specific responsibilities in terms of recapturing health.[32] The physician did not work against nature in the sense of staving off the pressures and forces of the milieu, but imagined the self-sufficiency and adequate functioning of the body under his (or rarely, her) care, reinforcing it against external threats and their internalization.

The origins of Cannon's reference to Hippocrates are unclear. It is possible that he imported it from Lawrence Henderson, who endorsed a *vis medicatrix naturae* despite his own biochemical focus. Henderson appears to have discovered it in 1926 while reading Vilfredo Pareto's socioeconomic theories of equilibrium and to have decided that it mattered for the treatment of both patients and societies in crisis.[33] The possibility of a sociological origin and hence a circular logic for the *vis medicatrix naturae* is tempting, particularly given economists' repeated use of the term and Henderson's Bernardian focus in the later 1920s, but even though Cannon had not yet articulated homeostasis in 1928, he was already evoking the *vis medicatrix* with an eye to the relationship between systemic regulation and medical practice. Delivering the annual oration of the Middlesex District Southern Medical Society in April of that year and offering, to his audience of physicians, "Reasons for Optimism in the Care of the Sick," Cannon became more concrete on the physician's relation to regulation:

> He should be well acquainted with the possibilities of self-regulation and self-repair in the body. . . . Then, he should realize that many of these remarkable capacities for adjustment require *time*—all the processes of repair belong to this category—and they can play an important rôle in restoring the organism to efficiency if only they are given the chance which time provides. The physician, therefore, should wisely assure the conditions which permit the damaged body to continue active until lost or injured parts have been rebuilt, strengthened, or compensated for. . . . The physician plays his rôle in making effective the self-regulating adjustments of the organism that have been disarranged or that are in need of reinforcement.[34]

With minor modifications, this passage would be repeated in *The Wisdom of the Body* with much the same emphasis: the physician was encouraged to observe and wait, to be attentive to "possibilities of self-regulation and self-repair." The physician's activity was not one of intervention but one of identifying homeostatic processes in action, and then at most facilitating the body's corrective tendencies. Without suggesting that the physician was useless, Cannon ignored any priorities of localized diagnosis or intervention and cast the doctor in the role of an aide replacing the orchestra conductor: it was up to the orchestra to correct itself.

This argument that the doctor aids an equilibration already under way in the individual was significant for three major reasons. First, it asked why the organism's self-regulation and self-repair might be out of sync. Second, it identified the patient as a self-healer, and it reversed the physician's objective from correction or intervention to aiding the natural return of a well-ordered system. In so doing, it opened the question of the doctor and patient as a duet.

One of the farthest-reaching versions of this neo-Hippocratism emerged out of Cannon's influence on theories of organismic adaptation and especially on stress as a form of alarm. Its most visible exponent was a self-proclaimed inheritor of Cannon's legacy, the Hungarian-born Canadian endocrinologist Hans Selye, who elaborated a model of biological stress (a term he popularized) in the "hypothalamic-pituitary-adrenal axis," a system of bodily regulation to cope with myriad forms of stress and insult to the organism.[35] Crucially, Selye posited both a local and a general adaptation syndrome as the organic mechanism of response and repair to inflammation at either a single site of injured tissue or a systemic assault on the organism, the latter working by means of alarm, resistance, and exhaustion.

To adopt Cannon as his direct precursor, Selye argued that Cannon spoke only in general terms about stresses and strains,[36] although he himself "was primarily inspired by Cannon's discovery of the 'emergency reaction.'"[37] And although the much-younger Selye asserted that the older Cannon "showed particular interest" in his approach to adaptation, he later admitted that he "could never convince Cannon of some of the basic tenets" of the general adaptation syndrome.[38] As Mark Jackson writes in *The Age of Stress*, Selye's 1946 article was dedicated to Cannon's memory and marked an awareness of the clinical implications and conceit of his work, but Cannon was deeply critical of the single-minded approach to laboratory experimentation to which Selye seemed committed.[39] In his 1947 *Textbook of Endocrinology*, Selye mentioned Cannon only once, and only as one in a long list of historical precursors to his own theory of the general adaptation syndrome.[40]

Selye's tug-of-war between Cannon as master and Cannon as stepping-stone would continually undergird many of his claims.[41] The goal was to standardize the points of homeostatic theory so it might be wrought into the general adaptation syndrome as the central axis of modern patienthood and medicine. In his popular and highly influential *The Stress of Life*, Selye portrayed *stress* as the central term in understanding the regulatory phenomena of the organism. Here he placed Cannon in a thread of conceptual inheritance that began with Bernard and ended with himself.[42] This created an opening for him to conveniently and vaguely define stress as "the nonspecific response of the body to any demand," and to forget the long genealogy of stress metaphors in earlier shock research: "There must be some element of stress here, at least in the sense in which the engineer speaks of stress and strain in connection with the interaction of force and resistance."[43]

Stress, in this analogy, meant a particular general response to the destabilization of organism-milieu relations; as the latter became distressed, the internal environment responded in kind, attempting an adaptation that would help obviate the external stressors. This internal mirroring is what would be named *stress*, and it offered a general form of systemic deregulation that required attention. For Anne Harrington, Selye's stress, expanding Cannon's concepts of fight or flight and homeostasis, provided a solid pillar—perhaps the most important one—for the rise of mind-body medicine. It completed the proto–New Age translation of Cannon's and others' physiological attempt to incorporate psychological factors, to render them corporeal, and to treat this physiological body as the basis for every natural medicine, on the assumption of a *vis medicatrix naturae*. Painting Cannon's definition of homeostasis in broad strokes allowed Selye to speak of the need to counter stress by actively reestablishing homeostatic regulation. With feigned surprise, he continued, "Apparently, disease is not just suffering, but a fight to maintain the homeostatic balance of our tissues, despite damage."[44]

In Selye's work, therapeutics would reinforce the *vis medicatrix* by focusing on the endocrine system's biological response to stress and the production of glucocorticoids. Cortisone injection therapies were designed to correct the "out-of-balance" cortisol response in the individual patient—that is, they put the "pros" and "antis" in alignment to address debilitation produced by the body's autonomic response to stress or alarm.[45] At issue was not the doctor's intervention so much as the restabilization that aided the individual.

By the mid-1950s, when Selye popularized the term, homeostasis had become a complex and problematic concept linked to—and often confused with—several others, including developmental homeorhesis, genotypical fitness, genetic homeostasis, and so on. Genetic and epigenetic attempts to account for stress and homeostasis only complicated things further.[46] Stress was especially viable and influential because of how it offered up the organism to the physician and physiologist as fundamentally disordered. The universalization of stress and the singularity of its experience went hand in glove. As a deregulation of the homeostatic body, stress posited most clearly that the neo-Hippocratic tendency was confronted with a problem of its own creation: a deregulation that belonged at once to modernity and to each individual body, a pathology without cause or proper treatment, yet claiming to explain the bulk of discomfort and suffering across human experience.

"The Art of the Healer": Transference, Individualized Care, and Self-Actualization

Roy Porter, in his canonical essay "The Patient's View," wrote that the clinical encounter is always a two-way interaction, yet its conceptualization has been one in which the physician directs its action.[47] The effort to examine this relationship by turning the expected dynamic on its head thus exposed two sets of concerns. First, the reappraisal of the doctor-patient relationship was an attempt to set aside the physician's concerns for the needs of the sufferer. Second, the retrained focus on the patient repaired a previous breach in the clinical encounter, which was figured as fundamentally detrimental to the sufferer.

Cannon's Hippocratism entered a complex terrain dominated by appeals to a new understanding of the physician-patient relationship, partly due to the mutual dependence of pathology and clinical medicine. Pierre Abrami, in his 1936 inaugural lecture as professor of pathology in Paris, theorized that clinical medicine could not do without pathology, but by the same token pathology had no raison d'être without clinical medicine. The two joined in a "science of the particular," and the relationship between the physician and the functional disorders from which the patient was suffering was perhaps the foundational factor affecting the latter's pathological picture.[48] A second part of the shift was due to the new recognition (partly thanks to Cannon, partly to Freud) that emotional factors played a significant role in the course of a disease: the simple encounter of the patient with the physician—made especially demanding in soldiers whose injuries were accompanied by elaborate psychological problems—played a role because the emotional component of the disorder could be decisively influenced by that encounter.

Walter Langdon-Brown's short article "Return to Aesculapius" (1933) explicitly cited quantum indeterminism to support his claim that patients' encounter with doctors changes them. He proposed the establishment of a relationship based on mutual faith and the physician's support of the patient's self-knowledge.[49] Fear, Langdon-Brown insisted, was decisive for the patient overwhelmed by an incomprehensible mass of data regarding his condition. Like Goldstein's catastrophic anxiety, like Cannon's fear (also presented in Abrami's derivative account as capable of generating functional disturbances[50]), this fear of being confronted by the doctor could be crippling to a patient. Langdon-Brown proclaimed, like Abrami, that a fear-hope binary was what physicians as much as psychiatrists needed to appreciate in a systematic fashion. The psychoanalytic discovery of transference fulfilled this role: in it resided a major reason for the spread of psychoanalysis and a third factor in the shift of the physician-patient relationship.

Freud had been explicit on the problems and possibilities offered by transference since his early work, particularly in the loving and hostile feelings a patient could exhibit toward the doctor. In 1911, Freud noted them in the patient's resistance to "the doctor"—by which he meant the psychoanalyst, but the choice of term matters:

> The patient regards the products of the awakening of his unconscious impulses as contemporaneous and real; he seeks to put his passions into action without taking any account of the real situation. The doctor tries to compel him to fit these emotional impulses into the nexus of the treatment and of his life-history, to submit them to intellectual consideration and to understand them in the light of their psychical value. This struggle between the doctor and the patient, between intellect and instinctual life, between understanding and seeking to act, is played out almost exclusively in the phenomena of transference. It is on that field that the victory must be won—the victory whose expression is the permanent cure of the neurosis.[51]

Triumph is born in this account from the heroic chaperoning of the patient to nervous health through a successful struggle over the direction of instinctual impulses. In his *Introductory Lectures to Psycho-Analysis* in 1917, Freud made the psychoanalyst somewhat more transient in the deployment of the patient's attachment, noting that by transference "we mean a transference of feelings on to the person of the doctor. . . . The whole readiness for these feelings is *derived from elsewhere.* . . . [T]hey were already prepared in the patient and, upon the opportunity offered by the analytic treatment, are transferred on to the person of the doctor."[52] Transference transformed the original disorder under treatment into a new class, "transference neurosis," which meant that not only did the relationship with one's physician realign one's condition; also, basic emotional components were recalibrated through the treatment.[53]

Langdon-Brown, for his part, conceded that psychoanalysis proper was too complex for most doctors to adopt as a praxis of hope and care; nevertheless, its insights for the doctor to engage the patient's state of mind and way of being were crucial. Others were explicit on the importance of transference. Rivers's goal of reinstilling the patient's autonomy in the face of battlefield trauma led him to object to transference by attempting to all but remove himself and oblige the patient to carry out as much of the work as possible independently.

The unorthodox approaches of René Allendy and Georg Groddeck followed suit, even if Allendy and Groddeck were not scientists in the same sense of the term.[54] Allendy, a homeopathic physician and founding member of the Société psychanalytique de Paris in 1926, is mostly remembered as the analyst of Antonin Artaud, Anaïs Nin, and René Crevel. Nin was at once resentful of and deeply attracted to Allendy's practice, in a way perhaps typical of transferential resistance to psychoanalysis. She wrote of her encounters with him as a "laboratory of the soul."[55] Allendy also became the leading figure in Paris for the adoption of a holistic and Hippocratic understanding of the doctor's role, which he explicitly privileged against the then-leading analytical approach. His 1929 *L'orientation des idées médicales*, as Weisz has indicated, marked something of a paradigm shift in that it signified the mainstreaming of different holistic approaches in France.[56] Canguilhem, then a twenty-five-year-old philosopher with no connection to medicine, reviewed and celebrated Allendy's anti-Pasteurianism by declaring: "The individual reappears. The day when we finally perceive that science points to the individual as its proper object, the philosophers who love generalities will be thrown into a panic. All the worse for them."[57] A few years later the Austrian physician Bernard Aschner, in an attempt to theorize medical care away from laboratory physiology, would cite Allendy as a contemporary contributor to the use of "historical and practical medicine" and to the "renaissance movement" that he declared emergent in the present.[58]

Like Allendy in Paris, the Baden-born "wild" psychoanalyst Groddeck incorporated somatic, psychological, and psychoanalytic practices into a broader theory of naturist care for the individual patient. Influenced by the nineteenth-century physician Ernst Schweninger as much as by Freud, in 1913 Groddeck published *Natura sanat, medicus curat: The Healthy and the Sick Man*. The title's Latin axiom, often abbreviated to *Nasamecu*, means "nature heals, the doctor cures."[59] However odd, this quasi-Hippocratic observation pervaded his understanding of the relation of illness to health. Circumscribing the role of the physician, Groddeck noted: "One can—to use the image of struggle—be defeated by the forces of life and death. But that one should believe himself to have no more weapons, that one should despair and give up the struggle, is an aberration of the mind. Where there's life there's hope, and where no hope remains, duty still remains. All the same we believe that the holy, the noble, the great work of the physician begins with the hopelessness of the incurable, of the dying."[60] The doctor's place was thus one of submitting to the becoming that is life, while attempting to facilitate the patient's return to hope. In 1923, his *Das Buch vom Es* (*The Book of the It*) elaborated this argument by taking up the use of psychoanalysis for more than merely psychological or even functional disease—indeed, even for the care of organic conditions.

In *Das Buch vom Es*, Groddeck went into great detail describing his work in a war hospital, treating wounded soldiers by psychoanalytic means. In one lengthy case study, Groddeck described a soldier-patient, "Herr D.," who presented recurrent fevers and rigors. After an inconclusive Wassermann test prompted by suspicion of syphilis, Groddeck treated him with hot baths, a strict diet, and massage before returning him to duty.[61] Sixteen months later, in the summer of 1916, his symptoms returned, and Herr D. came to Groddeck again. Groddeck resolved to abandon any attempt to elaborate on the physiological cause of Herr D.'s condition and treated him through psychoanalysis.[62] Far exceeding Freud's expectations of analysis and transference, but keeping with his own sense of the doctor aiding nature, Groddeck posited that "every sickness of the organism, whether it is physical or psychic, is to be influenced by analysis."[63] This was because in addition to having a *causa externa*, every disease was conditioned by a *causa interna* that went far beyond the *causa externa*.[64] Groddeck could even claim that "illness does not come from without; a man creates it for himself, uses the outer world merely as the instrument with which to make himself ill, selects from that inexhaustible supply to be found in the wide world, now the spirochete of syphilis, today a piece of orange peel, tomorrow the bullet of a revolver, the day after a chill, so that he may pile on his woes."[65] In his treatment of Herr D., Groddeck was uninterested in the spark; in this he was radicalizing Cannon and even Goldstein, with his theory of self-actualization.

Two months later, Herr D. died in battle. Groddeck continued to pronounce the case a success, despite the brevity of treatment, and numbly asked if the man's death could really undermine his claim to cure. Irrational as it seems, this claim followed stronger assertions on the capacity of transference to cure, and on the at once highly active and merely natural place of the physician whose role it was to force the patient back into his healthy life by overcoming the unnatural and culture-driven resistances that conditioned the disease. Disease, even organic disease, was not unnatural and was not imposed from without, Groddeck claimed; instead the sick person was sick because of the person's resistance.[66] The sick individual was always both the cause and cure of disease: "Recovery is brought about not by the physician, but by the sick man himself. He heals himself, by his own power, exactly as he walks by means of his own power, or eats, or thinks, breathes or sleeps."[67] Over and over, disease and resistance were identified with each other and feminized: "A sick man wishes to be sick, and he struggles against healing, much as a spoiled little girl who in her heart would like to go to the ball will nevertheless do everything she can to put obstacles in the way of going."[68]

Groddeck and Allendy were not alone in blending proto-psychosomatic medicine with transference and a theory of disease, but they were visible, controversial, and committed to an integrative physiology that would emphasize the restorative and natural force of the physician. This privileging of the doctor-nature-patient relationship, this affair in which no others partook, came at a cost—namely, forgetting the social fabrics into which both doctor and patient were woven. Psychoanalytic treatment was possible as a one-on-one relationship, but it had limits.[69] Nevertheless, neo-Hippocratism postulated a nature and a health that were spared institutional and marketplace complications. At times it brought in transference, at times it did not need more than the quasi-homeostatic theorization of internal stability. With Groddeck and Allendy, as with Goldstein, this health could be indulged as an ideal to be identified with individual healing.

Advances in psychoanalytic theory were significant for the spread of Goldstein's ideas. The therapeutic influence of his concept of self-actualization reveals the cross-fertilization between his writings and psychoanalysis—an influence that he generally rejected but that was central to his post-1934 career. In anticipation of Goldstein's fiftieth birthday on November 6, 1928, his collaborator Adhémar Gelb sent a letter to colleagues noting that his "students and friends" wanted to thank him by giving him as a gift the collected works of Sigmund Freud.[70] One of the recipients of this communication was Sigmund Fuchs, who had been a student of Goldstein's in the early 1920s and was training at the time as a psychoanalyst. In 1939 Fuchs would review *The Organism* in *Imago*, noting the close relation of some of Goldstein's positions in "dynamic biology" to psychoanalytic views.[71] Fuchs remained interested in the question of the doctor's position, as in the problem of transference. Under the Anglicized surname Foulkes, he would become known as the founder of group therapy, with its particular claims on the dynamic relationships within the group.

Other psychoanalysts concerned with pathological psychology—for example, Daniel Lagache in France in the mid-1930s[72]—also found

The Organism to be a pioneering and significant work on the relations between normal and pathological organisms that exhibited clearly the goals of therapeutic efforts. (Canguilhem first encountered Goldstein while following a course taught by Lagache in Clermont-Ferrand in 1941–1942.)[73] Ludwig Binswanger and other analysts similarly praised and adopted Goldstein's ideas,[74] and so did the Adlerian *Journal of Individual Psychology*, which later devoted a Festschrift to him for his eightieth birthday.[75]

Goldstein arrived in the United States in 1934, thanks to the Rockefeller Foundation's Emergency Committee in Aid of Displaced Foreign Scholars. Upon his arrival he was appointed university assistant in psychiatry with duties at the Psychiatric Institute at Columbia University and, as of 1935, at the Neurophysiological Laboratories at Montefiore Hospital.[76] Goldstein developed a following with the publication of *Der Aufbau des Organismus* soon after his arrival.[77] *The Organism* appeared in English in 1939, with a foreword by Karl Lashley, who apotheosized him as "one of the world's greatest authorities" on "neural and behavioral organization."[78] But even though Goldstein rose quickly at Montefiore, it was a decisive step down from his earlier position in Berlin.[79] Thus, Goldstein began developing a private practice in psychiatry to supplement his income, which brought him closer to analysts like Abram Kardiner and his mentee Frieda Fromm-Reichmann.[80] Teaching in the 1950s at Kardiner's Clinic for Psychoanalytic Training and Research at Columbia University, Goldstein highlighted the proximity of his thought to Freud's, all the while relishing the contrast elsewhere.[81]

In his post-1935 psychiatric career, Goldstein interacted with psychoanalytic concepts and began interpreting his treatment of the doctor-patient relationship in the context of contemporary pathophysiological and transferential tendencies in clinical and medical thought. In *The Organism*, he had disparaged Freud's theory of the "completely unintelligible" drives,[82] which he countered with the clearest expression of his monism: the "only" drive was toward self-actualization.[83] Even though Goldstein's fundamental experimental and therapeutic arsenal differed from that of the talking cure, for him treatment was a matter of restoring to patients something that belonged to them but that they could not quite articulate; this was self-actualization. Goldstein thus set the sights of his therapeutic experiments on the recognition and tentative restoration of order and self-actualization. Again, experimentation conceived as "the *means* of discovering the capacities of the individual" reconceived therapeutics as a mixture of experimental retraining—helping individuals to once again grasp for themselves behaviors forced and expected by biological and social milieus—and healing, in the specific sense of a restoration of adequacy.[84] Experimentation and therapeutic intent combined in Goldstein's dream so that by demonstrating through systematic experimentation patients' shrunken milieus and their contours, limits, and new geographies, he could collaborate with his patients and grant them anew a version of this relationship to nature and life.

To spotlight long-term individualized care and the doctor-patient relationship as means of returning to self-actualization, Goldstein again took on the problem of norms—norms that were not already established or "supra-individual,"[85] norms that did not await new preferred, ordered behaviors but were always achieved in them. The statement that "well-being consists of an individual norm of ordered functioning" meant that for both the patient and the attending physician, disease did not involve so much a "change of content" as a *disturbance in the course of the life processes.*"[86] No broad, singular norm, whether essential or statistical, could apply to a singular patient, whose individuality took priority; the physician took on the responsibility of not allowing the "individual" norm he developed for the patient to be wholly subjective, and hence to come unthreaded from concrete facts.[87] Noting that for the restoration of the patient's ordered behavior the physician might need to march the patient deeper into the canyons of disease, perhaps even convince the patient to suffer further in order to recover more fully, Goldstein wrote:

> The physician-patient relationship is not a situation depending alone on the knowledge of the law of causality, but . . . it is a coming to terms of two persons, in which the one wants to help the other to gain a pattern which corresponds, as much as possible, to his nature. This emphasis on the personal relationship between physician and patient marks off, impressively, the contrast between the modern medical point of view and the mere natural-science mentality of the physicians at the turn of the century. Although it may often seem as if the physician were interfering only with the bodily or mental event, he must always keep in mind that any effective interference, no matter how apparently superficial, must affect the patient's essential nature. He must remember that any interference, since it springs from freedom, affects the freedom of another person.[88]

In the sense that the physician must affect patients' essential nature and aid the restoration of their character to its normal order, Goldstein redrew the alliance of the physician and patient along pedagogical lines.[89] The doctor's place was to enable new norms, even if in so doing the physician "affects the freedom of another person," namely the patient, and even if the patient will in turn do this to others.

What others identified as a return to nature, Goldstein identified as a return to the capacity to self-actualize. Self-actualization emerged out of the organism's constancy mechanisms and its self-regulation but gave the sense of a different medicine that was concerned not only with the body but also with healing as a capacity to realize one's options and desires.[90] As the single underlying drive of each whole individual organism,[91] self-actualization was also the object of the experimenter's purpose and the physician's aim. With his 1940 book *Human Nature in the Light of Psychopathology*, as well as with a series of essays and lectures on anxiety and the emotions, Goldstein would popularize this argument toward a holistic theory of patienthood situated perilously close to an existentially minded rejection of health-care institutions as markets of organized physical soundness rather than health.

The Paroxysm of Medical Hope in the Individual: Antimedicine and Psychological Existentialism

We are faced with a paradox. Physiological and neurological integration, backed by a blend of therapeutically intended experimentation, provided a scenario that was open to a holistic and neo-Hippocratic approach to care at the same time that psychosomatically attuned theories of transference came to insist on the effect of the doctor. Goldstein and Cannon (perhaps with Leriche and Langdon-Brown on their side), Freud in a different way, and more radically still Groddeck and Allendy established a new set of scruples for long-term and individualized care. The individual patient whom Cannon and Goldstein described and whose care they demanded was just as holistic as the psychoanalytically treated analysand, but given their attention to neurological and physiological integration, it was supposedly "properly" scientific—indeed, this is what Goldstein and Lashley thought set Goldstein's holism apart from the thought of vitalist cranks. Yet Goldstein's and Cannon's approaches served brilliantly certain individualistic positions that persisted to the point that their motif of the patient came at the expense of the legitimacy of the health-care systems instituted in the twentieth century.

By the midpoint of the twentieth century, the individualist position did less to offer a new humanistic philosophy of the body than to leverage its promise as a response to a crisis in motion: the ever-increasing institutionalization of medicine that was supposedly erasing the unique quality of the individual in favor of a depersonalized and standardized image of the patient. Central to the development was the elaboration of plans for state-run medical systems, most famous among them the 1942 Beveridge Report (or Report on Social Insurance and Allied Services) in Britain, which placed health squarely in the scope of services rendered by the state.[92] As Michel Foucault would remark: "With the Beveridge Plan, health was transformed into an object of State concern, not for the benefit of the State, but for the benefit of individuals. Man's right to maintain his body in good health became an object of State action. . . . With the Beveridge Plan, health entered the field of macroeconomics."[93]

If this was an immense achievement for social equality and the welfare state, and no lesser an improvement to the lives of millions, critics of state-directed health came to see medicine representing two things: first a homogenizing and corrosive environment that failed to offer holistic healing and probably even produced sickness; and second, a set of impediments disallowing self-actualization especially for people in a diminished state of sickness. Put another way, the pressure to redefine norms and attend to the individual became, alongside integration, mottos for a rejection of the welfare state. The dynamic character of the environment was "rediscovered," René Dubos explained, through a "conceptual simplicity, almost pedestrian in obviousness."[94] The interplay of individual and environment in Selye's general adaption syndrome was slowly becoming the current version of the body's struggle in the milieu, and this milieu now extended to encompass the centers of medical activity as potential enemies. Dubos's writings on medical humanism and utopianism, however, should be read not as an embrace of Selye's picture of organism and environment, but as a cautionary note that biological life is far too easily refashioned through a notion of "social changes and ecological equilibria."[95] The English psychoanalyst Michael Balint, meanwhile, proposed that the doctor had an "apostolic function" that should fundamentally affect "the ways in which the doctor can administer *himself* to his patients," and limit prescriptions, especially pharmaceutical ones.[96]

The most forceful among the critics was Illich, who during the 1950s and 1960s grew from an anticolonial commentator on Western development schemes to a strident enemy of Western institutions. He broke into mainstream fame (or infamy) with his 1971 book *Deschooling Society*. The mutation of the doctor-patient relationship and the homogenization of differences between ailing bodies presented for Illich an opening not unlike those offered up by the educational system. Illich's 1975 *Medical Nemesis: The Expropriation of Health*, a book whose title declared war on its subtitle, demanded the personal recapturing of health. He argued that this expropriation, long in the making, found its landmark event in Beveridge's creation of the National Health Service in Britain, in part because its humanism was so obvious yet deceptive.[97] Illich concluded his accusation by declaring that a complete depersonalization of care occurred in standardized health care:

> Man's consciously lived fragility, individuality, and relatedness make the experience of pain, of sickness, and of death an integral part of his life. The ability to cope with this trio autonomously is fundamental to his health. As he becomes dependent on the management of his intimacy, he renounces his autonomy and his health must decline. The true miracle of modern medicine is diabolical. It consists in making not only individuals but whole populations survive on inhumanly low levels of personal health. Medical nemesis is the negative feedback of a social organization that set out to improve and equalize the opportunity for each man to cope in autonomy and ended by destroying it.[98]

Like so many others who took up the mantle of individualist humanism, Illich saw the central feature of life as a singular human drama. The individual character of suffering had been for too long trivialized by medicine as simply subjective, to the point that, citing Balint, Illich could claim that "medicalized addiction in 1975 has outgrown all self-chosen or more festive forms of creating well-being."[99] But rephrased through Illich's personalism, this subjectivity designated a form of personal suffering against the very institutions that deprived the person of meaning—religious, spiritual, or simply individual.

Medical Nemesis did not betray its Christian undercurrent, despite the occasional use of personalist language. Instead, Illich argued on the basis of traditions of medical health and showed himself to be a wide and able reader in public health, biomedicine, history of medicine, anthropology, sociology, and philosophy, all of which he held up against the medical system in order to prove its failure, or rather, to prove the danger posed by its practitioners. He quoted Hippocrates approvingly, argued against practices causing the perversion of Hippocratic thought, and placed the stressful impact of the doctor-patient encounter front and center.[100]

Stress and control of one's body are major—and frequently opposed—tropes in *Medical Nemesis*. But unlike Selye and Dubos, who placed themselves in traditions we have discussed, Illich did not use the ideas of medical thinkers so much as employ a wild citational strategy to adapt their thoughts to his thesis, drawing conclusions that often have little resemblance to the sources. When he pronounced, for example, that "some old people seek institutionalization with the intention of shortening their lives," and further that "the old have always obliged to die upon request," Illich cited Cannon's late article "Voodoo Death." This is presumably to turn a basic criticism of troubled institutions for care of the elderly into an attribution to the old and the sick of a primal ability to direct their own lives and deaths, to manage their own physiology by way of some kind of suggestion, standing beyond the powers of institutional control or threat. This is distant from Cannon's thesis.[101]

But not all his claims were as distant. Illich suggested that patients were capable of self-healing (or, in the case of the elderly, of psychic suicide), but these same patients were forever embattled by doctors and medical institutions, which seemed to have a near-infinite reach and a liquid, corporate control over them. The basic problem of Illich's antimedicine, framed as a custodianship of individualist systems of thought on the doctor-patient relationship, became evident in his attention to pain. Pain helped reinforce Illich's call for Hippocratic sensibilities in which suffering demanded a "deeper experience" to counter the biological functionalism he saw most doctors supporting.[102] For some of the figures we have looked at—Leriche, for example—such perspectives amounted to cheap humanisms that aimed to give moral significance to pain. Illich favored a metaphysical quality in pain.[103] This echoed Goldstein's position on self-actualization as freedom: even the highly restricted patient needed to find meaningfulness in the pained life he or she followed, to recover a measure of freedom.

Illich's conception of the subjective experience of sickness involved a contradiction. His "sick man" was caught in an endless cycle that somehow privileged his singularity but found value only when he participated in a broader human drama.[104] Although Illich did not cite Goldstein, he implicitly used his ideas of bodily control and self-actualization. The American existential psychologist Rollo May, in contrast, was entirely explicit about his connection to Goldstein.[105] May shared Illich's anxiety regarding the depersonalization of the patient, and while his approach was less polemical, it was no less an indictment of the medical establishment: "The medical model had turned out to be a dead end."[106] In May's translation of self-actualization, the potentials of the human subject needed to be realized for a true self to be discovered. Self-actualization and human freedom took on a character that shared a neurophysiological framework drawn, in part, from Goldstein. May wrote: "Human freedom involves our capacity to pause between the stimulus and response and, in that pause, to choose the one response toward which we wish to throw our weight. The capacity to create ourselves, based upon this freedom, is inseparable from consciousness or self-awareness."[107]

Not surprisingly, the adoption of self-actualization in the terms laid out by May presented serious theoretical problems, partly because of the number of theoretical matrices he cobbled together to fit a different philosophical purpose from Goldstein's.[108] Goldstein's self-actualization, as we have seen, resonated closely with Ernst Cassirer's concept of freedom, but it was also one of his most monistic and holistic concepts.[109] For May, the axis of self-actualization and freedom was the basis of a new psychological existentialism. For Goldstein, therapeutics veered between, on the one hand, suggesting that there existed a narrative of the patient's organismic life that affected how such a patient processed threats, a narrative the patient could grasp, and, on the other hand, guiding patients to become all but responsible for their reaction to his totalizing treatment for rehabilitation, and hence also for their healing. The problematic second option destabilized the first (which was clearly closer to Goldstein's intention), but for May it became the principal one. Individuals were required to wrangle meaning from illness, to have the "courage to create," or else to become subsumed by a medico-institutional machine bent on erasure, all the while possessing the power to harness their own vitality.

Our goal in this chapter has not been to offer an exhaustive account of practitioners in the mid-twentieth century who claimed a close kinship with the medical individualism of previous decades, but instead to hint at the way an entire system of thought has drawn a line back through Cannon, Goldstein, and others to Hippocrates and Aesculapius, then forward again in order to operationalize quasi-political attitudes posing as positions on humanistic, patient-centered problems. All throughout the book we have resisted the lure of the isolated clinical concern, even though for all their foibles Goldstein, W. H. R. Rivers, Head, and Cannon, like Freud, were physicians and psychiatrists in the best sense. When they wrote of their concern for patients, their cares derived from clinical situations, and the question of how to handle these clinical situations caused them evident anguish. But declaring that a patient is in need of individualized and specialized care is not free from the risk of imposing a biological monadism on the medical subject.

More than clinicians, these thinkers were intellectuals. Their therapeutic politics helped bring the utility of case histories to the fore and made the individual capacities of their patients so central as to demand a full accounting of the disease concept and the parameters of therapeutic offerings.[110] They nevertheless refused the impulse to explode the disease category for the sake of a so-called humanistic sensibility of the single patient. Some of their followers, Canguilhem for example, incredulously dismissed as a mirage the humanism attached to the hyperprivileged patient. Canguilhem reserved special ire for the assertions of Illich, while keeping precursors Groddeck, Allendy, and Goldstein closer to heart.[111] But for others, like May and Illich, the neo-Hippocratic thinkers had been heroes struggling against modernity—Starling's hopeful medical modernity—and pointing toward the new age they themselves were heralding: a new age of recovering and becoming oneself.

10 Closure: The Individual

What kind of construct is the individual? How does one establish the individual's place in thought and life? What scope, what kind of framework, can reveal the purview of a concept so slippery, woolly, and seemingly ever-present? How do individuals ground themselves, and to what end does such grounding operate? To what degree can one speak of individuality as a category of thought, of medicine and medical care, of politics? How would the understanding of individuality be defined by medical care and biological theory?

In this book we have sought to demonstrate that individuality was rearranged in medical thought during and after World War I thanks to a particular and largely new engagement with each body as single, whole, integrated, and always imperiled because of its integratedness. This rearrangement inflected political metaphors and economic, technological, and anthropological theory in ways that intellectual history and the history of political thought have ignored.

Our central claim is straightforward. For a while during and after World War I, in neurophysiological theories and related conceptions of medical care, human bodies became precisely individual at the moment when individuality—and indeed, individuals' very corporeality, their wholeness—was almost constantly collapsing. Before, individuality in biology and physiology had indexed the fundamental isomorphism of each example in a species and had ascertained that each one—a body, a patient, a subject—belonged to that species by sharing components with other members of the species and by reacting in the same manner to stimuli probing it. These components and their coexistence in the body could be explained as a result of evolutionary and phylogenetic developments. Even in physiology: from Starling's *Elements of Physiology* (1892) through Charles Sherrington's *The Integrative Action of the Nervous System* (1906) and up to Cannon's *Bodily Changes in Pain, Hunger, Fear, and Rage* (1915), the construction of the individual began with gestures affirming evolution's responsibility for the unification of the individual body out of its elements.[1]

Not so from World War I onward. The perception of wound shock and aphasia as whole-body disorders, alongside the proliferation of case studies in administration and therapeutics and a thoroughly new, neo-Hippocratic attention to the individual patient and his norms (we wish we could write her norms, but the gendering is clear), provided guidance to the new science of hormonal regulation, leading to its radicalization into theories of homeostasis and to the invention of autopharmacology. The new logic cast individuality in the framework of integration and disintegration. For Cannon, Head, Dale, Goldstein, Starling, and others, the structure of the organism was not a compilation of its biological bases (including race and gender), not an agglomeration or composition of lesser elements, but the conjoining of coextensive systems, functional goals, and regulatory and defensive behaviors that amounted to a single whole. As Goldstein put it, this unity was revealed in and through pathology. The organism's preferred, aberrant, and troubled behaviors, as well as its veering between order, disorder, and catastrophe were reconceptualized in terms of the decomposition, collapse, and reintegration of these systems. Integration and collapse, regulated interiority, and brittleness went hand in hand in Cannon's adage that "our bodies are made of extraordinarily unstable material," Alexander Luria's theory about the disintegration of behavior, and Dale's account of the pathologies of histamine response. The body was the brilliant strategist that assiduously deployed its defenses, and it was also the traitor within the walls.

Fragility was no novelty; other biologists had imagined the body in similar terms. Still, as a theoretical construct, a necessary component, a rationale, it cannot but be dated to the Great War. Nothing had prepared physiologists and physicians for the sight, feel, and meaninglessness of bodies torn asunder or perishing invisibly and counterintuitively.

This was not because violence of this order had never been experienced before, but because the surgeons, medical arts, and regimes of knowledge that had handled such patients had been relegated to medical prehistory. Its successes notwithstanding, the more recent and positivistic parts-to-whole mechanism had already been decried as simplistic and counterproductive, but no laboratory equivalent, no decorticated cat hissing and kicking and wailing could tacitly stand in for shaking, mute patients, so severely distorted in what they could do, their unity of mind broken, their bodies compounding their pain, their conditions similar but their suffering so divergent. Neurology, physiology, and medicine increasingly proposed the body as a singular whole threatened with a disintegration that was experienced in a highly particular manner.

We have traced the rise of a new conceptual architecture that offered a new epistemology of the body, a new ontology, notably of patienthood, and a new medical art, especially toward the fundamental medical experience for doctors and patients. Equilibrium, homeostasis, integration, autopharmacology, and holistic neurology—but also shock, collapse, fear, pathology, and disintegration. The body made itself, kept itself working, then, given some demand or other, it burned itself away and worked its own demise. As we have also seen, the resulting emphasis on "the individual" and on disintegration was also metaphorized, from politics to economics.

The conceptual architecture of this individuality was pluralistic, fragile, and tense. Its individual did not break out fully grown and clad in armor like Athena from Zeus's head, but was negotiated in often-contradictory terms, even by authors who worked together or clinician-theorists who largely agreed.[2] The tensions in and between their works compressed the joints of the overall epistemological construction and filled the halls of methodology—the laboratories, clinics, textbooks, and lecture halls that realigned, reworked, and retested this knowledge. Other tensions splintered the structure by trying to explain its connection with its metaphorical or metonymic possibilities in the political and economic worlds—in anthropologies at times medical, at times cybernetic, but always also philosophical.

In this regard, the brittleness of an integrated body succumbing because of the complexity of the systems guarding it was mirrored in the frailty of the conceptual structure of individuality. The same theories that delineated the individuality of the patient also traced, with unrelenting assurance, its conceptual instability. The new concepts of the individual were brittle too: they could pursue and celebrate individuality but could bend it only so far before conceptual tensions broke it apart. At the conceptual level the pressure to sustain an individual whole, to protect that whole, to distribute the tensions that threatened to tear it apart was a genuine problem, especially after a half century of mismatch between the credit philosophy and liberalism accorded to the individual and the downgrading of that individual in biology. The individual at the corporeal and experiential moment of threatened disintegration found its analogue in scientists' refusal to abandon the category of the individual altogether and to work within it to recover a more multifarious complex of systems. Dale and Cannon's systems accorded individuality less to the body than to the integrated and spectral totality of systems that worked at times independently and even against the body's integrity and survival. Goldstein worried about anxiety signaling or leading to catastrophic collapse. Rivers insisted on the restitution of an autonomy that predated the psychic wound. The individual was impossible to displace even if it was insufficiently plastic to ensure recovery.

This was also the case for medical therapeutics, as the obsessive attention to individuality and to a natural form of care clashed with the needs of large-scale hospital care and the health of whole populations. The warmer we feel toward Goldstein's or Cannon's promises regarding patient treatment, the more the incommensurability of those promises with what states and doctors could actually do for the sick comes into focus. Some of the resulting critiques of therapeutic narrowness, though forceful and convincing, led straight to New Age illusions of Romantic science and natural cures, and laid the groundwork for the antimedicine that followed some decades later. Ivan Illich and Rollo May's excesses reworked precisely these critiques and promised a new health *against* medicine by exploiting this incommensurability. Today, "health" and "illness" are not simply mitigated by the limitations of and unfairly distributed access to health care; they are defined against competing rubrics of personalized medicine, complementary medicine, internally conflicted naturalistic approaches to healing, and pharmaceutically and technologically centered treatments. Our abjuring of agency on biomedical matters clashes starkly with an otherwise steadfast commitment to individuality and political agency.

: : :

Between the terror of organismic collapse and the fantasy of self-actualization, this concept of the individual did its work. Systematically constructed, harmonized, and destroyed from within, the subject now became as distant from liberalism as it was from fascism and communism. The social models pursued by thinkers like Goldstein (an almost desperate democratic socialism) and Cannon (an at-times Soviet-friendly but mostly reserved and uncommitted welfare-statism) were thus compatible with their conceptions of individuality, of the individual as deserving of protection and restoration. In a sense, this insistence on protection and restoration inverted liberalism's dream of a self-grounded self with astonishing confidence: the self was constructed, grounded, normalized, upturned, and destroyed thanks to a totality that it possessed and with which it could be identified but which it could never control. Influential philosophers like Cassirer, Canguilhem, and Maurice Merleau-Ponty pursued philosophical approaches to individuality that retained, even accentuated, these ambiguities.[3] Individuality as a concept or a category of thought lacked givenness and permanence, and it mattered for just that reason.

This feedback loop between political and corporeal wholeness amounted to an awareness of the frailty of organized society and a fear of disintegration, perhaps clearest in Starling's account of deprivations in Germany in 1919 and in Cannon's description and worry over biocracy; again, philosophers like Merleau-Ponty and Canguilhem accentuated this awareness. Merleau-Ponty, in his 1947 engagement with the Soviet terror, worked on contrasting history, which he called "a holistic system moving toward a state of equilibrium," with the individual, whose experience of the world (and of history) was laced with ambiguity and was profoundly misunderstood by liberalism as well as Marxism.[4] At most what was on offer were the often irresolvable ambiguities of returning to balance, and righting a capsized body and mind: these offered the chances to rethink symbolic experience, the ambiguity written into phenomenological perception and existence, and the norms and normalities of life.

Such philosophical quests of a new individuality were only a small part of the story. Its political translation reached much farther. On the one hand, the world of political metaphors could replicate the structure of the organism as patient at the national, social, and international levels: the integration and wholeness of a human being could be mirrored in the imagination of a humanity that needed to be healed. Zimmern's description of a Europe in convalescence could be reimagined in terms of a humanity in convalescence; indeed, the hopes that Wiener, Mauss, and Lévi-Strauss toiled toward (namely a rational administration of communication, a socialist internationalism, and a UNESCO-like defense of the other) pointed precisely in this direction.

The other hand, though less promising, was perhaps more powerful. Once it was translated to social theory—whether in communication or in economics—the encoding and suspension of this individuality that occurred by way of regulated systems internal to the organism was obverted into an encoding and suspension of the individual by external systems. These replicated the mostly closed and regulated structure of the body at the level of social, economic, or linguistic wholes, but suspended the human individual from these wholes, even when they sought or pretended to protect him or her. From this external social or economic perspective, individuals seemed either threatened or capable of maintaining their position among others, but they no longer actively or fundamentally participated as irreducible units of the social or the linguistic. This is why the integrated individual diverged from the traditional model of the liberal individual. It was so individuals could be tinkered with in ways that often ignored their wants or goals.

From Cannon's terrified victim of bone pointing in "Voodoo Death" through Head's anxious aphasic patient who failed to keep up with the symbolic order, the individual was depicted as singular and consistently threatened. Goldstein concluded that "if it is true, . . . that these catastrophes are the expression of a clash of the individuality of the organism with the 'otherness' of the world, then the organism must proceed from catastrophe to catastrophe."[5] Merleau-Ponty translated this into political terms as clearly as Lévi-Strauss and Wiener, and the fear is unmistakable despite his political hope:

> Consciousnesses have the strange power to alienate each other and to withdraw from themselves. . . . They are outwardly threatened and inwardly tempted by absurd hatreds, inconceivable with respect to individuals. . . . If men are one day to be human to one another and the relations between consciousnesses are to become transparent, if universality is to become a fact, this will be in a society in which past traumas have been wiped out and the conditions of an effective liberty have been realized from the start. Until that time, the life of society will remain a dialogue and a battle between phantoms—in which real tears and real blood suddenly start to flow. . . . We are in the world, mingled with it, compromised with it.[6]

In this forceful expression of the antihumanism that in the 1960s would decry illusions of progress and humanist redemption, the integrated, fragile body was not a self-sovereign agent, or even a force, but merely an enchanted, hopeful, menaced, compromised perseverance. The vulnerable human being was suspended from the broader theories that promised only a small patch of a place within society, communication, and the universe of symbols. Easily subsumed into the broader systems proposed in economic theory, cybernetic control of information, and structural anthropology, and as easily dominated by their constitution as by internal systems, individuals lost the capacity to be the pilots of their lives while remaining perishable right to the end.

Acknowledgments

This project would not have been possible without the extremely generous funding offered us by an American Council of Learned Societies Collaborative Research Fellowship (2013–2015), supported by the Andrew W. Mellon Foundation, which gave each of us a year of leave to work together on the project. We are very grateful to Nicole Stahlmann, then the director of fellowships at ACLS, for guiding us through the fellowship. Fellowships from the University of Michigan's Eisenberg Institute for Historical Studies (Todd), the Max Planck Institute for the History of Science (Stefanos), and New York University's Global Research Initiative (Stefanos) made further research and work possible and thoroughly enjoyable. For publication support, we are also grateful to the Office of the Dean of the Humanities and the History Department at New York University and the Office of the Provost for New York University–Shanghai.

Archival research for this project was carried out at institutions in Britain, France, Germany, Austria, Australia, and the United States. In London, at the Wellcome Institute Library (Papers of William H. Bayliss, E. M. Cowell, Henry Head, Henry Dale, Charles Lovatt Evans, Ernest H. Starling, and Royal Army Medical Corps), the National Archives at Kew (the Medical Research Council archive), University College London (the W. H. R. Rivers

Papers in the Perry Papers), and the British Psychoanalytic Institute (Ernest Jones Papers). In Edinburgh, at the National Library of Scotland (J. S. Haldane Papers) and the University of Edinburgh Library's Special Collections (J. S. Haldane Letters). In New York, at the Columbia University Rare Book and Manuscript Library (Kurt Goldstein Papers) and in Tarrytown, at the Rockefeller Archive Center (Theodore Malinin Collection of the Papers of Alexis Carrel and Charles Lindbergh, and the archive of the Emergency Committee in Aid of Displaced Foreign Scholars). In Paris, at the Bibliothèque du Collège de France (Fonds Marcel Mauss), the Centre d'archives de philosophie et d'histoire des sciences at the École normale supérieure (Georges Canguilhem Archive), and the Muséum national d'histoire naturelle (Papiers Marcel Mauss). In Cleveland, at Case Western Reserve University (for the George W. Crile archive). In College Park, Maryland, at National Archives II (the archive of the Dijon Laboratory of American Expeditionary Force in World War I). In Boston, at the Francis A. Countway Library of Medicine at Harvard (for the Walter B. Cannon Papers) and the Baker Library of the Harvard Business School (for the Lawrence J. Henderson Papers). In Berlin, at the Free University, Humboldt University, and Max Planck Institute for the History of Science libraries. Last but certainly not least in Vienna, at the Sigmund Freud Museum. We are very grateful to librarians and archivists at all of these institutions for their help, and especially to Emily Holmes and Jennifer Lee at Columbia who helped with the lengthy process involved in digitally transferring of films we discuss in chapter 3; James Cox, archivist at Gonville and Caius College, Cambridge; Margaret Hogan at the Rockefeller Archive Center.

With extraordinary generosity, Stephen T. Casper offered us his primary research notes from L. J. Henderson's archive in addition to very kind advice. Valuable research assistance was provided by Richard Baxstrom, James Dunk, Jacob Krell, and Jamie Phillips. Nathaniel Boling's exemplary master's thesis on W. H. R. Rivers's theory of the unconscious and Rivers's and Freud's understandings of borders (New York University, 2014) provided a smart and useful counter to existing approaches to Rivers, and we benefited considerably from it. A number of our colleagues took the time to read the beast in its first—indeed, longer—version, and we thank them for their criticisms and support: Deborah Coen, Myles Jackson, Zac Levine, Julie Livingston, Emily Martin, Jamie Phillips, John Protevi, Anson Rabinbach, Camille Robcis, Leif Weatherby. Thanks also go to Stephanos J. Geroulanos and Nikos Geroulanos (Stefanos's uncle and father) for help with classical references; to Natasha Wheatley who offered a world of directions, including on case law, integration, and internationalism; to Jamie Martin on economic matters in chapter 8; as well as to Maria Muhle, Hannah Leffingwell, Lotte Houwink ten Cate, Ruth Leys, Thomas Hoffmann, and Frank Stahnisch for specific and very useful recommendations. Nasser Zakariya read the entire manuscript with tremendous care and brilliance, and helped us to make links we had not anticipated. Nuala Caomhanach helped establish the index. Finally we would like to thank David W. Bates, Richard Baxstrom, Stephen Chrisomalis, Katherine Fleming, Talia Gordon, Jeremy Greene, Dagmar Herzog, Nancy Rose Hunt, Céline Lefève, Paul Rabinow, Eugene Raikhel, Andrés Romero, Joan W. Scott, Anthony Stavrianakis, John Tresch, Larry Wolff, and Allan Young, and above all Rania Ajami and Fanny Gutiérrez-Meyers.

This project emerged from our collaborative work on Georges Canguilhem and the two translations of his work in which we participated (*Knowledge of Life*, 2008; *Writings on Medicine*, 2012). It matured when we were invited to publish a book on Kurt Goldstein with August Verlag, Berlin, which became *Experimente im Individuum: Kurt Goldstein und die Frage des Organismus*. That book also addresses in more detailed form a number of problems in Goldstein's thought that we also engage here.

Parts of this work have been presented by either one or both of us at several venues. For the invitation to the *Plasticity and Pathology* conference at the University of California, Berkeley, we thank Nima Bassiri and David Bates, and besides them also Katja Guenther, Catherine Malabou, and Tobias Rees for suggestions. For the invitation to speak at the University of California at San Diego's Science Studies Program, we are grateful to Cathy Gere. Peter Slezkine and Mark Mazower kindly invited us to present chapter 1 at the International History Workshop at Columbia University, and we are grateful to them, as well as to Marwa Elshakry and Deborah Coen for pressing us to persist with a discussion of a changing ontology of the soldier at war. Earlier still, we presented a first account of Cannon and Selye at the University of Amiens, thanks to the gracious invitation of François Delaporte, Sandra Laugier, and Paola Marrati, and we presented a version of early work on Kurt Goldstein's epistemology at the conference *History and Epistemology: From Bachelard and Canguilhem to Today's History of Science* at the Max-Planck-Institut für Wissenschaftsgeschichte in honor of Hans-Jörg Rheinberger. We acknowledge Jean-François Braunstein and Henning Schmidgen for their invitation, and Cornelius Borck for his helpful comments. Sara Guyer and Richard Keller kindly extended the invitation to present at the John E. Sawyer Seminar on Biopolitics at the

University of Wisconsin's Humanities Center, with additional thanks to Jason Smith, Donna Jones, and Timothy Campbell. Uffe Juul Jensen and Nikolas Rose brought us to Skagen, Denmark, for their joint (London School of Economics–Aarhus) workshop, "Canguilhem and the Contemporary Sciences of Life." Many thanks to Sandra Laugier, Sabine Arnaud, John Carson, and others at the workshop "Language, Citizenship, Forms of Life" at Max-Planck-Institut für Wissenschaftsgeschichte and Centre Marc Bloch, Berlin (May 2015). We also had the chance to share the work and thoughts with David Napier, Stephen Jacyna, and Roger Cooter during a presentation at University College London, and are very grateful to them as well.

At the University of Chicago Press, we would like above all to thank Priya Nelson. She has handled this manuscript as only a true intellectual could, and it is difficult to explain the difference she has made to it—her care for the ideas in this work and her attention to the shape those ideas should take has been masterful and extraordinarily generous. We have also benefited a great deal from the assistance of Meg Wallace and Katherine Faydash in editing and shortening the text.

Where works originally published in French or German are available in English, we have provided the English reference, frequently retranslating the original.

Abbreviations

Abbreviations Used in Notes

AKD   Henry Head, *Aphasia and Kindred Disorders of Speech*, 2 vols. (1926; London: Hafner Publishing, 1954).

BER   Claude Bernard, *Introduction à l'étude de la médecine expérimentale* (1865; Paris: Delagrave, 1912). Translated as *An Introduction to the Study of Experimental Medicine*, introduction by L. J. Henderson (New York: Schuman and Macmillan, 1927).

BMJ   *British Medical Journal.*

BPP   Sigmund Freud, *Beyond the Pleasure Principle*. Citations are, first, to *The Standard Edition of the Complete Psychological Works of Sigmund Freud*, vol. 18 (London: Hogarth Press, 1955), then to the commonly used Norton edition, and finally the *Critical Edition*, in *Psychoanalysis and History* 17, no. 2 (July 2015), 151–204 (prepared by Ulrike May and Michael Schröter, and presented in English by Matt ffytche, John Forrester, and Michael Molnar). We use the translation from the Revised Standard Edition (RSE) by Mark Solms as republished in the *Critical Edition*.

CBC   Walter B. Cannon, *Bodily Changes in Pain, Hunger, Fear, and Rage* (1915; New York: Appleton, 1929).

CKL   Georges Canguilhem, *Knowledge of Life*, trans. Stefanos Geroulanos and Daniela Ginsburg, ed. Paola Marrati and Todd Meyers (New York: Fordham University Press, 2008).

CNP   Georges Canguilhem, *The Normal and the Pathological* (New York: Zone, 1991).

CWB   Walter B. Cannon, *The Wisdom of the Body* (1932; New York: Norton, 1963).

IAN   Charles S. Sherrington, *The Integrative Action of the Nervous System* (1906; New Haven, CT: Yale University Press, 1961).

FSE   Sigmund Freud, *The Standard Edition of the Complete Psychological Works of Sigmund Freud*, trans. and ed. James Strachey (London: Hogarth Press, 1953–1974), followed by volume and page numbers.

HNP   Kurt Goldstein, *Human Nature in Light of Psychopathology* (New York: Schocken, 1940).

HRE   John Scott Haldane, *Respiration* (New Haven, CT: Yale University Press, 1922).

LSP   René Leriche, *The Surgery of Pain* (Baltimore: Williams and Wilkens, 1939).

MRC   Walter B. Cannon, E. M. Cowell, John Fraser, and A. N. Hooper, *Reports of the Special Investigation Committee on Surgical Shock and Allied Conditions, no. 2, Investigation of the Nature and Treatment of Wound Shock and Allied Conditions* (London: RAMC Medical Research Committee, December 25, 1917).

MSS   National Archives (Kew), FD 1/5262, Minutes of the Special Committee on Shock and Allied Conditions.

MTT   *Medical Research Committee Reports of the Special Investigation Committee on Surgical Shock and Allied Conditions*, no. 8, *Traumatic Toxaemia as a Factor in Shock* (March 14, 1919).

NHC   Alexander R. Luria, *The Nature of Human Conflicts, or Emotion, Conflict, and Will*, trans. W. Horsley Gantt (New York: Liveright Publishers, 1932).

OEB   John Scott Haldane, *Organism and Environment as Illustrated by the Physiology of Breathing* (New Haven, CT: Yale University Press, 1917)

RTS   Alfred Cort Haddon, W. H. R. Rivers, C. G. Seligman, Charles S. Myers, William McDougall, Sidney Herbert Ray, and Anthony Wilkin, *Reports of the Cambridge Anthropological Expedition to Torres Straits*, 6 vols. (Cambridge: Cambridge University Press, 1901–1935).

SWB   Ernest H. Starling, "The Wisdom of the Body: The Harveian Oration Delivered before the Royal College of Physicians on St. Luke's Day, 1923," *British Medical Journal* (October 20, 1923), 685–90.

RIU   W. H. R. Rivers, *Instinct and the Unconscious* (Cambridge: Cambridge University Press, 1920).

TO   Kurt Goldstein, *The Organism* (1933, 1939; New York: Zone Books, 1995).

Archives

United Kingdom
Wellcome Library, University College London
Papers of William H. Bayliss
Papers of E. M. Cowell
Papers of Henry H. Dale
Papers of Charles Lovatt Evans
Papers of Henry Head
Papers of Ernest H. Starling
Royal Army Medical Corps Archive

National Archives (Kew), London
Medical Research Council Archive
University College London Archives, London
Papers of William James Perry (W. H. R. Rivers Papers)
Papers of William M. Bayliss
British Psychoanalytical Society, London
Ernest Jones Papers
National Library of Scotland, Edinburgh
John Scott Haldane Papers
University of Edinburgh Special Collections, Edinburgh
Letters of J. S. Haldane

United States
Columbia University Rare Book and Manuscript Library, New York
Kurt Goldstein Papers

Rockefeller Archive Center, Sleepy Hollow, New York
Papers of Alexis Carrel
Papers relating to Grafton Elliot Smith
Theodore Malinin Collection of the Papers of Alexis Carrel and Charles Lindbergh
Archive of the Emergency Committee in Aid of Displaced Foreign Scholars

National Archives II, College Park, Maryland
Archive of the Dijon Laboratory of American Expeditionary Force in World War I

Case Western Reserve University, Cleveland, Ohio
George Washington Crile Archive

Francis A. Countway Library of Medicine, Harvard University, Cambridge, Massachusetts
Walter B. Cannon Papers

Baker Library, Harvard Business School, Cambridge, Massachusetts
Lawrence J. Henderson Papers

France
Bibliothèque du Collège de France, Paris
Fonds Marcel Mauss

Muséum national d'histoire naturelle, Paris
Papiers Marcel Mauss

Centre d'archives de philosophie et d'histoire des sciences, École normale supérieure, Paris
Georges Canguilhem Archive

Australia
State Library of New South Wales, Sydney
W. H. R. Rivers Archive

Austria
Sigmund Freud Museum, Vienna

Notes

PROLOGUE

1. Walter B. Cannon, draft of "Some Tentative Postulates Regarding the Physiological Regulation of Normal States," Walter B. Cannon Papers (hereafter Cannon Papers), 63:837, Francis A. Countway Library of Medicine, Harvard University, later published as "Physiological Regulation of Normal States: Some Tentative Postulates Concerning Biological Homeostatics," in *À Charles Richet: Ses amis, ses collègues, ses élèves*, ed. Auguste Pettit (Paris: Les éditions médicales, 1926), 91.

2. Paul Veyne, *Comment on écrit l'histoire, suivi de Foucault révolutionne l'histoire* (Paris: Seuil, 1978), 236.

CHAPTER ONE

1. Sigmund Freud, *Drei Abhandlungen zur Sexualtheorie* (1905), in *Gesammelte Werke* (London: Imago, 1991), 5:67; in English as *Three Essays on Sexuality* in *FSE* 7:167, translation amended. On the book being published around the beginning of June, see James Strachey, "Editor's Preface," in Freud, *Jokes and Their Relation to the Unconscious*, in *FSE* 8:5. See also Ernest Jones, *Sigmund Freud* (New York: Basic Books, 1955), 2:12, 2:335.

2. This was far from Starling's only contribution; he had recently been elected to the Royal Society of Medicine on the basis of his work on the formation of the lymph.

3. Joseph Needham, *Order and Life* (Cambridge: Cambridge University Press, 1936), 80, cited in John Henderson, "Ernest Starling and 'Hormones,'" *Journal of Endocrinology* 184 (2005), 9.

4. Ernest Henry Starling, "The Croonian Lectures on the Chemical Correlation of the Functions of the Body" (June 20, 22, 27, and 29, 1905), *The Lancet* 166 (1905), 339–40.

5. Freud, *Drei Abhandlungen*; Henry Liddell and Robert Scott, *A Greek-English Lexicon* (London: Oxford University Press, 1940), s.v. ορμάω, A1, B3.

6. *BER* 108–9/63–64.

7. William Bayliss, *An Introduction to General Physiology* (London: Longmans, Green, 1919), 2.

8. Erwin H. Ackerknecht, *A Short History of Medicine* (1955; Baltimore: Johns Hopkins University Press, 1982), 228–29.

9. On functional disturbances actually causing organic lesions, see Pierre Abrami, "Les troubles fonctionnels en pathologie," *La presse médicale* (December 23, 1936).

10. George L. Mosse, *Fallen Soldiers* (New York: Oxford University Press, 1990); Joanna Bourke, *Dismembering the Male* (London: Reaktion, 1996); and Alfredo Bonadeo, *Mark of the Beast* (Lexington: University Press of Kentucky, 1989).

11. *IAN*.

12. *NHC* xi.

13. On the difference between these sciences, see François Dagognet, *Philosophie biologique* (Paris: Presses universitaires de France, 1955), 3.

14. "Styles of thought" is Jonathan Harwood's phrase, from his *Styles of Scientific Thought: The German Genetics Community, 1900–1933* (Chicago: University of Chicago Press, 1993).

15. On the neurological patient, see L. Stephen Jacyna and Stephen T. Casper, eds., *The Neurological Patient in History* (Rochester, NY: University of Rochester Press, 2012).

16. *TO* 48, 49, and passim.

17. J. S. Mill, *On Liberty* (Oxford: Oxford University Press, 1985), 223–24.

18. E. H. Starling, *Elements of Human Physiology*, 3rd ed. (London: Churchill, 1897), 1.

19. Volker Roelcke correctly notes that the division between supposedly acceptable animal experimentation and ethically unacceptable human experimentation was at best frail and at most a matter of utility. Nevertheless, it is worth noting that Victorian and American cultural norms—alongside the epistemological possibilities offered by phylogenetic recapitulation—worked against the effacement of the distinction. See Roelcke, "Tiermodell und Menschenbild," in *Kulturgeschichte des Menschenversuchs im 20: Jahrhundert*, ed. Birgit Griesecke, Marcus Krause, Nicolas Pethes, and Katje Sabisch (Frankfurt: Suhrkamp, 2009), 19.

20. The process of abstracting from animals to human beings would be criticized by, for example, Goldstein. See his criticism of decortication (*TO* 74), or his sense that animal experiments are of a lower quality than human ones (*TO* 82, 149, 159). On derivations of the normal from pathology, see *CNP*.

21. Walter Fletcher, "Aims and Boundaries of Physiology," *Science* 55, no. 1430 (May 26, 1922), 552.

22. Bayliss to Cannon (September 18, 1915), box 62.F.811.Bayliss, Cannon Papers.

23. As Karl Rothschuh notes, Britain "displaced Germany as the leading country in physiological research, after the deaths of Brücke, Helmholtz, Ludwig, and Du Bois-Reymond. This successful development was achieved by a group of young British physiologists who by and large had trained in French and German laboratories." See Rothschuh, *History of Physiology* (New York: Krieger, 1973), 307. Cannon established his physiology lab at Harvard College in his first year of medical studies, citing the inadequacy of facilities for experimental work; see Saul Benison, Clifford Barger, and Elin L. Wolfe, *Walter B. Cannon: The Life and Times of a Young Scientist* (Cambridge, MA: Belknap Press of Harvard University Press, 1987); W. Bruce Fye, *The Development of American Physiology* (Baltimore: Johns Hopkins University Press, 1987; and E. M. Tansey, "Charles Sherrington, E. D. Adrian and Henry Dale," in *Cambridge Scientific Minds*, ed. P. Harman and S. Mitton (Cambridge: Cambridge University Press, 2002), 187–201.

24. An abbreviated list of general physiology textbooks and their multiple editions published in Britain and the United States includes the following: Starling, *Elements of Human Physiology*, 8th ed. (1892; London: Churchill, 1907); Starling, *A Primer of Physiology* (London: Murray, 1904); Starling, *Principles of Human Physiology* (London: Churchill, 1895)—from 1918 on this work was revised by Charles Lovatt Evans, with the fourteenth edition published in 1968 (London: J. A. Churchill); Bayliss, *Principles of General Physiology*, 3rd ed. (1915; London: Longmans, Green, 1920); Cannon, *A Laboratory Course in Physiology*, 5th ed. (1910; Cambridge, MA: Harvard University Press, 1926); Leonard Hill, *Manual of Human Physiology*, 4th ed. (1899; London: Arnold, 1935); Joseph H. Raymond, *Human Physiology, Prepared with Special Reference to Students of Medicine*, 2nd ed. (1894; Philadelphia: Saunders, 1901)—a third revised edition was published in 1905 with 443 illustrations; Albert Brubaker, *Textbook of Human Physiology, Including a Section on Physiologic Apparatus*, 8th ed., with 367 illustrations (1904; Philadelphia: Blakiston's Son, 1925); Friedrich Schenck and August Gürber, and W. Zoethout, trans., *Outlines of Human Physiology* (1900; New York: Henry Holt, 1901); Robert Tigerstedt, *A Textbook of Human Physiology*, trans. John Murlin from the third German edition (New York: Appleton, 1910); Diarmid Paton, *Essentials of Human Physiology for Medical Students*, 5th rev. ed. (1903; Edinburgh: William Green, 1920); William H. Howell, *A Textbook of Physiology for Medical Students and Physicians*, 10th ed. (1905; Philadelphia: Saunders, 1926); John Thornton, *Human Physiology*, 4th ed. (1894; London: Longman's, Green, 1901), with a new edition, revised by William Smart, with 281 illustrations, published in 1926; William S. Furneaux, *Human Physiology* (London: Longmans, Green, 1878), the twenty-first impression was released in 1920, and a new edition, revised by William Smart, in 1924; Luigi Luciani, *Human Physiology*, 5 vols. (London: Macmillan 1911–1921); M. Foster, *A Text-Book of Physiology*, 6th ed., 5 vols. (1878; London: Macmillan, 1900); and Albrecht Bethe, *Handbuch der normalen und pathologischen Physiologie*, 18 vols. (Berlin: Springer, 1925–1932). "Almost every great teacher of the

subject has written a treatise on physiology" noted an exhausted-sounding Fielding Garrison in 1913. See Fielding Hudson Garrison, *An Introduction to the History of Medicine*, 2nd ed. (1913; Philadelphia: Saunders, 1917), 601.

25. Garrison, *Introduction to the History of Medicine*, 732.

26. Sharpey-Schafer, "History of the Physiological Society, 1876–1926: The First Fifty Years," *BMJ* (February 25, 1928), 313.

27. *SWB* 685. See a similar language in Fletcher, "Aims and Boundaries of Physiology," 554.

28. Garrison, *Introduction to the History of Medicine*: on Bayliss, Starling, and Pavlov, see 583, 714; on Cannon, see 580–81.

29. Charles Singer, *A History of Biology to about the Year 1900*, 3rd ed. (London: Abelard-Schuman, 1959), 573.

30. Bayliss, *Introduction to General Physiology*, viii; and Cannon, "The Effect of Adrenal Secretion on Muscular Fatigue," *American Journal of Physiology* 32, no.1 (1913), 44.

31. For example, Cannon and Henry Lyman, "The Depressor Effect of Adrenalin," *American Journal of Physiology* 31, no. 6 (March 1913), 376, where "the literature on the adrenal glands" consists of works from 1897, 1899, 1900, 1903, 1905, 1906, 1911, 1912, and 1913. Cannon's "An Explanation of Hunger," coauthored with his student A. L. Washburn, did cite pre-1900 work but self-servingly: to demonstrate that the older theories lacked experimental validation. See Cannon, "An Explanation of Hunger," *American Journal of Physiology* 29, no. 5 (March 1912), 441–54.

32. Starling to Evans, October 4, 1918, PP/CLE/D1, Charles Lovatt Evans Papers, Wellcome Library.

33. *Walter Bradford Cannon: Exercises Celebrating Twenty-Five Years as George Higginson Professor of Physiology* (Cambridge, MA: Harvard University Press, 1932), 77–90.

34. That the extended affair, like the trial, also played "in Bayliss's favor"—which is to say, in the coherence and commitment of physiology in Britain—is emphasized by Charles Lovatt Evans in his *Reminiscences of Bayliss and Starling* (Cambridge: Cambridge University Press, 1964), 4.

35. Singer, *History of Biology*, 433.

36. Andrew D. Evans, *Anthropology at War* (Chicago: University of Chicago Press, 2010), 27. On craniometry and anthropology in France, see Jennifer Hecht, *The End of the Soul* (New York: Columbia University Press, 2003), 92–93 and passim; and Andrew Zimmermann, *Anthropology and Antihumanism in Imperial Germany* (Chicago: University of Chicago Press, 2001), chs. 3–4. For Britain, see the work of George W. Stocking Jr.

37. Stefanos Geroulanos, "The Plastic Self and the Prescription of Psychology," in *Republics of Letters* 3, no. 2 (2014), http://arcade.stanford.edu/rofl/plastic-self-and-prescription-psychology.

38. Franz Boas, *The Measurement of Differences between Variable Quantities* (New York: Science Press, 1906); and Boas, "Changes in the Bodily Form of Descendants of Immigrants," *American Anthropologist* 14, no. 3 (1912), 530–62.

39. Marcel Mauss, *Oeuvres II* (Paris: Minuit, 1969), 421–22 and 440.

40. Bronisław Malinowski, *Argonauts of The Western Pacific* (1922; London: Routledge & Kegan Paul, 1932).

41. W. H. R. Rivers, ed., *The Depopulation of Melanesia* (Cambridge: Cambridge University Press, 1922), 95, 104.

42. Gustave Le Bon, *Psychologie des foules* (Paris: Alcan, 1895), translated as *The Crowd* (London: Unwin, 1976); William McDougall, *The Group Mind* (New York: G. P. Putnam's Sons, 1920); and Sigmund Freud, *Massenpsychologie und Ich-Analyse* (1920), translated as *Group Psychology and the Analysis of the Ego*, in *FSE* 18:65–144.

43. Rothschuh, *History of Physiology*, xviii.

44. Garrison, *Introduction to the History of Medicine*, 711.

45. Singer, *History of Biology*, viii.

46. Ackerknecht, *Short History of Medicine*, 238.

47. For Canguilhem's reliance on Singer, see *CKL* ch. 2; for his vitalism, see ch. 3. See also our "Georges Canguilhem's Critique of Medical Reason," in *Writings on Medicine*, ed. and trans. Stefanos Geroulanos and Todd Meyers (New York: Fordham University Press, 2012), esp. 5.

48. See the critique of physicochemical reductionism in Starling, *Elements of Human Physiology*, 27–28.

49. *IAN* 3–4. In chapters 4 and 5, we will see how this "internal environment" was resuscitated in the 1920s to serve a new integrative paradigm while retaining the mechanistic premises.

50. *BER* 162/93.

51. Sherrington, "Physiology," in *Lectures on the Method of Science*, by T. B. Strong (Oxford: Clarendon, 1906), 59.

52. For a similar critique of turn-of-the-century mechanistic thought, see Arturo Castiglioni, "Neo-Hippocratic Tendency of Contemporary Medical Thought," *Medical Life* 41 (1934), "It was no more necessary to think about the problem itself and man had become a mere engine, though slightly more complicated" (118).

53. Bayliss, *Introduction to General Physiology*, 1.

54. Bayliss, *Principles of General Physiology*, viii; see also p. xi for a direct discussion of vitalism.

55. *NHC* 2–6.

56. Karl Lashley, foreword to *The Organism*, by Kurt Goldstein (New York: American Book Co., 1939), v–vi.

57. *OEB* 89–122.

58. Haldane and J. B. Priestley, "The Regulation of Lung-Ventilation," *Journal of Physiology* 32 (1905), 225–66; also *HRE* 395–96.

59. *OEB* 91, 99, 111, 11.

60. *HRE* 391, also 389.

61. Haldane, "The New Physiology," *Science* 44, no. 1140 (November 3, 1916), 629–30.

62. Cannon, *The Way of an Investigator* (New York: Norton, 1945), 91.

63. *HRE* 387.

64. Ibid., 392.

65. *SWB* 689.

66. Ibid., 690.
67. Christiane Sinding, *Utopie médicale* (Arles, France: Actes Sud, 1989).
68. Anne Harrington, *Reenchanted Science: Holism in German Culture from Wilhelm II to Hitler* (Princeton, NJ: Princeton University Press, 1996), ch. 5.
69. *TO* 54–55, 66–67, 333, 338, 40.
70. *NHC* 8.
71. See, for example, the work of the French physiologist Camille Soula in bridging integration and individuality: Soula, "L'expérimentation en physiologie," in *Somme de la médecine contemporaine*, ed. René Leriche (Paris: Diane française, 1951), 142.
72. *TO* 108–9.
73. *IAN* 1.
74. Ibid., 237–38.
75. Ibid., 3–4.
76. Canguilhem, "La constitution de la physiologie comme science," in *Études d'histoire et de philosophie des sciences*, 5th ed. (Paris: Vrin, 1983), 270. On Jackson, see Nima Bassiri, "Epileptic Insanity and Personal Identity," in *Plasticity and Pathology*, ed. Nima Bassiri and David Bates (New York: Fordham University Press, 2016), 65–111.
77. Cannon, in *The Way of an Investigator* (New York: W. W. Norton and Co., 1945), 19, noted the influence of Davenport on his studies, but Davenport is more notable for his absence in the later works, especially *The Wisdom of the Body*. Notice also Starling's dismissal of eugenics in his own "The Wisdom of the Body," in *SWB* 690: "If longevity is our goal it is not medical science we must look to but eugenics, and I doubt whether the question is one with which we are concerned." See also S. Benison, "Walter B. Cannon and the Politics of Medical Science, 1920–1940," *Bulletin for the History of Medicine* 65, no. 2 (1991), 234–51; and Owsei Temkin, *The Double Face of Janus* (Baltimore: Johns Hopkins University Press, 2006), 345.
78. Starling, *The Fluids of the Body* (London: Constable, 1909), 41.
79. On positivism and "primitive" medicine, see Erwin H. Ackerknecht, *Medicine and Ethnology*, ed. H. Walser and H. Koelbing (Baltimore: Johns Hopkins University Press, 1971), 114.
80. W. H. R. Rivers, *Psychology and Ethnology* (New York: Harcourt, Brace, 1926), 61.
81. For example, *CBC* 1929, ch. 13.
82. Arthur Hall, "The Wall between the Clinic and the Laboratory," *The Lancet* (October 7, 1933), 822–23. On this separation and the prioritization of institutionalized over clinical medicine, see George Weisz, *Divide and Conquer: A Comparative History of Medical Specialization* (Oxford: Oxford University Press, 2005); and Joel D. Howell, *Technology in the Hospital* (Baltimore: Johns Hopkins University Press, 1996).
83. Haldane, "New Physiology," 630, 631.
84. *AKD* 1:63.
85. *CWB* 19.
86. Walter Langdon-Brown, "The Return of Aesculapius," *The Lancet* (October 7, 1933), 821–22.

87. Thus, whereas physicians elsewhere (notably in Vienna) were recognized for their attention to internal secretions and the internal environment (Garrison, *Introduction to the History of Medicine*, 714–16), the shift toward the dual engagement with integration and pathology was quite new to British-led physiology.

88. To offer a methodological note on proximities between the disciplines and analogies beyond them, let us consider the approaches of three authors: Canguilhem, Hans Blumenberg, and Paul Veyne. In his study of Greek mythology, Veyne proposed: "Let us suppose that explanation is reduced to envisaging a polygon of minor causes that do not remain constant from one set of circumstances to the next and that do not fill the specific places that a pattern would assign to them in advance. . . . [The task is] to reveal the unpredictable contours of this polygon, which no longer has the conventional forms or ample folds that make history into a noble tragedy, and to restore their original silhouette to events, which has been concealed under borrowed garments." Veyne, *Did the Greeks Believe in Their Myths?* (Chicago: University of Chicago Press, 1988), 33. Also citing Greek philosophy, Canguilhem asks: "In Aristotle, the hierarchy of freedom and servitude, theory and practice, nature and art parallels an economic and political hierarchy—the hierarchy, within the city, of free men and slaves. A slave, says Aristotle in the *Politics*, is an animate machine. Does the Greek conception of the dignity of science engender contempt for technology and thereby a paucity of inventions, thus leading, in a certain sense, to a difficulty in transposing the results of technical activity to the explanation of nature? Or, rather, does the concept of the eminent dignity of a purely speculative science, a contemplative and disinterested knowledge, translate the absence of technological inventions? Is a contempt for work the cause of slavery, or does an abundance of slaves, in connection with military supremacy, engender contempt for work? Must we here explain ideology by the economic structure of society or, rather, that structure by the orientation of ideas? Is it the ease with which man exploits man that leads to disdain for techniques of man's exploitation of nature—or is it the difficulty of man's exploitation of nature that necessitates justification of man's exploitation of man? Is there a causal relation here, and if so, in which direction? Or are we faced with a global structure of reciprocal relations and influences?" Canguilhem, *Knowledge of Life* (New York: Fordham University Press, 2008), 80–81. Hans Blumenberg, finally, proposes: "The relationship of metaphorology to the history of concepts . . . [is] an ancillary one: metaphorology seeks to burrow down to the substructure of thought, the underground, the nutrient solution of systematic crystallizations; but it also aims to show with what 'courage' the mind preempts itself in its images, and how its history is projected in the courage of its contractures." Blumenberg, *Paradigms for a Metaphorology* (Ithaca, NY: Cornell University Press, 2010), 5. Similarly, we are concerned here with a "polygon" or with a "global structure of reciprocal relations and influences," with an often metaphorical substructure of thought concerning the (re)emergence and (re)unification of the body as a single entity, and the conception of this unified body as reemerging at the same time as the body was most threatened and damaged. These are epistemological as well as onto-

logical concepts, matters of knowing the body and of being a body. And so we do intend to look at how concerns of integration and disequilibrium arose and transformed in interconnected fields (neurology, psychiatry, endocrinology, and so on), fields that at the time were often less or variously differentiated from each other and often also pointed beyond their immediate reach, without insisting on a causalist model of explanation. We are concerned with establishing such a constellation of thinkers around a particular problem: not with telling the reader the whole state of physiology at the time, not with giving a history of all its problems or all its conceptions of the body, but with considering one particular trend, brought about partly because of problems in the understanding of the body that crossed disciplinary and experimental barriers, partly because of epistemological and experimental frontiers, partly because of opposition to existing and previous research, and partly because of demands of war and hospital care. This, to be clear, was just one set of approaches, and it could be contrasted to other approaches emerging at the same time. It was a way of thinking that not only characterized, and even defined, a significant array of influential thinkers—at once original, institutionally well placed, and philosophically and medically consequential—but also ranged in its scope from anthropology to cybernetics, from social theory to medico-psychological therapeutics.

89. "Economic integration" began appearing in earnest in English at this point, and discussions of political integration became central, with internationalism, especially of the functionalist school, and with the Paris Peace Conference and the institution of the League of Nations.

90. Ferdinand de Saussure, *Course in General Linguistics* (New York: McGraw-Hill, 1966), 9.

91. Ibid., 9, 13, 15, and passim.

92. In "Two Aspects of Language and Two Types of Aphasic Disturbances," *On Language* (Cambridge, MA: Harvard University Press, 1995), 115–33, the structural linguist Roman Jakobson used Head and Goldstein as his starting points.

93. George Crile, *An Autobiography* (Philadelphia: Lippincott, 1947), 86–87.

94. Harvey Cushing, *From a Surgeon's Journal* (Boston: Little, Brown, 1936), 79–80.

95. Mrs. William Vanderbilt, "Miracles of Surgery on Men Mutilated in War," *New York Times*, January 16, 1916.

96. Peter C. English, *Shock, Physiological Surgery, and George Washington Crile in the Progressive Era* (Santa Barbara, CA: Greenwood, 1980), 208.

97. Eric Rutkow and Ira Rutkow, "George Crile, Harvey Cushing, and the Ambulance Américaine," *Archive of Surgery* 139 (2004), 678–85; and George Crile, "A Composite Report of the Three Months' Service of the Lakeside Unit at the American Ambulance," *Cleveland Medical Journal* 14 (1915), 421–39. Innovations included the expanded role of triage for the stretcher bearer, the standardization of field medical kits, the training of regimental medical officers, building ambulances on Ford Model T chassis for rapid transport, and the training of civilian surgeons by Crile and others; these drew on a vast re-

source of research and previous experience in war (notably, for Crile, his time during the Spanish-American War).

98. From a handwritten note between leaves of an essay on wound shock, box 13, folder 1, George Washington Crile Archive (hereafter Crile Papers). All further citations to the Crile Papers use a colon to distinguish box and folder; for example, 13:1 refers to box 13, which holds folder 1.

99. Crile Papers, 13:1; Ackerknecht to Sigerist, in *Henry E. Sigerist: Correspondences with Welch, Cushing, Garrison, and Ackerknecht* (Frankfurt: Lang, 2010), 336. See also Ackerknecht's review of Crile's autobiography, in which Ackerknecht sneers, "His unrestricted admiration for George Crile carries him often to attribute to himself work accomplished decades before his birth (e.g. in the case of shock experimentation, CO poisoning)." Ackernecht, "George Crile: An Autobiography," *Quarterly Review of Biology* 24, no. 1 (1949): 46.

100. Canon [sic] to Crile [Crile's note], February 26, 1902, Crile Papers, 20:61, file 1.

101. Notes for "Brain Cell Symposium" (1915), Crile Papers, 3:13, files 54–55. After compiling keywords from the titles of papers published in *American Journal of Physiology* over the previous twenty-five years, Crile strikingly concluded that "the study of the brain is almost shunned by the physiologist" and asked not to be quieted by those "who cry metabolism." Crile was as much responding to what he saw as competing disciplinary priorities in surgery as asserting his stance as a fierce and unrepentant interventionist.

102. It is hard not to think here of Thomas Kuhn's 1957 paper "Energy Conservation as an Example of Simultaneous Discovery," republished in Kuhn, *The Essential Tension* (Chicago: University of Chicago Press, 1977), 66–104.

CHAPTER TWO

1. "Conference in Boulogne of Surgeons and Pathologists of the B.E.F. on Treatment of Shock and Haemorrhage, held on November 15th 1918," GC/223/A.1/3-5, William H. Bayliss Papers. A copy is kept at the National Archives (Kew), FD 1/5262.

2. Ibid., 2. It is unclear from this report whether Crile found the application of gum to be unreliable during the colder months only or thought it unreliable in general; on other occasions his belief that gum arabic was harmful was obvious.

3. The quote is from William M. Bayliss, *Intravenous Injection in Wound Shock* (London: Longman's, Green, 1918), 1.

4. No consensus would be reached before World War II, although the work of the London Shock Committee discussed here would establish something of an Anglophone paradigm. See Medical Research Council, Committee on Traumatic Shock and Blood Transfusion, *The Treatment of Wound Shock* (London: H. M. Stationery Office, 1940).

5. George W. Crile and William E. Lower, *Surgical Shock and the Shockless Operation through Anoci-Association* (Philadelphia: Saunders, 1920).

6. *Webster's Revised Unabridged Dictionary* (Springfield, MA: Merriam, 1913), 603.

7. Allan Young cites William Gowers and Hermann Oppenheim as emphasizing that functional disturbance was specifically not to be understood in psychological terms. Young, *The Harmony of Illusions* (Princeton, NJ: Princeton University Press, 1995), 53. But among American physicians and physiologists, this was not clearly so: in his *Shell-Shock and Other Neuropsychiatric Diseases, Presented in Five-Hundred and Eighty-Nine Case Histories for the War Literature, 1914–1918* (Boston: W. M. Leonard Publisher, 1919), E. E. Southard uses *functional* to refer quite specifically not to whole-body conditions (which he calls organic) but to psychic conditions whose organic basis is unclear or inexistent. William Whitla's *Dictionary of Treatment* similarly uses "functional" and "purely functional" to refer to psychological conditions or psychogenic effects. William Whitla, *A Dictionary of Treatment, including Medical and Surgical Therapeutics*, 6th ed. (Toronto: Hartz, 1920).

8. Research on wound shock in recent years has been carried out principally in reference to nursing. Christine Hallett, *Containing Trauma* (Manchester: Manchester University Press, 2009); M. M. Manring, Alan Hawk, Jason H. Calhoun, and Romney C. Andersen, "Treatment of War Wounds," *Clinical Orthopaedics and Related Research* 467, no. 8 (2009), 2168–91. For the (repeated) reference to a "vicious circle," see Bayliss, *Intravenous Injection in Wound Shock*, 4.

9. On brutalization, see George Mosse, *Fallen Soldiers* (Oxford: Oxford University Press, 1990), 172 and passim; and Daniel Pick, *War Machine* (New Haven, CT: Yale University Press, 1993).

10. Young, *Harmony of Illusions*, 50–52.

11. E. H. Starling, *Elements of Human Physiology* (London: Churchill, 1897); and Starling, *Principles of Human Physiology* (Philadelphia: Lea & Febiger, 1915). On physicians' and surgeons' acceptance of the research value of the war, see, for example, V. Warren Low, "Presidential Address on Surgery during the War, Delivered before the Medical Society of London," *The Lancet* (February 21, 1920), 419: "But from it all will emerge observations, technique, and conceptions which will, I believe, have a far-reaching influence on medicine and surgery." Cannon speaks of the "recognition of the importance of utilizing the opportunities offered by warfare to obtain further knowledge of the shock state," in *Traumatic Shock* (New York: Appleton, 1923), viii.

12. Crile to Samuel Mather, "Diary of Dr. Crile for American Ambulance Service," Crile diary, vol. 1, November 3, 1914, Crile Papers, 64:254. Psychiatrists, too, treated the war as a ready-made laboratory. See Stefanie Caroline Linden and Edgar Jones, "German Battle Casualties," *Journal of the History of Medicine and Allied Sciences* 68, no. 4 (2013), 628.

13. Alexis Carrel to Simon Flexner, December 17, 1914, RU450:C232:2 (1906–1916), Papers of Alexis Carrel, Rockefeller Archive Center (hereafter Carrel Correspondence).

14. Crile lectured on the matter before a group of French, English, and American surgeons in early 1915, claiming that "the people as a whole were showing as a result of the severe and long-continued strains to which they had been subjected precisely the same effects that follow the exhaustion of an individual's central nervous system when he is long under the knife. . . . Dr. Crile

summed up what had happened by calling it the vivisection of a race.... For them there has been none of the 'blocking off' of the afferent nerves by which surgical shock can now be prevented, for what may be called the 'operations' of the Germans on this unhappy people have been performed with no such precautions and with no anesthetics." "Surgical Shock to Belgium," in "Diary of Dr. Crile for American Ambulance Service, 1914–1915," vol. 2, Crile Papers, 64:254. Crile also wrote in detail of "Belgium's vivisection" in Crile, *A Mechanistic View of War and Peace* (New York: Macmillan, 1915), 79, 84–86. Crile's book title demonstrates his unwillingness to be branded a Romantic.

15. Crile, *Mechanistic View of War and Peace*, 69.

16. Ibid., 79. Crile's research on Belgium relied largely on interviews with medical personnel who had either fled Belgium or visited hospitals or refugee camps. His diary includes transcriptions of several such interviews in which he repeatedly asks about physiological functions among refugees and those who had witnessed atrocities. Crile diary, vol. 2, Crile Papers, 64:254.

17. Humphrey Cobb, *Paths of Glory* (New York: Viking, 1935), 3–4.

18. Sigmund Freud, "Charcot," in *FSE* 3:13.

19. Notorious is the example of gas masks: that they were hard to use and dysfunctional was an open secret among researchers as much as soldiers, who appealed over and over for improved masks and training. See Charles Lovatt Evans to Ernest Starling, May 25, 1916, PP/CLE/A.3, Lovatt Evans Papers.

20. E. M. Cowell, "The Initiation of Wound Shock," in *MRC* 59.

21. Crile, *Mechanistic View of War and Peace*, 19–20.

22. Later reports on shock would focus on both the activation and the distention and exhaustion. On emotional influence, see also the reference to the centrality of fear and pain in Captain E. M. Cowell, "Wound Shock," *Military Surgeon* (June 1926), 625.

23. William Townsend Porter, *Shock at the Front* (New York: Atlantic Monthly, 1918); Editorial, "A New Hypothesis Concerning Traumatic Shock," *New York Medical Journal* (February 23, 1918), 389–40.

24. See the discussion of prewound factors and the conclusion that "the psychology and physiology of the average healthy 'veteran' soldier living in the fighting zone under 'peace' conditions are for practical purposes neutral," in Cowell, "Initiation of Wound Shock," in *MRC* 59, 66. Cowell raises questions of temperament and emotion in a note on page 63, praising temperamental stability, and in Cowell, "Arris and Gale Lecture on the Initiation of Wound Shock and Its Relation to Surgical Shock," *The Lancet* (July 26, 1919), 137.

25. See the discussion of anxiety (as opposed to fright) in *BPP* 12/11/156, and *Inhibitions, Symptoms, and Anxiety* in *FSE* 20:164–66. See also *Introductory Lectures on Psycho-Analysis*, in *FSE* 16:394–95. Of course, anxiety is not an unambiguously positive tool for protecting the organism; consider its role in anxiety neuroses.

26. Crile diary (January 29, 1915), 1:399, Crile Papers, 64:254.

27. Christine Debue-Barazer and Sébastien Perrolat, "1914–18: Guerre, chirurgie, image," *Sociétés & représentations* 25, no. 1 (2008), 236.

28. Low, "Presidential Address on Surgery," 419. See also Carrel on the dangers of surgery itself causing infection and sepsis: "Losses from

surgical infections are terrible. It is the more important problem we have to deal with." Carrel to Flexner (December 4, 1914), RU450:C232, 2, Carrel Correspondence.

29. See Low's celebration of the prevention of sepsis by way of the early treatment of infection by the Carrel-Dakin solution, in Low, "Presidential Address on Surgery," 419–20.

30. Hallett traces this with regard to practices of nursing in Hallett, *Containing Trauma*, 35. In contrast to the "sophisticated scientific response" that shock required, Hallett writes, the one for hemorrhage was "simple, decisive, mechanical."

31. H. M. W. Gray, *The Early Treatment of War Wounds* (London: Frowde, 1919), 80.

32. Low, "Presidential Address on Surgery," 419.

33. On the disordered action of the heart, see Young, *Harmony of Illusions*, 53, 60–61.

34. Crile to Winchester du Bouchet, Crile diary, 1:32, Crile Papers 64:254. See the similar note from Crile to Samuel Mather (November 3, 1914), in the diary at 1:28. See also Crile and W. E. Lower, *Anoci-Association* (Philadelphia: Saunders, 1915), 19.

35. Crile, *Mechanistic View of War and Peace*, 23. Ambulance in the caption refers to the Ambulance américaine in Paris, a hospital funded privately by Americans in Paris since its founding during the Franco-Prussian War. Eric I. Rutkow and Ira M. Rutkow, "George Crile, Harvey Cushing, and the Ambulance Américaine," *Archives of Surgery* 139 (2004), 678–85.

36. For Crile on exhaustion, see John B. Roberts, *War Surgery of the Face* (New York: William Wood, 1919), 106.

37. Crile, *Mechanistic View of War and Peace*, 26.

38. For the English, the category of "neurasthenia" would come to cover symptoms of both exhaustion and shell shock; but as Young has shown, these designations were in part political considerations concerning the class status of the sufferer. Young, *Harmony of Illusions*, ch. 2.

39. The insets after pages 20, 22, and 24 in Crile's *Mechanistic View of War and Peace* are reproduced in Crile and Lower, *Surgical Shock and the Shockless Operation*, at 43, 47, and 45, respectively, where their captions point out that they are photographs from rabbits. No such reference accompanies the photographs in *Mechanistic View*; rabbits are not referenced in the book, and the only reference to animal research addresses its humaneness compared to the battle that makes possible the study of humans (84).

40. Anson Rabinbach, *The Human Motor* (New York: Basic Books, 1992).

41. For the complications involved in naming "soldier's heart," see the editorial "'Soldier's Heart,'" *The Lancet* (April 14, 1917), 580–81.

42. Ibid., 580.

43. On the history of "irritable heart" and "soldier's heart," see Charles F. Wooley, *The Irritable Heart of Soldiers and the Origins of Anglo-American Cardiology* (Burlington, VT: Ashgate, 2002).

44. Whitla, *Dictionary of Treatment*, 394.

45. Adolphe Abrahams, "Soldier's Heart," *The Lancet* (March 24, 1917),

442–45. Studies on soldier's heart include James Mackenzie, "Soldier's Heart," *BMJ* 1, no. 2873 (January 22, 1916), 117–19; Mackenzie, R. M. Wilson, Philip Hamill, Alexander Morrison, O. Leyton, and Florence A. Stoney, "Discussion of the Soldier's Heart," *Proceedings of the Royal Society of Medicine: Therapeutical and Pharmacological Section* 9 (1916), 27–60; Mackenzie, "A Lecture of the Soldier's Heart and War Neurosis," *BMJ* 1, no. 3094 (1920), 530–34; R. D. Rudolf, "The Irritable Heart of Soldiers," *Canadian Medical Association Journal* 6, no. 9 (1916), 796–810.

46. J. S. Haldane, J. C. Meakins, and J. G. Priestley, "Effects of Shallow Breathing," *Journal of Physiology* 52, no. 6 (1919), 433–53.

47. See, for example, D.D.M.S. XVII Corps, Headquarters of the Third Army, OC, 18 FA, November 6, 1917, "The Initiation of Wound Shock with Suggestions for Its Early Treatment," RAMC 466/3-4, Royal Army Medical Corps Archive, Wellcome Library.

48. *MSS*, August 17, 1917.

49. See Crile diary, 2:531 (February 6, 1915), Crile Papers 64:254. Crile, very impressed, called the Balkan splint "a unique but scientific contribution born of Balkan poverty but so efficient that I shall adopt it at Lakeside."

50. T. K. Rowlands and J. Clasper, "The Thomas Splint," *Journal of the Royal Army Medical Corps* 149, no. 4 (December 2003), 291–93.

51. Cowell, "Front Line Application of Thomas's Splint," GC/116/1, E. M. Cowell Papers, Wellcome Library. Cowell also credits General Harry Neville Thompson with the idea of a "drill by numbers" approach to teaching the splint's application, which facilitated its usage.

52. The particularity of the blankets in World War I becomes clearer when we consider that both the original splint by Hugh Owen Thomas and later manuals of military surgery that told its story emphasized the immobilization technique. The blankets and their rationale were a particular invention in the Great War and specifically designed for shock. See Hamilton Bailey, ed., *Surgery of Modern Warfare* (Edinburgh: Livingstone, 1941), 379–86.

53. "As the result of the widespread adoption and practice of the method, the mortality of gunshot wounds of the femur was reduced by over 30% in forward areas. French gunshot femurs still arrived at the forward hospitals, dead from bleeding or wound shock. This was one of the great advances in front-line surgical treatment." Cowell, "Front Line Application of Thomas's Splint," GC/116/1, Cowell Papers; Gray, *Early Treatment of War Wounds*, 59 and passim. See P. M. Robinson and M. J. O'Meara, "The Thomas Splint," *Bone and Joint Journal* 91B, no. 4 (2009), 540–44.

54. Cowell, denying that shock could be distinguished from collapse, noted, "At the present time the term 'collapse' is used to describe the symptom-complex resulting from the effects of more or less prolonged low blood pressure (Bayliss)." Cowell, "Arris and Gale Lecture," 143. See also the more detailed discussion in H. H. Dale, "The Nature and Cause of Wound Shock," in *Harvey Society Lectures, 1919–1921* (Philadelphia: J. B. Lippincott, 1921), 30–31. Dale emphasizes that Crile had inverted the terminology of shock and collapse; before him, the latter had generally referred to both rapid and slow "depression of vitality."

55. Crile to Harvey Cushing (January 30, 1915), Crile diary, 2:418, Crile Papers 64:254.

56. Shock was so multifaceted that the American Expeditionary Force did not identify it as a category among causes of death. Records of the American Expeditionary Forces WWI, History of the Central Medical Department Laboratory, Dijon (1918–19), box 1, entries 2840 and 2841 (NM-92HM1999), Records Group (RG) 120, Archive of the Dijon Laboratory of American Expeditionary Force in World War I, National Archives II.

57. Cowell, "Arris and Gale Lecture," 138, quoted in Cannon, *Traumatic Shock*, 5.

58. Cowell, "Wound Shock," *Military Surgeon* (June 1926), 624.

59. John Fraser and Cowell, "Clinical Study of Blood Pressure in Wound Conditions," in *MRC* 14–15. See also Porter's insistence on thighs, bones, and fractures (*Shock at the Front*, 120 and passim). Leriche, lamenting the extreme danger of thigh fractures, wrote in 1917 that those who arrive with a smashed hip "are always under the influence of extreme shock, may survive, but quite exceptionally." René Leriche, *The Treatment of Fractures* (London: University of London Press, 1918), 118, also 6, 125, 129.

60. But see the report by Wallace from December 1915 that notes that in the abdomen, too, the occasion and effects of shock were unexpected: "A man may have a portion of the belly blown away and the viscera exposed without profound or even marked shock. At the same time, a few perforations of the intestine may produce so much shock as to make it doubtful if the patient will stand an operation." Wallace, "The Treatment of Abdominal Wounds in War," *Journal of the Royal Army Medical Corps* 25, no. 6 (December 1915), 594. Wallace notes that in peritoneal wounds "the main causes of death are shock, hemorrhage, peritonitis and septic infection of the retroperitoneal tissue."

61. See in general both Crile's work and H. Tyrell Gray and Leonard Parsons, "The Arris and Gale Lectures of 1912 on the Mechanism and Treatment of Shock," *BMJ* 1, no. 2680 (May 11, 1912), 1065–72.

62. Théodore-Marin Tuffier, "Practical Problems from the Viewpoint of a Field Inspector," in Crile Papers 64:254, pp. 502–3. Tuffier was convinced by Crile's explanation concerning anoci-association and the importance of the patient's morale (he thought the morale issue would later be played down somewhat). Tuffier's statement about shock also appears in modified form as his "Practical Problems from the Viewpoint of a Field Inspector," *Surgery, Gynecology, & Obstetrics*, 20 (1915) (Franklin Henry Martin, ed.; Surgical Publishing Company of Chicago), 713–15.

63. Cited in Sophie Delaporte, *Les médecins dans la grande guerre, 1914–1918* (Paris: Bayard, 2003), 27.

64. On discoloration, see *Military Surgeon* (June 1926), 226.

65. Donald Macrae Jr. "Prevention and Treatment of Wound Shock in the Theatre of Army Operations," *Illinois Medical Journal* 38 (1920), 108–14, 110.

66. Cannon, Fraser, and Cowell, "The Preventive Treatment of Wound Shock," in *MRC* 93; and Fraser and Cowell, "A Clinical Study of the Blood Pressure in Wound Conditions," in *MRC* 8.

67. Bayliss, *Intravenous Injection in Wound Shock*, 14.
68. Dale, "Nature and Cause of Wound Shock," 40.
69. See, for example, Cannon, Fraser, and Hooper, "Some Alterations in the Distribution and Character of the Blood," in *MRC* 27. Bayliss, *Intravenous Injection in Wound Shock*, 1. See also the characterization "the great unresolved riddle of military surgery," in Gray, *Early Treatment of War Wounds*, 80, quoted in Hallett, *Containing Trauma*, 30.
70. Crile and Lower, *Surgical Shock and the Shockless Operation through Anoci-Association*, 24.
71. Porter, *Shock at the Front*, 55, 98. Cannon, in *Traumatic Shock*, objected: "All the evidence, both clinical and experimental, proves that the stagnant blood is not in the large veins in secondary shock." As blood was "lost" neither in the arteries nor in the veins, Cannon considered the capillaries, which he sought to show could indeed pool a considerable amount. Cannon, "A Consideration of the Nature of Wound Shock," in *MRC* 80, 73–76. See also the discussion of high blood count in the capillaries and low count in the veins, in Cannon, Fraser, and Hooper, "Some Alterations in the Distribution and Character of the Blood," in *MRC*, 31–32, 34; Cannon and McKeen Cattell, "Studies in Experimental Traumatic Shock," *Archives of Surgery* 4, no. 2 (1922), 300; and Dale, "Nature and Cause of Wound Shock," 34. See also the discussion of the "missing blood" in Frederick Heaton Millham, "Brief History of Shock," in *Surgery* 148, no. 5 (2010), 1031ff.
72. See Cannon's critique of Yandell Henderson's pre–World War I acapnia theory of shock, as well as of Henderson's modifications of it, in Cannon, "Consideration of the Nature of Wound Shock," in *MRC* 69 and passim. See also *MSS*, November 9, 1917, p. 2.
73. Henry James to Carrel, March 16, 1916, in Carrel Correspondence, RU450:C232:2 (1906–1916). The term *nervous shock* is found as early as Edwin Morris, *A Practical Treatise on Shock after Surgical Operations . . .* (London: Hardwicke, 1867).
74. Cannon, "Consideration of the Nature of Wound Shock," in *MRC* 81. Millham suggests the identification to be false ("Brief History of Shock," 1027), but even though Cannon was inventing the genealogy to suit his ends, Hippocrates did use ἔξαιμος, at least in *De capitis vulneribus*, pt. 16. See also the approving notes in Bayliss, "Intravenous Injection in Wound Shock: Abstract of the Oliver-Sharpey Lectures," in *BMJ* (May 18, 1918), 553, and in Macrae, "Prevention and Treatment of Wound Shock."
75. Medical Research Committee, "Memorandum upon Surgical Shock and Some Allied Conditions," *The Lancet* (March 31, 1917), 502. But see also H. Dale to Cannon, October 11, 1918, box 62, folder 816. Dale, Cannon Papers.
76. Cannon, *Traumatic Shock*.
77. Bayliss, *Intravenous Injection in Wound Shock*, 1.
78. Ibid., 1. See also Porter, *Shock at the Front*, 3, 7 and passim.
79. In "Du shock dans les blessures de guerre," *La presse médicale* (February 8, 1918), 69. E. Quénu did not distinguish shock from commotion. Similarly, Fernand Masmonteil discussed an "état de dépression très marquée" in his "Le shock chez les blessés de guerre," *Paris médical* 27 (1918), 419.

Secondary sources agree: "Le choc traumatique et le choc hypo-volémique étaient mal différenciés. Il n'existait pas de consensus pour le traitement de l'un ou de l'autre," claimed Frédéric Chauvin, Louis Paul Fischer, Jean-Jacques Ferrandis, Édouard Chauvin, and François-Xavier Gunepin in their "L'évolution de la chirurgie des plaies de guerre des membres en 1914–1918," in Société française d'histoire de la médecine histoires des sciences médicales 26, no. 2 (2002), 165; and Sophie Delaporte's account in Les médecins dans la grande guerre, 26–30, where psychological factors and general suffering appear crucial for shock.

80. This distinction often remains unclear in the secondary literature on the matter, perhaps because of the dominance of shell shock. To give an example from one of the ablest of interpreters, Young, in discussing Cannon's "Voodoo Death" article in Harmony of Illusions (24), conflates wound and shell shock, citing Cannon on wound shock but in a context that suggests psychic trauma.

81. Bayliss, Intravenous Injection in Wound Shock, 1.

82. Cannon, Traumatic Shock, 2.

83. Sophie Delaporte, Les médecins dans la grande guerre; Annette Becker and Stéphane Audoin-Rouzeau, 14–18 (Paris: Gallimard, 2000).

84. Bayliss "paid a visit to the Western front and was arrested as a spy, but was promptly released when recognized." Charles Lovatt Evans, Reminiscences of Bayliss and Starling (Cambridge: Cambridge University Press, 1964), 13.

85. "University Lecture" notes, Crile Papers 3:13.

86. Crile diary, 2:463 (February 4, 1915), Crile Papers, 64:254.

87. Henry James to Carrel, March 16, 18, and 29, 1916 (for Cannon), and June 13, 1916 (for Porter), RU450:C232:2 (1906–1916), Carrel Correspondence.

88. Cannon to Sherrington, February 14, 1916, box 63, folder 836.Sherrington, Cannon Papers.

89. Gaston Bachelard, The Psychoanalysis of Fire (London: Routledge and Kegan Paul, 1964), 22.

90. Cannon, The Way of an Investigator (New York: Norton, 1945), 130–45; Cannon, Traumatic Shock, viii–x. See also Cannon's discussion of his clinical work in Béthune in Cannon and Cattell, "Studies in Experimental Traumatic Shock," 303, 306–7. Cannon's competition with Porter dated to the early years of the century: in his autobiography, The Way of an Investigator, Cannon wrote his colleague out altogether.

91. Carrel forcefully disparaged French bureaucracy and competence, and he was not alone in comparing them negatively to American operations in particular. Carrel to Flexner, December 4, 1914, and December 17, 1914. RU450:C232:2 (1906–1916), Carrel Correspondence. William van der Kloot, in "William Maddock Bayliss's Therapy for Wound Shock" in Notes and Records of the Royal Society 64 (2010), 283, claims that the Medical Research Committee in London "had no exchanges with French scientists whereas the British and Americans worked shoulder to shoulder." This should not be overstated, as the Medical Research Committee discussed French reports (e.g., MSS, December 2, 1918, pp. 2–3), and Cannon cited at length from French sources in Traumatic Shock and lectured to French surgeons on shock at an

interallied conference in 1918: Cannon, "'Le choc,' première séance interalliée consacrée à la biologie de guerre, 19 octobre 1918," *Paris médical* 29 (1918), 371–72.

92. Leriche, *Treatment of Fractures*, 147. See also Delaporte, *Les médecins dans la grande guerre*, 27.

93. Dale, "Nature and Cause of Wound Shock," 37. See also Quénu's contribution to Cannon's presentation to the interallied meeting of October 19, 1918. Cannon, "Le choc," 372.

94. Leriche, *Treatment of Fractures*; Carrel and Georges Dehelly, *The Treatment of Infected Wounds* (New York: Hoeber, 1917); Angelo May and Alice May, *Two Lions of Lyons: Alexis Carrel and René Leriche* (Rockville, MD: Kabel, 1992).

95. Van der Kloot, "William Maddock Bayliss's Therapy for Wound Shock," 277.

96. Cannon, *Traumatic Shock*, vii, and *MRC* passim.

97. Anthony Bowlby and Cuthbert Wallace, "Development of British Surgery at the Front," in *BMJ* (June 2, 1917), 705.

98. Leriche, *Treatment of Fractures*, 116. See also Quénu, "Du shock dans les blessures de guerre," 69. Cannon, in contrast, followed Soubeyran in arguing that immediate shock is infrequent or specific to head injuries. Differently put, primary wound shock follows soon after injury. See Cannon, *Traumatic Shock*, 4; and Soubeyran, "Les projectiles du foie (Onze observations d'ablation)," *Bulletin et mémoires de la Société de chirurgie de Paris* 45 (1919), 1523.

99. Dale, "Nature and Cause of Wound Shock," 31.

100. D.D.M.S. XVII Corps, Headquarters of the Third Army, OC, 18 FA, November 6, 1917, "The Initiation of Wound Shock with Suggestions for Its Early Treatment" (signed by Cowell), RAMC 466/3-4, RAMC Archive. See also "Suggestions for Prevention and Early Treatment of Wound Shock," RAMC 466/5, H.Q. First Army, November 2, 1917, RAMC Archive.

101. Dale, "Nature and Cause of Wound Shock," 31.

102. Wallace, "Introductory," in *MTT* 5.

103. Cowell and Cannon both described secondary shock as occurring "some" or "several" hours after the injury. Cowell, in "Arris and Gale Lecture," 138; Cannon, *Traumatic Shock*, 5. Secondary shock is called "insidious" in Medical Research Council, *The Treatment of Wound Shock* (London: H. M. Stationery Office, 1940), 1.

104. Dale, "Nature and Cause of Wound Shock," 31.

105. Ibid. See also Cuthbert Wallace's similar description in "Introductory," *MTT* 5.

106. See, for example, the section "Importance of the Time Factor in Secondary Shock" in Cannon and Cattell, "Studies in Experimental Traumatic Shock," 322–23. Earlier sections of the article make it clear how protracted shock affects the vasomotor center and central nervous system. Cannon, Fraser, and Cowell had specifically noted in 1917 that "secondary wound shock is proportionate to the length of time the pernicious factors are allowed to work—in other words, to the period during which the wounded man is 'lying

out'"(Cannon, Fraser, and Cowell, "Preventive Treatment of Wound Shock," in *MRC* 89). See also Gray, *Early Treatment of War Wounds*, 81.

107. Masmonteil, "Le shock chez les blessés de guerre," 420–21. Masmonteil relies at length on Quénu and Paul Santy. The categorization of shock into immediate, primary, and secondary is echoed throughout French publications on the matter, with terms *immediate*, *prompt*, and *belated*. It is unclear whether French researchers took up the distinction from Anglophone ones or were the first to propose it.

108. See, for example, Carrel, "Traitement abortif de l'infection des plaies" (September 18, 1915), 5, RU450:C232:2 (1906–1916), Carrel Correspondence: "Seuls les blessés atteints d'hémorragie grave ou de choc sont soignés sur place." Tuffier wrote: "The wounded man must not suffer—he is in a state of shock. If during transportation he suffers great pain his resistance is diminished." Tuffier, "Practical Problems from the Viewpoint of a Field Inspector" in Crile diary, 2:505, Crile Papers 64:254. Earlier, and in a similar context, Tuffier had told Crile (as Crile recorded in his diary) that "they did not attempt to move the cases of highest-up compound fractures of the femur" and that these patients were treated with great care at the base hospital. Crile diary, 2:399 (January 29, 1915), Crile Papers 64:254.

109. Leriche, *Treatment of Fractures*, 129–30, 118–19. On gangrene, see also Cowell, "Initiation of Wound Shock," 139; and on symptomatic similarities, see Cannon, *Traumatic Shock*, 53–54.

110. Quoted in Chauvin et al., "L'évolution de la chirurgie des plaies de guerre des membres en 1914–1918," 165.

111. Leriche, *Treatment of Fractures*, 122, 125, 135.

112. Ibid., 44.

113. See also Gray, *Early Treatment of War Wounds*, 100. Gray describes surgical intervention and anesthetics as concerns of "great difficulty."

114. See, for example, Crile and Lower, *Anoci-Association*, ch. 1. Also see the draft essay, "Are Traumata of All Parts of the Body Equally Capable of Producing Shock?," Crile Papers 3:13.

115. Draft essay, "The Kinetic System," Crile Papers 3:13. When Crile turned his attention to neurological function, he did so to distinguish "traumatic shock" from "psychic shock" as a surgical concern with histological brain lesions.

116. Notes on "The Phenomena of Exhaustion" (1915), Crile Papers 3:13.

117. Carrel's research on shock was limited and unexceptional. Cf. May and May, *Two Lions of Lyons*, 173.

118. Porter, *Shock at the Front*, 44, 120.

119. Ibid, 95.

120. *MSS*, August 23 and September 3, 1917.

121. Medical Research Committee, "Surgical Shock and Some Related Conditions," *BMJ* (March 24, 1917), 381–83; Medical Research Committee, "Memorandum upon Surgical Shock and Some Allied Conditions," *The Lancet* (March 31, 1917), 502–5.

122. In declining the Rockefeller Institute's early invitation to work on shock under Carrel, Cannon cited his experimental commitments and lack

of clinical experience; see James to Carrel, March 29, 1916, RU450:C232:2, Rockefeller Archive Center; and Saul Benison, A. Clifford Barger, and Elin L. Wolfe, "Walter B. Cannon and the Mystery of Shock," *Medical History* 35 (1991), 217.

123. "The Study of Wound Shock," *The Lancet* (July 26, 1919), 159, 160. See also Fraser and Cowell, "Clinical Study of Blood Pressure in Wound Conditions," in *MRC* 1.

124. "Study of Wound Shock," 160.

125. See Fraser and Cowell, "Clinical Study of Blood Pressure in Wound Conditions," in *MRC* 1–2; "Wound Shock," *Military Surgeon* (June 1926), 624; Bayliss, "Intravenous Injection in Wound Shock," 553.

126. Cannon, Fraser, and Hooper, "Some Alterations in Distribution and Character of Blood in Shock and Hemorrhage," 27; Cannon, "Acidosis in Cases of Shock, Hemorrhage, and Gas Infection," in *MRC* 48.

127. See the short reference to a four-and-a-half-year-old girl in Gray and Parsons, "Arris and Gale Lectures on the Mechanism and Treatment of Shock," 942.

128. Cannon, Fraser, and Hooper, "Some Alterations in Distribution and Character of Blood in Shock and Hemorrhage," in *MRC* 27. Bayliss discussed, in *Intravenous Injection in Wound Shock*, the trouble he had in replicating earlier animal experiments (1918). See also Cannon and James McKeen Cattell's critique of Henderson for not correlating animal experiments to the clinical picture ("Studies in Experimental Traumatic Shock," 308), in comparison to their general praise of Robert Gesell (300–301).

129. Cannon, "Consideration of the Nature of Wound Shock," in *MRC* 80.

130. Cannon, Fraser, and Hooper, "Some Alterations in Distribution and Character of Blood in Shock and Hemorrhage" in MRC 28, 30, 31–32, 34.

131. Cannon, "Consideration of the Nature of Wound Shock," in *MRC* 75.

132. Ibid., 77.

133. Cannon, Fraser, and Hooper, "Some Alterations in Distribution and Character of Blood in Shock and Hemorrhage," in *MRC* 34–35.

134. Cannon, "Acidosis in Cases of Shock, Hemorrhage, and Gas Infection," 55. From 1918 onward, this would become a major research subject, and Cannon would cite several studies of the complex effects of the drop in alkali reserve. For example, Cannon and Catell, "Studies in Experimental Traumatic Shock," 304. Dale, in 1919, would insist that the drop in alkali reserve was either impossible to maintain consistently or was a side effect. Dale, "Nature and Cause of Wound Shock," 35.

135. Cannon and Catell, "Studies in Experimental Traumatic Shock," 309. See also Bayliss, *Intravenous Injection in Wound Shock*, where Bayliss doubts the importance of acidosis as such (35–37), and insists that only coupled with reduction in oxygen (66) would it affect the overall condition of the blood so strongly.

136. Cannon, "Acidosis in Cases of Shock, Hemorrhage, and Gas Infection," in *MRC* 55.

137. Ibid., 54. Bayliss also argued specifically against exhaustion as a cause or major factor in surgical and wound shock. See Bayliss, *Intravenous Injection in Wound Shock*, 2–3.

138. Cannon and Catell, "Studies in Experimental Traumatic Shock," 310.
139. Cannon, "Consideration of the Nature of Wound Shock," in *MRC* 80.
140. Ibid.
141. Ibid., 82. Bayliss, *Intravenous Injection in Wound Shock*, 14. See also Wallace, "Introductory," in *MTT* 3.
142. Robert Gesell, "Studies on the Submaxillary Gland," *American Journal of Physiology* 47, no. 4 (January 1919), 438–67.
143. Cannon and Catell, "Studies in Experimental Traumatic Shock," 301–2.
144. Editorial, *The Lancet* (July 27, 1919), 160.
145. E.g., Bayliss, *Intravenous Injection in Wound Shock*, 5.
146. Cowell, "Initiation of Wound Shock," in *MRC*, 60; see also Cowell, "Arris and Gale Lecture," 137.
147. Cannon, "Consideration of the Nature of Wound Shock," in *MRC* 80.
148. Cowell, "Arris and Gale Lecture," 137.
149. Ibid.
150. Bayliss, *Intravenous Injection in Wound Shock*, 50.
151. See also Dale, "Nature and Cause of Wound Shock," 36, 37, 38 ("either histamine itself or some substance having a close chemical relation to it").
152. Bayliss, *Intravenous Injection in Wound Shock*, 11.
153. Ibid., 41.
154. Ibid., 66.
155. Cannon and Catell, "Studies in Experimental Traumatic Shock," 304. But see the 1919 criticism of this position by Dale in "Nature and Cause of Wound Shock," 35 and 36.
156. Cannon and Catell, "Studies in Experimental Traumatic Shock," 318, 320–21.
157. Ibid., 320–21.
158. For example, Gray, *Early Treatment of War Wounds*, 82; Dale, "Nature and Cause of Wound Shock," 32; and N. M. Keith, "Blood Volume Changes in Wound Shock and Primary Haemorrhage," in *Medical Research Committee Reports of the Special Investigation Committee on Surgical Shock and Allied Conditions: Traumatic Toxaemias a Factor in Shock, Special Report no. 26* (London: His Majesty's Stationary Office, 1919), 36–43.
159. *MTT*, passim. See also Dale, "Nature and Cause of Wound Shock," 37–38.
160. See Dale's notebooks from experiments in Hampstead during the war: "Experiments with Gas Gangrene and Its Toxins," PP.HHD.7, Wellcome Library. Dale's conviction that during shock toxins were released, perhaps from the wound, but in any case carried through the body (emphasized as inexplicable and wrong by William van der Kloot in "William Maddock Bayliss's Therapy for Wound Shock," 283) came from his studies of both histamine and gas gangrene. A further parallel between gas gangrene and shock, namely that gangrene required an anaerobic environment, which shock also produced, was emphasized by several authors. Chauvin and coauthors note that at the end of the war, the French surgeons Édouard Quénu and Pierre Delbet also came to believe that a toxemia resulting from the wound was responsible for shock

in much the same way as Dale understood; see Chauvin et al., "L'évolution de la chirurgie des plaies de guerre des membres en 1914–1918," 165; and Dale, "Nature and Cause of Wound Shock," 37.

161. Cowell, "Arris and Gale Lecture," 137.
162. Bayliss, *Intravenous Injection in Wound Shock*, 14.
163. Gray, *Early Treatment of War Wounds*, 82.
164. Ibid., 102.
165. Dale, "Nature and Cause of Wound Shock," 39.
166. Clifford Allchin Gill, "The Immediate Treatment of Extensive Wounds on Field Service," *The Lancet* 167, no. 4317 (May 26, 1906), 1467.
167. Leriche, *Treatment of Fractures*, 120.
168. Cowell, "Initiation of Wound Shock," in *MRC* 61n.
169. Cowell, "Front Line Application of Thomas's Splint," GC/116/1, Wellcome Library. On the standardization of the treatment of shock in 1917 and the "universal use of the Thomas splint by the casualty clearing stations and field ambulances of every army," see Bowlby, "A Sketch of the Growth of the Surgery of the Front in France," *BMJ* (August 2, 1919), 129.
170. Hallett, *Containing Trauma*. For hot-air bottles, see also *BMJ* (August 18, 1917), 224.
171. Bowlby and Wallace, "Development of British Surgery at the Front," 709.
172. N. M. Keith, "Blood Volume Changes in Wound Shock and Primary Haemorrhage," 13.
173. Wallace, "Introductory," in *MTT* 4.
174. Chauvin et al., in "L'évolution de la chirurgie des plaies de guerre des membres en 1914–1918," 165–66, write that transfusion was rarely used in the French army before 1917. They count only 192 "certain" cases of blood transfusion in the French army between July 1917 and November 1918. On the lack of blood typing in transfusions, see van der Kloot, "William Maddock Bayliss's Therapy for Wound Shock," 280.
175. *MSS*, September 10, 1917.
176. For the logic behind the use of colloids, see van der Kloot, "William Maddock Bayliss's Therapy for Wound Shock."
177. Bayliss, *Intravenous Injection in Wound Shock*, 137. Recall that Dale also considered this drop insignificant.
178. Wallace, "Introductory," *MTT* 6.
179. Bayliss to Cowell, October 25, 1917, RAMC/466/2/5, Wellcome Library; in a separate letter of February 17, 1918, Bayliss asks about the adequacy of sodium bicarbonate without gum, thereby indicating that he and the other researchers were not heavily prejudiced against trying out alternatives.
180. Cannon to Dale, October 2, 1918, Box 62.F 816. Dale, Cannon Papers.
181. See Bayliss, "Further Observations on the Results of Muscle Injury and Their Treatment," in *MTT* 149–51. Van der Kloot, "William Maddock Bayliss's Therapy for Wound Shock," 280–81.
182. Keith, "Blood Volume Changes in Wound Shock and Primary

Haemorrhage," 14. See also the discussion of reasons for its occasional failure in Macrae, "Prevention and Treatment of Wound Shock," 111–12.

183. Wallace, "Introductory," in *MTT* 7; Keith, "Blood Volume Changes in Wound Shock and Primary Haemorrhage," 9–13.

184. "Report Made by Dr. Carrel, March 27, 1919: Experimental Surgery," p. 3, 450/C232/2, Rockefeller Archive Center.

185. Peter C. English, *Shock, Physiological Surgery, and George Washington Crile* (Westport, CT: Greenwood, 1980), 205.

186. *MSS*, November 9, 1917, where Crile's theory was discussed at length and dismissed on several points, but especially for its focus on the vasomotor center and the claim that veins remained full of blood. See also Low, "Presidential Address on Surgery," 420, and Cannon's discussion of Crile's work, which includes his agreement on matters of anesthesia and critique of the treatment of shock, in Cannon, "Consideration of the Nature of Wound Shock," in *MRC* 70. Dale, "Nature and Cause of Wound Shock," 33.

187. Cannon, "Consideration of the Nature of Wound Shock," in MRC 69. "Porter has declared that shock occurs mainly in cases of fractured femur or multiple wounds of subcutaneous tissue. He ascribes shock to the effects of fat thus brought into the circulation—an inference for which there exists as yet very little evidence. In our experimental cases, there was no noteworthy amount of subcutaneous fat, and examination of the lungs after death revealed no indication of Pulmonary Emboli." Bayliss and Cannon, "Note on Muscle Injury in Relation to Shock," in *MTT* 19. Dale, in a letter to Cannon, also noted that in discussions at the Shock Committee, "we were unable ... to trace any logical connection between the phenomenon of fat embolism and the symptoms of wound shock." Dale to Cannon, October 4, 1918, box 62.F 816. Dale, Cannon Papers. Cannon's reply listed no fewer than six reasons that the incidence of fat in the lungs was not significant for shock. Cannon to Dale, October 10, 1918. *MSS*, December 7, 1917. This is not to say that Porter should be altogether dismissed: the *Index Medicus* of 1917, for example, cited no fewer than four of his articles, by comparison to one by Cannon and no major report by the Medical Research Committee. See Fielding H. Garrison and Frank J. Stockman, *Index Medicus: War Supplement* (Washington, DC: Carnegie, 1917), 136.

188. Yandell Henderson's "acapneic" theory of shock had been the main competitor to Crile's approach; it was put to the test by the Shock Committee and found to be generally right on the symptoms but wrong on their hierarchization. See *MSS*, November 9, 1917, for the discussion and critique of Henderson; Cannon's critique of Henderson (Cowell, "Initiation of Wound Shock," in *MRC* 69); Cowell's critique of Lockhart-Mummery's Hunterian lectures (in "Arris and Gale Lecture," 143); and Bayliss's opening of *Intravenous Injection in Wound Shock* (2–4), where he lists and rejects prewar theories, including Crile's and Henderson's. Dale, in "The Nature of Wound Shock," recalls that the committee tested Crile's theory against Henderson's, discarded Crile's, and found Henderson's ordering of the symptoms to be misguided (33). Dale's contrast of Crile's and Henderson's works is stylized to mark off the starting point of his and the committee's own successful research.

189. Anson Rabinbach, *The Human Motor* (New York: Basic Books, 1992).

190. A partial but detailed description of the history of shock is offered in Millham, "Brief History of Shock," 1026–37.

191. Henri-François LeDran, *Traité ou réflexions tirées de la pratique sur les playes d'armes à feu* (Paris: Osmont; 1740); Millham, "Brief History of Shock," 1027–28.

192. Lapointe, "Simples réflexions sur le choc et autres états de collapsus traumatique," *Paris médical* 29 (1918), 480–83. "Mitchell, Morehouse and Keen, in their report to the Surgeon General of the United States Army in 1864, described cases occurring after wounds of grave nature, in which the patient, immediately after being wounded, suffered from a state of depression which continued and which was marked by great weakness, feeble circulation, pallor, etc. This is immediate shock." Cannon, *Traumatic Shock*, 1923, 3, 17. See Cannon's quotation from a case reported by Hermann Fischer in 1870: Hermann Fischer, "Ueber den Shok," in *Sammlung klinischer Vortrage*, ed. Richard Volkmann (Leipzig: Breitkopf und Härtel, 1870), 10:70, in Cannon, *Traumatic Shock*, 1. Fischer's text was frequently cited in the literature on wound shock, including in Yandell Henderson and Marvin McRae Scarbrough "Acapnia and Shock," in *American Journal of Physiology* 26 (1910), 260–86.

193. During the Boer War, the surgeon Watson Cheyne defined traumatic shock much in the way wound shock is discussed here. Delaporte, *Les médecins dans la grande guerre*, 27.

194. Gill, "Immediate Treatment of Extensive Wounds on Field Service," 1467. Clifford Allchin Gillwas a lieutenant in the British army and a member of the Royal College of Surgeons and the Royal College of Physicians, stationed in Jullundur, India. Occasional reports of shock and psychiatric effects on soldiers also came from the Balkan Wars.

195. Morris, *Practical Treatise on Shock*, 30.

196. Ibid., 42.

197. Ibid., 13, 17.

198. Crile, *An Experimental Research into Surgical Shock* (Philadelphia: Lippincott, 1899).

199. George James Guthrie, *Treatise on Gun-Shot Wounds . . .* (London: Burgess and Hill, 1820); John Hennen, *Principles of Military Surgery* (1818; Philadelphia: Carey & Lea. 1830), quoted in Morris, *Practical Treatise on Shock*, 2–3.

200. Ibid., 16.

201. This is not to suggest that Morris cast too wide a net in terms of bodywide diagnosis—identifying misdiagnosis was a priority. Indeed, he echoed his contemporaries and prefigured a number of World War I–era anxieties when he stated that syphilis "may closely resemble a spinal concussion," but reminded his readers that "it must not be forgotten that a patient with syphilis may also sustain an injury of the spine." The concern about misdiagnosing syphilis as (shell) shock was shared by Southard and Myers nearly half a century later. Morris, *Practical Treatise on Shock*, 261.

202. John Eric Erichsen, *On Concussion of the Spine, Nervous Shock, and Other Obscure Injuries of the Nervous System, in Their Clinical and Medico-Legal Aspects*, 2nd ed. (1866; New York: Bermingham & Co., 1882).

203. Morris, *Practical Treatise on Shock*, 19. Morris also attends to malingering or imposture, and what he describes as "the rise of compensation for alleged cases."

204. See his citation of John Eric Erichsen, "Mr. Erichson's Work 'Railway Spine and Other Injuries of the Nervous System," *BMJ* 2 (1866), 612.

205. Morris, *Practical Treatise on Shock*, 69–70.

206. Letter from F. H. Pike to Crile, February 1, 1916, Crile Papers 20:61.

207. Myers, *Shell Shock in France, 1914–1918* (New York: Cambridge University Press, 2012), 25.

208. Dale, "Nature and Cause of Wound Shock," 40 and 29.

209. Cannon, *Way of an Investigator*, 130. But see also his publications until 1923 or so, including *Traumatic Shock*, and Cannon and McKeen Cattell, "Studies in Experimental Traumatic Shock," *Archives of Surgery* 4, no. 2 (1922), 1–22. In a letter to Edward Sharpey Schafer dated March 1, 1929 (PP/ESS/A.6/7, Wellcome Library), Cannon noted: "When I came home in 1919 I had two jobs on my hands—one, the completion of the work on shock begun during the war and the other the controversy with Stewart and Rogoff. . . . In consequence of these two intrusive interests I did not get back to an interest in the thyroid until about two years ago."

210. Dale, "Nature and Cause of Wound Shock," 35. Dale was not the first to speak before the Harvey Society on shock; Porter did so on October 27, 1917. See *Harvey Lectures 1917–1919*, ser. 13–14 (Philadelphia: Lippincott, 1919), 21–43. Among the society's officers in 1919, when Dale lectured, ranked Pike as treasurer (9).

211. *CWB* 39.

212. George Gould, *The Practitioner's Medical Dictionary* (Philadelphia: Blakiston's Son, 1910), 1035.

213. Thomas L. Stedman, *A Practical Medical Dictionary*, 6th ed. (New York: W. Wood, 1920), 1119.

214. Whitla, *Dictionary of Treatment*, 1045–50.

215. Anthony Bowlby, "Wounds in War," *The Lancet* (December 25, 1915), 1385.

216. Gould's, Stedman's, and Whitla's dictionaries do not mention surgical or wound shock.

217. Fredrick W. Mott, *War Neuroses and Shell Shock* (London: Oxford University Press, 1919).

218. But see Young, *Harmony of Illusions*, 51–52.

219. Tracey Loughran, "Shell Shock, Trauma, and the First World War," *Journal of the History of Medicine and Allied Sciences* 67, no. 1 (2010), 112.

220. Crile, *Mechanistic View of War and Peace*, 27.

221. Crile diary, 2:535 (February 6, 1915), Crile Papers 64:254.

222. Cannon to J. B. Cleland, January 8, 1935, box 67.F.901.AUS, Cannon Papers.

223. F. W. Mott, "The Microscopic Examination of the Brains of Two Men

Dead of *Commotio cerebri* (Shell Shock) without Visible External Injury," *Journal of the Royal Medical Corps* 29 (1917), 662–77; Mott, "Discussion of Shell Shock without Visible Signs of Injury," *Proceedings of the Royal Society of Medicine London* 9 (1915–1916), 1–44; Mott, "Punctiform Hemorrhages of the Brain in Gas Poisoning," *BMJ* 1 (1917), 637; Mott, "The Lettsomian Lectures on the Effects of High Explosives upon the Central Nervous System," *The Lancet* 1 (1916), 190, 331–38.

224. Mott, *War Neuroses and Shell-Shock* (London: Frowde, Hodder & Stoughton, 1919), 4–5.

225. On sound and injury, see J. Gordon Wilson, "Effects of High Explosives on the Ear," *BMJ* (May 5, 1917), 353–55.

226. James to Carrel, March 16, 1916, RU450:C232:2 (1906–1916), in Carrel Correspondence; Cannon and Catell, "Studies in Experimental Traumatic Shock," 318, 320–21.

227. Mott, *War Neuroses and Shell-Shock*, 16–22.

228. Mott, "Changes in the Central Nervous System Occurring in Various Forms of Shock," in *MTT* 44–47.

229. Mott, "Microscopic Examination of the Brain in Cases of 'Surgical Shock,'" 25–40.

230. Myers, "Hearing," in *RTS* 2:141–68.

231. Myers, *Shell Shock in France, 1914–1918*, 11, 12, 13, 16, 28, 30, and passim.

232. Ibid., 13.

233. Ibid., 34–35; see also his unstable distinction of shell shock from shell concussion on 29, 31.

CHAPTER THREE

1. "Neurasthenia and Shell Shock," *The Lancet* (March 18, 1916), 627.

2. In his excellent study of hospital case files, "The Case of the Archive," *Critical Inquiry* 39, no. 3 (2013), 532–47, Warwick Anderson contrasts case histories, which he identifies as the short, fragmented "illness trajectories" collected by physicians and functioning as administrative case files, to case studies, which he understands as modernist works that are "obviously distinct" (536) in their conscious construction. This seems to us correct, if imagined in a contrast of Freud's case studies to hospital case files. Yet the binary is hardly satisfactory when one considers that both extremes—narration, synthesis, and conscious reconstruction of a trajectory on the one hand, and fragmentation and bureaucracy on the other—play a role, including in World War I–era accounts. How one would distinguish raw data from the final account is not often clear; one might also wonder to what degree case histories, in the fragmentation, standardization, and pedagogical exemplarity that Anderson attributes to them, really escape modernist bias (as he indicates on p. 540). We use the two terms *case study* and *case history* interchangeably, and respectively suggest that the difference portrayed by Anderson does exist but should not be seen as a given qualitative contrast.

3. Charles S. Myers, "The Revival of Emotional Memories and Its Therapeutic Value (II)," *British Journal of Medical Psychology* 1 (1920), 20–22;

William McDougall, "The Revival of Emotional Memories and Its Therapeutic Value (III)," *British Journal of Medical Psychology* 1 (1920), 23–29; Stefanie C. Linden, Edgar Jones, and Andrew J. Lees, "Shell Shock at Queen Square: Lewis Yealland 100 Years On," *Brain* 136, no. 6 (2013), 1976–88; Edgar Jones, "Shell Shock at Maghull and the Maudsley," *Journal of the History of Medicine and Allied Sciences* 65 (2010), 368–95, 376.

4. Emily Martin, "Toward an Ethnography of Experimental Psychology," in *Plasticity and Pathology: On the Formation of the Neural Subject*, ed. David Bates and Nima Bassiri (New York: Fordham University Press, 2015), 12.

5. Myers, *Shell Shock in France, 1914–1918* (New York: Cambridge University Press, 1940), 11.

6. Ibid, 11.

7. Ibid, 12–13.

8. Ibid., 13.

9. Ibid.

10. Ibid.

11. Ibid. Indeed, for the possibility of physical effects from shock vibrations, see Myers, "A Final Contribution to the Study of Shell Shock," *The Lancet* 193, no. 4976 (January 11, 1919), 51.

12. Myers, *Shell Shock in France*, 26.

13. Ibid., 26.

14. Ibid., 96.

15. Ibid., 40–41.

16. The late John Forrester's magisterial account of cases in psychoanalysis and beyond, *Thinking in Cases* (London: Polity, 2016), was published too late for us to fully take it into account here.

17. We are also not concerned with the case study as a mode of sociological and historical inquiry. For that approach see Charles Ragin, "Cases of 'What Is a Case?,'" in *What Is a Case?*, ed. Charles Ragin and Howard Becker (Cambridge: Cambridge University Press, 1992), 9.

18. A chapter of cases accompanies each of the seven books of the *Epidemics*. See Hippocrates, *Epidemics* 2.4–7, ed. Wesley D. Smith (Cambridge, MA: Loeb, 1994).

19. On Aristotle and infinite variation, see Angela Creager, Elisabeth Lunbeck, and M. Norton Wise, introduction to *Science without Laws*, ed. Angela Creager, Elisabeth Lunbeck, and M. Norton Wise (Durham, NC: Duke University Press, 2007), 13.

20. Owsei Temkin, *The Double Face of Janus* (Baltimore: Johns Hopkins University Press, 1977), 441–55; see Julia Epstein's detailed study of the narrative and textual aspects of the case, *Altered Conditions: Disease, Medicine, and Storytelling* (London: Routledge, 1994), 37.

21. Philippe Huneman, "Writing the Case," *Republics of Letters* 3, no. 2 (January 2014), 2–3.

22. Ibid.

23. Recall Breuer and Freud's *Studies on Hysteria* (1895, see *FSE* vol. 2) and Freud's *Dora* (1901–1905, see FSE vol. 7). On the case in French psy-

chiatry, see Jan Goldstein, *Hysteria Complicated by Ecstasy* (Princeton, NJ: Princeton University Press, 2011), 3.

24. Steven Shapin, *A Social History of Truth* (Chicago: University of Chicago Press, 1995), ch. 5.

25. Walter B. Cannon, "The Case System in Medicine," *Boston Medical and Surgical Journal* 142 (May 31, 1900), 564.

26. Ibid., 567, 578–79.

27. Cannon, *The Way of an Investigator* (New York: W. W. Norton and Co., 1945), 85.

28. Christopher Lawrence, "Incommunicable Knowledge: Science, Technology and the Clinical Art in Britain, 1850–1914," *Journal of Contemporary History* 20 (1985), 503–20.

29. Arturo Castiglioni, "Neo-Hippocratic Tendency of Contemporary Medical Thought," *Medical Life* 41 (1934), 120–21.

30. CNP 154–62.

31. Ruth Leys, "Types of One," *Representations*, 34 (1991), 10–11.

32. Jean-Luc Nancy, *Corpus* (New York: Fordham University Press, 2008), 53.

33. Cannon, "Case System in Medicine," 563–64. See also Creager, Lunbeck, and Wise, introduction to *Science without Laws*, 13.

34. Fredrick Schauer and Walter Sinnott-Armstrong, *Philosophy of Law* (New York: Oxford University Press, 1995).

35. Cannon, "Case System in Medicine," *Boston Medical and Surgical Journal* (May 31, 1900), 4.

36. Roy Porter, "The Patient's View," *Theory and Society* 14, no. 2 (March 1985); Porter identifies case histories with the doctor's point of view (182) but also allows for their use in "resurrecting the anxieties and tribulations of rural communities" (183–84).

37. L. Stephen Jacyna and Stephen T. Casper, introduction to *The Neurological Patient in History*, ed. L. Stephen Jacyna and Stephen T. Casper (Rochester, NY: University of Rochester Press, 2012), 4–6.

38. Michel Wieviorka, "Case Studies," in *What Is a Case?*, ed. Charles Ragin and Howard Becker (Cambridge: Cambridge University Press, 1992), 159.

39. Creager, Lunbeck, and Wise, introduction to *Science without Laws*, 13, citing John Forrester, "If p, Then What?" *History of the Human Sciences* 9, no. 3 (1996), 1–35.

40. Kathleen Frederickson, *The Ploy of Instinct* (New York: Fordham University Press, 2014), 87.

41. On the disruptiveness and promise of case studies (in the context of bioethics), see Marta Spranzi, "The Normative Relevance of Cases," *Cambridge Quarterly of Healthcare Ethics* (2012), 482, 490.

42. Lauren Berlant, "On the Case," *Critical Inquiry* 33, no. 4 (2007), 664; Frederickson, *Ploy of Instinct*, 66.

43. Carlo Ginzburg, "Clues," in *Clues, Myths and the Historical Method* (Baltimore: Johns Hopkins University Press, 1989), 106.

44. Julia Epstein, *Altered Conditions* (London: Routledge, 1995), 26, 35–36.

45. Michel Foucault, *Discipline and Punish* (New York: Vintage, 1995), 191.

46. Frederickson, *Ploy of Instinct*, 87, citing Forrester, "If *p*, Then What?" Anderson relies on this contrast too.

47. Arthur F. Hurst, *A Twentieth-Century Physician* (London: Edward Arnold, 1949); see Edgar Jones's detailed study, "*War Neuroses* and Arthur Hurst: A Pioneering Medical Film about the Treatment of Psychiatric Battle Casualties," *Journal of the History of Medicine and Allied Sciences* 67, no. 3 (2012), 345–73.

48. Arthur F. Hurst and J. L. Symns, "The Rapid Cure of Hysterical Symptoms in Soldiers," *The Lancet* 192, no. 4953 (August 3, 1918), 139–41; Hurst, "An Address on Hysteria in the Light of the Experience of War," *The Lancet* 194, no. 5018 (November 1, 1919), 771–75; Hurst, "Cinematograph Demonstration of War Neuroses," *Proceedings of the Royal Society of Medicine* (Neurology Section) 11 (1918), 39–42; Hurst, *Medical Diseases of War* (London: Arnold, 1918); and Hurst, "Observations on the Etiology and Treatment of War Neuroses," *BMJ* 2, no. 2961 (September 29, 1917), 409–14.

49. In 1906 Hurst traveled to the United States, where he met with Cannon and observed Cannon's experiments on the mechanisms of digestion using X-rays. He later wrote of his great indebtedness to Cannon, yet there seems to have been little or no correspondence between them during the war regarding shock or war neuroses. See Arthur F. Hurst, *Selected Writings of Sir Arthur Hurst (1879–1944)* (London: British Society of Gastroenterology, 1969).

50. Jones, "*War Neuroses* and Arthur Hurst."

51. Thomas Hunt, "History of the British Gastroenterology Society," *Gut* 1, no. 3 (1960), 2–5; Thomas Hunt, "Sir Arthur Hurst," *Gut* 20 (1979), 463–66.

52. G. Schaltenbrand, "Max Nonne," in *Grosse Nervenärzt*, ed. Kurt Kolle (Stuttgart: Thieme 1963), 3:164–73.

53. Arthur F. Hurst, "The Etiology and Treatment of War Neurosis," in Hurst, *Selected Writings*, 178–79, originally published in *BMJ* 2 (1917), 409–13.

54. Ibid., 180–81.

55. Ibid., 182.

56. Ibid., 185, 187.

57. Ibid., 185.

58. Jones, "War Neuroses and Arthur Hurst," 364–65.

59. E. E. Southard, "Syphilis and the Psychopathic Hospital," *Boston Medical and Surgical Journal* 174 (1916), 50–53. In many ways the continued dominance of syphilis in a larger arena of diagnostics and comorbid conditions during the war is no surprise. Only a few years prior, the Wassermann diagnostic assay and effective treatments for syphilis with Salvarsan (arsphenamine), thanks to the efforts of Paul Ehrlich, had been widely adopted. E. E. Southard and H. C. Solomon, *Neurosyphilis* (Boston: W. M. Leonard, 1917).

60. E. E. Southard, *Shell-Shock and Other Neuropsychiatric Problems, Presented in Five Hundred and Eighty-Nine Case Histories, from the War Literature, 1914–1918* (Boston: W. M. Leonard, 1919).

61. Ibid., 854.
62. Ibid., 831, 834.
63. Ibid., 855, italics in original.
64. Ibid., 834–35.
65. Southard and L. J. Henderson were invested in the regulation and standardization of medical school curricula before the war around the same time as Abraham Flexner was reporting on American and Canadian medical education. See Southard and Henderson, "Education in Medicine," *Harvard Bulletin* 12 (December 15, 1909); and Abraham Flexner, *Medical Education in the United States and Canada* (New York: Carnegie Foundation, 1910). Henderson and Southard are not cited in the Flexner Report.
66. Southard, *Shell-Shock and Other Neuropsychiatric Problems*, 835.
67. Ibid., 839.
68. Ibid., 848.
69. Ibid., ii, italics in original.
70. Ibid., preface, i–iii.
71. Ibid., 854, 849.
72. E. E Southard and Mary C. Jarrett, *The Kingdom of Evils* (New York: MacMillan, 1922).
73. Ibid., xi, ix, italics in original.
74. Ilana Löwy, in her study of cancer treatments, shows how these sites of laboratory and clinic are at odds not simply because they are opposites, but rather because they (their practitioners, their concepts) attempt to occupy the same space, object of intervention, practice, and framework. See Löwy, *Between Bench and Bedside* (Cambridge, MA: Harvard University Press, 1996).
75. W. H. R. Rivers, *Psychology and Ethnology* (New York: Harcourt, 1926), 61.
76. While "psychophysics" is the recognized basis of Rivers's work, he was clearly familiar with Gestalt theory, as suggested by the visual experiment images in his account. See Rivers, "Introduction," in *RTS* 2:128–29 and elsewhere.
77. W. H. R. Rivers, introduction to *Reports of the Cambridge Anthropological Expedition to Torres Straits*, by Alfred Cort Haddon, W. H. R. Rivers, C. G. Seligman, Charles S. Myers, William McDougall, Sidney Herbert Ray, and Anthony Wilkin (Cambridge: Cambridge University Press, 1901–1935), 2:1–2.
78. Martin, "Toward an Ethnography of Experimental Psychology," 12.
79. Charles S. Myers, "The Influence of W. H. R. Rivers," in Rivers, *Psychology and Politics, and Other Essays*, 165.
80. Partly as a result of Pat Barker's *Regeneration* trilogy, the attention on Rivers and Sassoon in the secondary literature has been extensive. Richard Slobodin, *W. H. R. Rivers* (New York: Columbia University Press, 1978); Carole Shelton, "War Protest, Heroism, and Shellshock," *Focus on Robert Graves and His Contemporaries* 1 (1992), 43–50; Karolyn Steffens, "Communicating Trauma," *Journal of Modern Literature* 37, no. 3 (2014), 36–55; Paul Fussell, *The Great War and Modern Memory* (Oxford: Oxford University Press, 1975), 101 and passim.

81. Rivers, "On the Repression of War Experience," *The Lancet* 191, no. 4927 (February 2, 1918), 173.
82. Rivers first discusses this instinct, as more fundamental than any other, in *RIU* 5.
83. Frederic Bartlett, "Obituary notice for W. H. R. Rivers," *The Eagle* (St. John's College, Cambridge, 1922), 12–14.
84. Rivers, "Repression of War Experience," in *RIU* 173.
85. Ibid., 175–76.
86. Ibid.
87. Rivers, "Psychiatry and the War," *Science* 49, no. 1268 (April 18, 1919), 367.
88. On illustrative value, see Rivers's preface to John T. MacCurdy, *War Neuroses* (Cambridge: Cambridge University Press, 1918), v.
89. Rivers, "Repression of War Experience," in *RIU* 173. Rivers was referring to Hurst, among others, when he claimed that "the patients are instructed to lead their thoughts to other topics, to beautiful scenery and other pleasant aspects of experience."
90. Ibid., 177.
91. See, notably, ibid., 176.
92. Ibid., 175.
93. Leriche, *Des résections de l'estomac pour cancer* (Lyon: Storck, 1906).
94. *LSP* 239; in her excellent study *The Story of Pain* (Oxford: Oxford University Press, 2014), Joanna Bourke places Leriche within the longer arc of pain research inside and outside of the laboratory: for Leriche, "pain was more than tissue damage. It was intrinsically affected by interactions with other people and the environment" (*LSP* 228). Unlike experimental pain, it occurred, in Leriche's words, as "a continuous phenomenon" in the everyday.
95. Leriche, *De la méthode de recherche dans la Chirurgie* (Paris: Masson, 1933).
96. *LSP* 22.
97. Interestingly, few individual case studies are actually described in the book, outside a short section on gunshot wounds and their treatment. Unlike Southard's aim, Leriche's goal was not to describe and aggregate cases but to build from them. *LSP* 190–91.
98. *LSP* translator's note, viii.
99. Ibid., 6.
100. Ibid., 3, italics in original.
101. Ibid., 27.
102. Ibid., 28.
103. Ibid., 29.
104. Ibid., 30.
105. Ibid., 7–8.
106. René Leriche, *The Treatment of Fractures* (London: University of London Press, 1918), 1:xxi.
107. Ibid., xx.
108. *LSP* 24.
109. Ibid., 25, italics in original.

110. Ibid., 178.
111. Ibid., 181, italics in original.
112. Joseph M. Hone, *The Life of Henry Tonks* (London: Heinemann, 1939); Joanna Bourke, *Dismembering the Male* (London: Reaktion, 1996); Sander Gilman, *Making the Body Beautiful* (Princeton, NJ: Princeton University Press, 1999); Harold Gillies, *Plastic Surgery of the Face* (London: Henry Frowde, 1920); Emma Chambers, *Henry Tonks* (London: University College London, 2002); Suzannah Biernoff, "Flesh Poems and the Art of Surgery," in *Visual Culture of Britain* 11 (2010), 24–47; Chambers, "Fragmented Identities," *Art History* 32 (2009), 578–607; François Delaporte, Emmanuel Fournier, and Bernard Devauchelle, *La fabrique du visage* (Turnhout: Brepols, 2010).
113. Frederic Manning, *Her Privates We* (London: Davies, 1930).
114. Roselyne Rey, *René Leriche (1879-1955): Une oeuvre controversée* (Paris: Editions CNRS, 1994).
115. Kurt Goldstein, *After-Effects of Brain Injuries in War* (New York: Grune & Stratton, 1942), 13. Goldstein argues for the need for long-term, close observation in *TO* 40–41.
116. Kurt Goldstein, *Die Behandlung, Fürsorge und Begutachtung der Hirnverletzten* (Leipzig: Vogel 1919); and Goldstein, *Über Aphasie* (Leipzig: Orell Füssli, 1927). See also the editors' introduction by Thomas Hoffmann and Frank Stahnisch to Kurt Goldstein, *Der Aufbau des Organismus* (Paderborn: Wilhelm Fink, 2014), xxi–xlvi.
117. *TO* 42, 40–41.
118. Ibid., 41. On the imperative of individualized longer-term care, see his *Die Behandlung, Fürsorge und Begutachtung der Hirnverletzten*, 211–12.
119. *TO* 34–35; Goldstein, *Selected Writings* (The Hague: Nijhoff, 1971), 155–56.
120. *TO* 41–42.
121. *TO* 34, 36.
122. Ibid., 28.
123. Ibid., 42, 46.
124. Gelb and Goldstein, "Zur Psychologie des optischen Wahrnehmungs- und Erkennungsvorganges," in *Psychologische Analysen hirnpathologischer Fälle* (Leipzig: Verlag von Johann Ambrosius Barth, 1919), 9.
125. Ernst Cassirer, *Philosophy of Symbolic Forms* (New Haven, CT: Yale University Press, 1957), vol. 3, ch. 6; Maurice Merleau-Ponty, *Phenomenology of Perception* (London: Routledge, 1962); and Anne Harrington, *Reenchanted Science* (Princeton, NJ: Princeton University Press, 1999).
126. Gelb and Goldstein, "Zur Psychologie des optischen Wahrnehmungs- und Erkennungsvorganges," 18, translation in *Source Book of Gestalt Psychology*, ed. W. D. Ellis (London: Routledge, 1938), 318–19.
127. *TO* 39, 43–44.
128. Goldstein brought five films with him when he emigrated from Germany, for the purposes of illustrating lectures and convincing American institutions and learned societies of the significance of his work. See Robert Lambert to Collector of Customs, October 15, 1934, RF/1.1/200A/78/939, Rockefeller Archive Center.

129. On pointing and past pointing, see Goldstein, "Über Zeigen und Greifen," *Der Nervenarzt* 4 (1931) 453–66, and *TO* 127, 177, 279, 335.

130. A longer description of Goldstein's film appears in our German-language book *Experimente im individuum, Kurt Goldstein und die Frage des Organismus*, translated by Nils F. Schott and Holger Wölfle (Berlin: August Verlag, 2014), 56–60.

131. *TO* 272–73.

132. *HNP* 120.

133. Ibid., 170–72.

134. *TO* 89, 128.

135. Gelb and Goldstein, "Über den Einfluss des vollständigen Verlustes des optischen Vortstellungsvermögens auf das tactile Erkennen," in *Psychologische Analysen hirn-pathologischer Fälle*, 157–250.

136. *CNP* 197.

137. Ibid, 39.

138. Bourke, *Dismembering the Male*; and Emily Mayhew, *Wounded* (London: Oxford University Press, 2014).

CHAPTER FOUR

1. Robert Graves, *Good-Bye to All That* (London: Anchor, 1958), 114.

2. Ernst Jünger, *Storm of Steel* (London: Penguin, 2007) 211, 251.

3. Paul Fussell has discussed the theatricality in Graves's writing in *The Great War and Modern Memory* (London: Oxford University Press, 1975), 203.

4. Lashley cites Goldstein and Head in, among other places, his crucial article "Basic Neural Mechanisms in Behavior," *Psychological Review* 37 (1930), 1–24. Lashley also wrote the foreword for the translation of Goldstein's *The Organism* (New York: American Book Company, 1939), v–vi. Luria cites Goldstein, Head, and Lashley in *NHC*.

5. *AKD* 1:140.

6. Ibid., 1:63.

7. Mitchell Ash, *Gestalt Psychology in German Culture, 1890–1967* (Cambridge: Cambridge University Press, 1998).

8. Cf. Canguilhem's different interpretation of the relationship of the Head and Goldstein: *CNP* 86.

9. *IAN* 7, italics in original.

10. Ibid., 313.

11. Ibid., 352.

12. See the biography by L. Stephen Jacyna, *Medicine and Modernism* (London: Pickering, 2008).

13. Goldstein either did not possess significant correspondence with Head or opted to leave it behind when he was forced to emigrate in 1933.

14. *AKD* 1:141 and 1:129–33. He cites Gelb and Goldstein, "Zur Psychologie des optischen Wahrnehmungs- und Erkennungsvorganges," *Psychologische Analysen hirnpathologischer Fälle* (Frankfurt, 1918); Gelb and Goldstein, "Über Farbenamnesie," in *Psychologische Forschung* 6 (1924), 127–86.

15. Goldstein to Head, July 23, 1926, PP/Hea/D8, Wellcome Library.

16. Cassirer to Goldstein, January 5, 1925, in Cassirer, "Two Letters to Kurt Goldstein," *Science in Context* 12 (1999), 663.

17. Undated lecture from 1950s, p. 5, box 3, Aphasia folder, Kurt Goldstein Papers, Rare Book and Manuscript Library, Columbia University.

18. Jacyna, *Medicine and Modernism*.

19. *AKD*, 1:viii. Head notes that he first encountered aphasia at Addenbrooke's Hospital at Cambridge in 1886 (a decade before he moved to London Hospital).

20. Goldstein's earliest essays on aphasia are a 1906 case study and a commentary on François Moutier's commentary on Broca's area: Goldstein, "Zur Frage der amnestischen Aphasie und ihrer Abgrenzung gegenüber der transcorticalen und glossopsychischen Aphasie," *Archiv für Psychiatrie und Nervenkrankheiten* 41, no. 3 (1906), 911–50; Goldstein, "Einige Bemerkungen über Aphasie im Anschluss an Moutier's 'L'aphasie de Broca,'" *Archiv für Psychiatrie und Nervenkrankheiten* 45, no. 1 (1908), 408–40.

21. Institut für die Erforschung der Folgeerscheinungen von Hirnverletzungen. See Frank W. Stahnisch and Thomas Hoffmann, "Kurt Goldstein and the Neurology of Movement during the Interwar Years," in *Was bewegt uns?*, ed. Christian Hoffstadt, Franz Peschke, Andreas Schulz-Buchta, and Michael Nagenborg (Bochum, Germany: Projekt, 2010), 288.

22. On Rivers's importance to the establishment of psychology in Britain, see John Forrester, "1919: Psychology and Psychoanalysis," *Psychoanalysis and History* 10, no. 1 (2008), 37–94; see Rivers's own articles "Psychiatry and the War," *Science* 49, no. 1268 (April 18, 1919), 367–69; "The Repression of War Experience," in *The Lancet* (February 1918), 513–33. For Rivers's part in the Torres Straits expedition, see *RTS* vols. 2, 5, 6.

23. *New York Evening Post*, September 8, 1912.

24. In the 1950s, Goldstein stepped back somewhat: "Head's utterances overstated the criticism." At box 3, Aphasia folder, p. 5, Goldstein Papers. It is not clear that Goldstein in his earlier years was much warmer toward his contemporaries than Head was.

25. There are significant exceptions to this: both Head and Goldstein praise Constantin von Monakow and Arnold Pick, in addition to Jackson, although Head oriented his argument toward criticizing them for what they did not manage.

26. Head republished Jackson's principal essays on aphasia in 1915, in *Brain* 38, nos. 1–2 (1915), 59–64. Note his admiration for Jackson in his obituary in *Nature* 87, no. 2190 (October 19, 1911), 524; and his (and Gordon Holmes's) 1911 Croonian lecture, published in Head, *Studies in Neurology* (London: Hodder & Stoughton, 1920), 2:533.

27. Head criticized the Spencerian trend in "Hughlings Jackson on Aphasia and Kindred Affections of Speech," *Brain* 38, nos. 1–2 (1915), 1–27; Allan Young, *The Harmony of Illusions* (Princeton, NJ: Princeton University Press, 1995), 47. Emily Martin has noted the anti-Spencerian effect of the Torres Strait expedition in her "Toward an Ethnography of Experimental Psychology," *Plasticity and Pathology: On the Formation of the Neural Subject*,

ed. David Bates and Nima Bassiri (New York: Fordham University Press, 2015), 12.

28. Thomas Kuhn, "Energy Conservation as an Example of Simultaneous Discovery," in Kuhn, *The Essential Tension* (Chicago: University of Chicago Press, 1977), 66–104.

29. Henry Head, "Hughlings Jackson on Aphasia and Kindred Affections of Speech," 1.

30. *AKD* 1:133.

31. Ibid., 1:x.

32. Ibid., 1:134 and ix.

33. Ibid., 1:133.

34. Given his polemical predilection, Head is remarkably kind to Gelb and Goldstein and does not target their use of terms and notions (such as word blindness) whose use by others (*AKD* 1:117) he strenuously disparages.

35. Head warned that even these case details were "drastically reduced versions of voluminous clinical records" (*AKD* 2:vii). The first seventeen cases, placed first to prioritize wartime injury, were all of men who came under Head's care during the war, suffering from brain injury due to gunshots or penetration by other fragments; the remaining nine cases, of seven men and two women, were almost all patients aged fifty-five to sixty-five, and dated with only one exception to the postwar period. The exception is case 21, dating to 1910; the patient of case 25 also visited Head before the war, but he attended to her, he said, with much greater care in 1920.

36. *AKD* quoted by Goldstein, box 3, Aphasia folder, p. 5, Goldstein Papers.

37. Ibid., 1:138.

38. See his refutation of approaches he deemed prevalent, in *AKD* 1:202.

39. Ibid., 1:218.

40. Ibid., 1:210–18 and passim.

41. Ibid., 1:203, emphasis ours.

42. Ibid., 1:323, 269.

43. Ibid., 1:220.

44. Ibid., 1:289, emphasis ours.

45. Ibid., 1:269.

46. Ibid., 1:270.

47. Ibid., 1:535.

48. Ibid., 1:166.

49. H. Charlton Bastian, *A Treatise on Aphasia and Other Speech Defects* (London: Lewis, 1898).

50. *AKD* 1:220. First discussed in Head, "Aphasia and Kindred Disorders of Speech," in *Brain* 43, no. 2 (July 1920), 119–57, and Head, "Disorders of Symbolic Thinking and Expression," *British Journal of Psychology* 11, no. 2 (January 1921), 179–93.

51. *AKD* 1:229.

52. Ibid., 1:239.

53. Ibid., 1:241.

54. Ibid., 1:257; see also ibid., 1:261, 268.

55. See the discussion of "the grammarians" in ibid., 1:120–21.
56. *AKD* 1:122.
57. Ibid., 2:xvi–xvii.
58. Ibid., 1:140–41. See Jacyna, *Lost Words: Narratives of Language and the Brain 1825–1926* (Princeton, NJ: Princeton University Press, 2000), 152.
59. On Head's approach to his experiments on nerve division (including later ones), see the critiques by E. G. Boring, "Cutaneous Sensation after Nerve Division," *Quarterly Journal of Experimental Physiology* 10 (1916), 1–95; and R. A. Hensen, "Henry Head," *British Medical Bulletin* 33 (1977), 91–96.
60. Head and W. H. R. Rivers, "A Human Experiment in Nerve Division," *Brain* 31 (1908), 323–450, republished in Head, *Studies in Neurology*, 1:225–328.
61. Head, "The Conception of Nervous and Mental Energy II: Vigilance," in *British Journal of Psychology* 14, no. 2 (October 1923), 133; *AKD* 1:486.
62. Ibid., 134–35; *AKD* 1:487.
63. *AKD* 1:539.
64. Head, "Conception of Nervous and Mental Energy II," 137; *AKD* 1:490.
65. *AKD* 1:535, 533.
66. Ibid., 1:538.
67. Goldstein's *Über Aphasie* is based on a lecture given in Bern on February 27, 1926. Goldstein wrote to Head about the latter's *Aphasia and Kindred Disorders of Speech* only that July (according to Jacyna, *Medicine and Modernism*, 143, it was published by March). It is unclear whether Head's work played a role in Goldstein's argument.
68. *TO* 203–7.
69. Ibid., ch. 2, esp. 70, 74, 76, 78.
70. Ibid., 69. Head says something quite similar, but not in critique of Sherrington. See Head, "Conception of Nervous and Mental Energy II," 139; *AKD* 491–92.
71. For the critique of reductionist physiology, see *TO* 108–9, 261.
72. Ibid., 336.
73. Ibid., 48.
74. Ibid., 54.
75. Ibid., 43–44.
76. Ibid., 301.
77. Goldstein and Martin Scheerer, *Abstract and Concrete Behavior*, *Psychological Monographs* 53, no. 2 (1941), 2, 3.
78. *TO* 45.
79. Ibid., 54–55. For the implications of this position, see also *CKL* 132.
80. Ibid., 29. See also *CNP* 184 and passim.
81. Ibid., 80–81, 177–78. L. Halpern, "Studies on the Inductive Influence of Head Posture," in *The Reach of the Mind*, by Marianne Simmel (New York: Springer, 1968), 75; Raoul Mourgue, "La conception de la neurologie dans l'oeuvre de Kurt Goldstein," *L'encéphale* 32, no. 1 (1937), 38–39.
82. See Kurt Goldstein, "Über den Einfluss der Motorik auf die Psyche," *Klinische Wochenschrift* 1924, no. 3, 1255–60; "Über induzierte

Tonus-Veränderungen beim Menschen. II," *Zeitschrift für die gesamte Neurologie und Psychiatrie* 89 (1924), 383–428; Goldstein and Walther Riese, "Über induzierte Veränderungen des Tonus. V," *Monatsschrift für Ohrenheilkunde* 58 (1924), 931–40; "Zum Problem der Tendenz zum ausgezeichneten Verhalten," *Deutsche Zeitschrift für Nervenheilkunde* 109 (1929), 1–61.

83. Goldstein's commitment to understanding systems of organismic functioning also distinguished his holistic method from others that focused on Gestalt psychology's figure-ground relationships to develop a theory of perception and cognition. Wolfgang Köhler, *The Place of Value in a World of Facts* (New York: Liveright, 1934); and Kurt Koffka, *Principles of Gestalt Psychology* (New York: Harcourt, Brace & World, 1935).

84. See also another surviving film, *Diegelman*, discussed in our *Experimente im individuum, Kurt Goldstein und die Frage des Organismus*, trans. Nils F. Schott and Holger Wölfle (Berlin: August Verlag, 2014), 79–83.

85. Isaac H. Jones and Lewis Fisher, *Equilibrium and Vertigo* (Philadelphia: Lippincott, 1918), 221, 218.

86. Ibid., 227.

87. Ibid., 225.

88. Goldstein, "Das Wesen der amnestischen Aphasie," *Schweizer Archiv für Neurologie und Psychiatrie* 15 (1924), 163–75.

89. Koffka, *Principles of Gestalt Psychology*, 50–52.

90. Goldstein, "Notes on the Development of My Concepts" (1959), in *Selected Papers* (The Hague: Nijhoff, 1971), 5–14.

91. *TO* 334–35.

92. Ernst Cassirer, *The Philosophy of Symbolic Forms* (New Haven, CT: Yale University Press, 1957), 3:243, 217, italics in original.

93. *TO* 270.

94. Ibid., 266.

95. *AKD* 1:544.

96. *TO* 56, 339; *CNP* 185, 186.

97. *AKD* 1: 220–21.

98. Ibid., 1: 535. See also, e.g., ibid., 1: 536.

99. Uta Noppeney, *Abstrakte Haltung* (Wurzburg: Konigshausen & Neumann, 2000).

100. Head depends on Gelb and Goldstein's early publications for his infrequent uses of the term *abstract*. For its derivation from Goldstein, see *AKD* 1:133. Head also swings as if on a pendulum between the position that there is a fundamental diminution (1:538–39) and the position that this diminution can perhaps be undone (1:535).

101. *TO* 163.

102. *CKL* 113.

103. *TO* 45.

104. Ibid., 54–55.

105. Ibid., 355.

106. On "disintegration," see also *TO* 33, 42, 44, 45, 115, 208, 341, 365, 370.

107. Head, "Disorders of Symbolic Thinking and Expression," 180. Head,

too, considers, in a manner largely derivative of Sherrington, the tonus and posture problem, but he does so unsystematically and certainly not with an aim to use it as central to the whole organism.

108. *AKD* 1:544.

109. On monism and dualism in Goldstein, see our *Experimente im Individuum*, 134–36.

110. Virginia Woolf, *On Being Ill* (Ashfield, MA: Paris Press, 2002), 17.

CHAPTER FIVE

1. *CBC* 1929:11.

2. Walter B. Cannon, "The Movements of the Stomach, Studied by Means of the Röntgen Rays" *Journal of the Boston Society of Medical Sciences* 2, no. 6 (February 15, 1898), 59–66.

3. *Walter Bradford Cannon: Exercises Celebrating Twenty-Five Years as George Higginson Professor of Physiology* (Cambridge, MA: Harvard University Press, 1932), 27.

4. J. P. Pawlow (Pavlov), *The Work of the Digestive Glands* (London: Griffin, 1902); and Ernest H. Starling, *Recent Advances in the Physiology of Digestion* (Chicago: Keener, 1906).

5. Edward Sharpey-Shafer, *History of the Physiological Society* (Cambridge: Cambridge University Press, 1927), 132.

6. Cannon, *The Mechanical Factors of Digestion* (New York: Longman's, Green, 1911).

7. *CBC* 1929: ch. 19.

8. *CWB* 35.

9. Cannon, *Digestion and Health* (New York: Norton, 1936); and Cannon and Arturo Rosenblueth, *Autonomic Neuro-Effector Systems* (New York: Macmillan, 1937). A book-length publication by Rosenblueth of collaborative work appeared after Cannon's death as *The Supersensitivity of Denervated Structures* (New York: Macmillan, 1949).

10. *CWB* 19.

11. In *The Wisdom of the Body*, he offered a sense of the movement of his work, from the phenomena of swallowing through motions of the stomach and intestines, then to the digestive canal as a single unit (with a growing interest in nervous control and the effects of the emotions on the canal), then to emotional effects on adrenal secretion, and finally to the autonomic nervous system (*CWB* xiii). This list does not even mention his work on the thyroid, his shifts of scale, or the fact that very different experimental, technical, theoretical work was required to handle different systems.

12. Cannon and John B. Blake, "Gastro-Enterostomy and Pyloroplasty," *Annals of Surgery* 41 (1905), 686–710; Cannon, "Stimulation of Adrenal Secretion by Emotional Excitement," *Proceedings of the American Philosophical Society* 1 (1911), 226–27. Cannon persisted in seeing his later experiments as complementary to the earlier ones, not merely developed out of them. "Preface to the First Edition," in *CBC* 1929: viii.

13. Cannon, "Movements of the Stomach," 59–66.

14. Cannon, "The Movements of the Intestines Studied by Means of the

Röntgen Rays," *American Journal of Physiology* 6, no. 5 (January 1902), 250–77, 251.

15. Ibid., 259, 261, 269, 271, 274, 276.

16. Cannon, *Digestion and Health*, 13.

17. For a complete list of Cannon's writings in defense of vivisection, see *Walter Bradford Cannon: Exercises*, 73–94. See also SA/RDS/K/2/15 and SA/RDS/K/2/17, Wellcome Library.

18. Cannon and J. B. Blake, "Gastro-Enterostomy and Pyloroplasty," *Annals of Surgery* 41, no. 5 (May 1905), 686.

19. Cannon, "The Influence of Emotional States on Functions of the Alimentary Canal," *American Journal of the Medical Sciences* 137 (1909), 480–87.

20. Cannon, "The Nature of Gastric Peristalsis," *American Journal of Physiology* 29 (1911–1912), 253, 255, fig. 2, and see 258. For the removal of the adrenal glands, see Cannon, "The Emergency Function of the Adrenal Medulla in Pain and the Major Emotions," *American Journal of Physiology* 33, no. 2 (February 1914), 356. Claude Bernard strongly opposed surgical removal of organs, as Cannon had early on. See François Dagognet, *Philosophie biologique* (Paris: Presses Universitaires de France, 1955), 32.

21. Cannon and A. L. Washburn, "An Explanation of Hunger," *American Journal of Physiology* 29, no. 5 (March 1912), 441 and 442 ("a dull persisting sensation"). For introspective psychology, see also pp. 444, 450. In *Digestion and Health*, Cannon would emphasize the role that observing himself "with a stethoscope" (27) had played in the study.

22. Cannon and Washburn, "Explanation of Hunger," 449–51, 449.

23. Ibid., 454.

24. Cannon, "The History of the Physiology Department of the Harvard Medical School," *Bulletin of the Harvard Medical School Alumni Association* 1 (1927), 18.

25. Cannon, "Movements of the Stomach," 66.

26. Cannon, "Movements of the Intestines," 250–77, 275.

27. Cannon, "Influence of Emotional States on Functions of the Alimentary Canal."

28. Cannon had a habit of reusing the same passages in different papers—especially when referring to cases and experiments. This practice establishes a sense of continuity and helps us see the direction of his experimentation around such types of evidence, in particular his movement toward using experiments to prove or confirm results or intuitions already gathered from cases or reports.

29. Even as astute a reader as Anne Harrington backdates the experimental system and epistemic questions of the 1910s to the early papers; we see not continuity here, but Cannon citing himself. Anne Harrington, *The Cure Within* (New York: Norton, 2008), 145–46.

30. Cannon, *Mechanical Factors of Digestion*, 220. See also his description of frustration with animals refusing digestion turning into interest in their emotional state in "The Career of the Investigator," *Science* 34, no. 864

(July 21, 1911), 68; also "Influence of Emotional States on Functions of the Alimentary Canal," 485.

31. Cannon, *Mechanical Factors of Digestion*, 210.

32. Cannon, "Stimulation of Adrenal Secretion by Emotional Excitement," 226–27; Cannon and H. Lyman, "The Depressor Effect of Adrenalin on Arterial Pressure," *American Journal of Physiology* 31 (1912–1913), 379–98; Cannon and L. B. Nice, "The Effect of Adrenal Secretion on Muscular Fatigue," *American Journal of Physiology* 32, no. 1 (May 1913), 44–60. Cannon went so far as to describe the adrenal glands as being "dominated" by the sympathetic nervous system. Cannon and Daniel de la Paz, "The Stimulation of Adrenal Secretion by Emotional Excitement," *Journal of the American Medical Association* 56, no. 10 (March 1911), 742. Cannon and de la Paz, "Emotional Stimulation of Adrenal Secretion," *American Journal of Physiology* 28, no. 1 (1911), 64.

33. See Cannon and Lyman's discussion in "The Depressor Effect of Adrenalin," *American Journal of Physiology*, 376, 379, and see his explanatory schema in 396–97. For further effects on the blood, see Cannon and Daniel de la Paz, "The Emotional Stimulation of Adrenal Secretion," 64–70.

34. Cannon and Nice, "Adrenal Secretion on Muscular Fatigue," *American Journal of Physiology* (1913), 45, 60. Cannon argued that adrenal secretions caused a "slight but distinct" change, which was not enough to claim that they caused muscular work or fatigue. Also, Cannon and de la Paz, "Emotional Stimulation of Adrenal Secretion," 67.

35. Cannon, *Digestion and Health*, 82.

36. Ibid., 126–27.

37. Cannon, "Influence of Emotional States" *American Journal of Medical Sciences* (1909), 480–87.

38. Cannon, "Movements of the Intestines," 275.

39. *CBC* 1929:358.

40. Ibid., 2.

41. Ibid., ch. 13.

42. Otniel E. Dror argues that Cannon's effort was principally to exclude pain from the category of the emotions. We approach the problem differently, to accentuate that pain was closer to fear and anger than these were to other, notably weaker emotions, because at stake in Cannon's book were first the bodily changes and the factors that forced them, and only secondarily those states called emotions. Dror, "Fear and Loathing in the Laboratory and the Clinic," in *Medicine, Emotion and Disease, 1700–1950*, ed. Fay Bound Alberti (Basingstoke, UK: Palgrave Macmillan, 2006), 131.

43. *CBC* 1915:277–78, 1929:342. See also *CBC* 1929:248.

44. *CBC* 1929:16–17.

45. Saul Benison, A. Clifford Barger, and Elin L. Wolfe, *Cannon: The Life and Times of a Young Scientist* (Cambridge, MA: Belknap Press of Harvard University Press, 1987), 2.

46. "Epistemic things—things embodying concepts," in Hans-Jörg Rheinberger, *Toward a History of Epistemic Things* (Stanford, CA: Stanford University Press, 1997), 8.

47. Usually only the reliance on evolutionary thought is pointed out. Tracey Loughran, "Shell Shock, Trauma, and the First World War," *Journal of the History of Medicine and Allied Sciences* 67, no. 1 (2010), 111; and Anne Harrington, *The Cure Within*. Allan Young sees Cannon's argument as a physiologization of Herbert Spencer's psychology. Young, *The Harmony of Illusions: Inventing Post-Traumatic Stress Disorder* (Princeton, NJ: Princeton University Press, 1995), 21.

48. On Darwin and Cannon, see also J. Scott Turner, "Homeostasis and the Forgotten Vitalist Roots of Adaptation," in *Vitalism and the Scientific Image in Post-Enlightenment Life Science, 1800–2010*, ed. Sebastian Normandin and Charles T. Wolfe (Heidelberg: Springer, 2013), 273.

49. CBC 1929:1.

50. CBC 1915; CBC 1929: 2.

51. Young, *Harmony of Illusions*, 22. Young connects Cannon to Crile on the issue of fear.

52. Cannon, "The Interrelations of Emotions as Suggested by Recent Physiological Research," *American Journal of Psychology* 25, no. 2 (April 1914), 256.

53. CBC 1929:194; see also CBC 1929:221.

54. Charles Darwin, *The Expression of the Emotions in Man and Animals* (London: Murray, 1872).

55. CBC 1929:195–96. See also Cannon, "The Emergency Function of the Adrenal Medulla," *American Journal of Physiology* 32 (1914) 356–72, 356.

56. Cannon, "Stimulation of Adrenal Secretion by Emotional Excitement," 227.

57. Ibid. On the idea that modernity imposes stress, see Harrington, *Cure Within*, and Andreas Killen, *Berlin Electropolis* (Berkeley: University of California Press, 2006).

58. Cannon, *Digestion and Health*, 135.

59. CBC 1929: ch. 13.

60. Ibid., 225.

61. Ibid., 225–26, original emphasis.

62. Cannon, "Interrelations of Emotions," 260, 259.

63. CBC 1929:193–94.

64. By *The Wisdom of the Body*, Cannon had established the sympathico-adrenal system as perhaps the central component necessary to keep the fluid matrix constant (see also *Digestion and Health*, 140).

65. CBC 1929:241.

66. On McDougall, see CBC 1915:187–88, 216–17, 276–77. Cannon mentions "fighting or flight" in CBC 1915:211.

67. William James, "Energies of Men," *Science* 25, no. 635 (1907), 321–32.

68. Cannon, "Interrelations of Emotions," 275; also CBC 1929: chs. 12, 14.

69. Cannon, A. T. Shohl, and W. S. Wright, "Emotional Glycosuria," *American Journal of Physiology* 29, no. 2 (1911), 280.

70. CBC 1915:277, 1929:343. This passage and overall discussion derive from his "Interrelations of Emotions," 277–80, italics in original.

71. CBC 1915:188–89, 1929:197. See also CBC 1915:211, 1929:219.

72. Cannon, "Studies on the Conditions of Activity in Endocrine Glands," *American Journal of Physiology* 72 (1925), 283–313, 283; and *CBC* 1929:246.

73. On the ease with which it becomes possible to slide into a causality argument, see the discussion of Cannon in the fine essay by Bettina Hitzer, "Healing Emotions," in *Emotional Lexicons*, ed. Ute Frevert (Oxford: Oxford University Press, 2014), 140–41.

74. *CBC* 1915: 211–12, italics ours.

75. Ibid., 44–45.

76. The recognition that Cannon used multiple sources of evidence is particularly important if we want to avoid the mistake of directly extrapolating from animal experiments to the overall theory. Even excellent studies like Dror's have retained an abstraction of this variety, which leads to a specific set of objections to Cannon's argument.

77. Cannon, *Way of an Investigator*, 48, 92, 130–33. Saul Benison, A. Clifford Barger, and Elin L. Wolfe, "Walter B. Cannon and the Mystery of Shock," *Medical History* 35 (1991), 249.

78. *CWB* 39.

79. Cannon, "Organization for Physiological Homeostasis," *Physiological Reviews* 9, no. 3 (July 1929), 399–431.

80. *CWB* 24.

81. Ibid., 19.

82. *CWB* 286. See also *CWB* 311: "extraordinarily labile material." The reference to instability recurs across his writings from the period: "We are made of the most unstable stuff," in Cannon, "Biocracy," *Technology Review* 35, no. 6 (March 1933), 203–6, 227, quote on 203.

83. *CWB* 263, 79, 176.

84. Ibid., 59.

85. Ibid., 288; on sentinels, see ibid., 57.

86. Ibid., 46–48, 49–51.

87. Ibid., ch. 14.

88. Ibid., 227.

89. Ibid., 309.

90. On overflow, see *CWB* 290, 293, 294; on inundation, see ibid., 85, 290, 88.

91. See, e.g., ibid., 296.

92. Ibid., 296ff.

93. In one instance Cannon all but cited Ernst Haeckel in writing that homeostasis confirms the idea of ontogeny recapitulating phylogeny. *CWB* 301.

94. Ibid., 231.

95. Ibid., 27, 263.

CHAPTER SIX

1. For a consideration of individuality among scientific categories in the nineteenth century, see Roger Smith, "The Background of Physiological Psychology in Natural Philosophy," in *History of Science* 11 (1973), 102–4.

2. Arturo Castiglioni, "Neo-Hippocratic Tendency of Contemporary Medical Thought," *Medical Life* 41 (1934), 115–16.

3. Claude Bernard, *Leçons sur les phénomènes de la vie* (Paris: J.-B. Baillière et fils, 1875) 1:113; quoted in *CWB* 38.

4. *BER* 110/64.

5. Bernard, *Leçons sur les phénomènes de la vie*, 121.

6. *BER* 147/85.

7. *BER* 130/76; translation modified.

8. Georges Canguilhem comments on the innovation involved in Bernard's passage: "So long as scientists conceived of organs' functions within the organism in the image of organisms' functions in the external milieu, it was natural for them to borrow the . . . guiding ideas of biological explication and experimentation from the pragmatic experience of the human living being." *CKL* 7–8.

9. Joseph Barcroft, "'La fixité du milieu intérieur est la condition de la vie libre,'" *Biological Reviews and Biological Proceedings of the Cambridge Philosophical Society* 7 (1932), 24–87. See also F. L. Holmes, "Joseph Barcroft and the Fixity of the Internal Environment," *Journal of the History of Biology* 2 (1961), 88–122; and the examination of Bernard by Pierre Vendryès in his *Vie et probabilité* (Paris: Albin Michel, 1942), and *CNP*, pt. 1, ch. 3, and pt. 2, ch. 3.

10. Starling, *Elements of Human Physiology* (London: Churchill, 1897), 17, 20, 26.

11. Ibid., 13.

12. Ernest H. Starling, *The Fluids of the Body* (London: Archibald Constable, 1909), 41.

13. Ibid., 134.

14. Ibid., 87, 134ff.

15. Starling, *Principles of Human Physiology* (Philadelphia: Lea & Febiger, 1912), 4.

16. Cited in Walter B. Cannon, "Organization for Physiological Homeostasis," *Physiological Reviews*, 9, no. 3 (July 1929), 399.

17. Cited in Cannon, "Reasons for Optimism in the Care of the Sick," *New England Journal of Medicine* 199 (September 27, 1928), 597.

18. *OEB* 89; see also *OEB* 114.

19. *HRE* 383. For citations of Haldane's phrase, see, for example, Cannon, "Organization for Physiological Homeostasis," 400.

20. Barcroft, "La fixité du milieu intérieur," 24.

21. Ibid., 86.

22. On Bernard's reception, see Mirko Grmek, *Le legs de Claude Bernard* (Paris: Fayard, 1997); E. H. Olmsted, "Historical Phases in the Influence of Bernard's Scientific Generalizations in England and America," in *Claude Bernard and Experimental Medicine*, ed. Francisco Grande and Maurice B. Visscher (Cambridge, MA: Schenkman, 1967), 24–34. Accumulating references without offering a clear reading, Mark Jackson, in *The Age of Stress* (Oxford: Oxford University Press, 2013), 63–64, sees physiological research and innovation in the early twentieth century as deeply indebted to Bernard. This overshoots the target, for physiologists only turned to Bernard belatedly and used him as an authority figure. He was not influential in advance (for example, on research protocols or conceptual organization), and they only

retroactively cited him from the middle of the war onward, and especially in the late 1920s. John Parascandola, "Organismic and Holistic Concepts in the Thought of L. J. Henderson," *Journal of the History of Biology* 4, no. 1 (1971), 72; and F. L. Holmes, "Claude Bernard's Concept of the Milieu Intérieur" (PhD diss., Harvard University, 1962), 195–98.

23. Henry H. Dale, "Progress in Autopharmacology," *Bulletin of the Johns Hopkins Hospital* 53 (1933), 297.

24. See the records of the Collège philosophique preserved in Dossier Collège philosophique, Fonds Jean Wahl, Institut mémoires de l'édition contemporaine. Canguilhem's manuscript for the talk is preserved at GC 12.1.8, Georges Canguilhem Archive, Centre d'archives en philosophie, histoire et éditions des sciences (CAPHES), École normale supérieure.

25. CKL 113, 110–11, 9. *Auseinandersetzung* is Goldstein's term.

26. Ibid., 113; see also 118.

27. Hans Spemann, *Experimentelle Beitrage zu einer Theorie der Entwicklung* (Berlin: Springer, 1936); and G. E. Coghill, *Anatomy and the Problem of Behavior* (Cambridge: Cambridge University Press, 1929). See also Coghill's "The Structural Basis of the Integration of Behavior," *Proceedings of the National Academy of Sciences* 16, no. 10 (1930), 637–43; and Sven Hörstadius, "The Mechanics of Sea Urchin Development," *Année biologique* 26, no. 8 (August 1950), 381–98.

28. Ludwig von Bertalanffy, *Modern Theories of Development* (London: Oxford University Press, 1933); von Bertalanffy, *Problems of Life* (New York: Harper, 1952).

29. Jakob von Uexküll, *Umwelt und Innenwelt der Tiere*, rev. ed. (1909; Berlin: Springer, 1921); von Uexküll, *Streifzuge durch die Umwelten von Tieren und Menschen* (Hamburg: Rowohlt, 1958). On Uexküll, see also Anne Harrington, *Reenchanted Science* (Princeton, NJ: Princeton University Press, 1996), ch. 2.

30. Fenton B. Turck, *The Action of the Living Cell* (New York: Macmillan, 1933). Turck made contributions to the study of shock, which he sought to reconstruct in the laboratory as early as 1891. Turck declared it to be an effect of the "autolysis" of cells and disparaged approaches that did not focus on cellular injury. Turck, "The Primary Cause of Shock," *Medical Record* 93, no. 22 (June 1, 1918), 927–39; and Turck, *Action of the Living Cell*, 13–46.

31. On Whitehead, see Lawrence J. Henderson, "A Philosophical Interpretation of Nature," *Quarterly Review of Biology* 1, no. 2 (April 1926), 289–94. Henderson's review of Whitehead—his endorsement of Whitehead's endorsement of "organicism" or "organic mechanism"—is perhaps the most difficult of his writings to square with his commitment to mechanism and reductionism.

32. L. J. Henderson, "The Fitness of the Environment," *American Naturalist* 47, no. 554 (February 1913), 106–7.

33. Henderson, *Blood: A Study in General Physiology* (New Haven, CT: Yale University Press, 1928), 8, 20.

34. Ibid., 10.

35. Ibid.

36. Ibid., 355–57.

37. Henderson, lecture before the Eastern Associate of Graduates of the Angle School of Orthodontia (March 1926), Lawrence J. Henderson Papers 1:2, pp. 143–44, Baker Library, Harvard Business School.

38. Henderson to David L. Edsall, July 27, 1929, Henderson Papers 1:46.

39. Henderson, "Preface," in *BER*/v–xii.

40. Henderson, *Blood*, 3–4.

41. Bernard, *Leçons de pathologie expérimentale* (Paris: Baillière, 1872), 87 and 123, 331.

42. *CKL* xx, 16, 87.

43. Replying privately to Haldane's critique of his understanding of Bernard, and in support of his own mechanism, Henderson cited at length from Bernard's *Leçons sur les phénomènes de la vie communs aux animaux et aux végétaux* (Paris: Baillière, 1878) 1:50. See Henderson to Haldane, March 9, 1929, Henderson Papers 1:60. For Frederic L. Holmes, only this "late" Bernard mattered to Henderson and Cannon on physiological constancy, not the Bernard of the *Introduction*. Holmes, "Claude Bernard and the Milieu intérieur," *Archives internationales d'histoire des sciences* 16 (1963), 376.

44. Henderson, *Blood*, 21.

45. Yandell Henderson, "Is This Science or Metaphysics?" *Science* 69, no. 1776 (January 11, 1929), 39–41.

46. James McKeen Cattell to L. J. Henderson, February 13, 1929, Henderson Papers 1:25.

47. Lawrence Henderson had his supporters, notably in *Science*, where a month later he was defended from Yandell Henderson's critique by Raymond Pearl and Donald Van Slyke: *Science* 69, no. 1780 (February 8, 1929), 161–63. It is against this background that Haldane's critique appeared two months later. Several others wrote to Lawrence Henderson personally—including Baird Hastings (January 18, 1929, Henderson Papers 1:65) and A. V. Hill (March 19, 1929, Henderson Papers 2:2).

48. Yandell Henderson to L. J. Henderson, January 31, 1929; L. J. Henderson to Yandell Henderson, February 5, 1929; and Yandell Henderson to L. J. Henderson, February 7, 1929—all in Henderson Papers 2:1.

49. J. S. Haldane to Henderson, February 23, 1929, Henderson Papers 1:60.

50. J. S. Haldane, *Organism and Environment as Illustrated by the Physiology of Breathing* (New Haven, CT: Yale University Press, 1917), 117.

51. Haldane, "Claude Bernard's Conception of the Internal Environment," *Science* 69, no. 1791 (April 26, 1929), 453–54; and Haldane, "Claude Bernard's Conception of the Internal Environment," *Journal of Physiology* 67 (1929), xxii–xxiv.

52. On the kidneys' role in regulating the blood, see *OEB* ch. 3.

53. Haldane, "Claude Bernard's Conception of the Internal Environment," *Science*, 453–54. Jackson situates the relationship of Haldane, Lawrence Henderson, Barcroft, and Cannon in the development of theories of stress, but underrates the disagreements and the context of coordination and integration. Jackson, *Age of Stress*, 65–66.

54. Haldane and J. G. Priestley, "The Regulation of the Lung-Ventilation," *Journal of Physiology* 32, nos. 3–4 (May 9, 1905), 224–66.
55. Haldane, J. C. Meakins, and J. G. Priestley, "The Effects of Shallow Breathing," *Journal of Physiology* 52, no. 6 (1919), 433–53.
56. *HRE* 31ff., 242.
57. Ibid., 386.
58. Henderson, review of *Mechanism, Life and Personality, Science* 42, no. 1081 (September 17, 1915), 378–82.
59. Haldane, "The New Physiology," *Science* 44, no. 1140 (November 3, 1916), 629–30.
60. Haldane to L. J. Henderson, February 23, 1929, Henderson Papers 1:60. This was Haldane's classic objection to mechanism: *HRE* 387.
61. Haldane was insistent on holistic tendencies in organs and systems, not just at the cellular level: *HRE* 386.
62. "For irreversible chemical reactions physiology has but little use," see *HRE* 386. On "irreversible," see 385.
63. *HRE* 56–59, 242, 288, 297.
64. See, among others, Parliamentary Paper C 8112 (1896).
65. *HRE* 128.
66. Ibid., 388, 52.
67. Ibid., 27.
68. *CKL* 70, see also 114, 117, and the citation of Haldane on 118.
69. *HRE* 387.
70. Ibid., 388.
71. Ibid., 390–92.
72. Ibid., 391.
73. *OEB* 114.
74. *HRE* 383.
75. Ibid., 387.
76. Ibid., 386.
77. Ibid., 117.
78. Ibid., 397.
79. Haldane to L. J. Henderson, February 23, 1929, Henderson Papers 1:60.
80. L. J. Henderson to Hill, March 28, 1929, Henderson Papers 2:2.
81. John Fulton to Hallowell Davis, March 17, 1929 (which Davis forwarded to Henderson on March 28), Henderson Papers 1:38.
82. L. J. Henderson to Yandell Henderson, January 16, 1929, Henderson Papers 2:1.
83. Cannon, "Organization for Physiological Homeostasis," 399–431.
84. Cannon to L. J. Henderson, January 21, 1929, Henderson Papers 2:1.
85. Cannon cited Yandell Henderson in "Organization for Physiological Homeostasis," 425.
86. Ibid., 400.
87. Cannon expanded on Bernard's *milieu intérieur* in *CWB* 37–40.
88. Cannon, "Some Tentative Postulates Regarding the Physiological Regulation of Normal States," Walter B. Cannon Papers 63:837 and "E. Starling and Harveian Lecture Notes."

89. *CWB* 290.

90. Cannon, "Some Tentative Postulates Regarding the Physiological Regulation of Normal States," 1.

91. *CWB* 231, 239, 243. S. J. Meltzer, "The Factors of Safety in Animal Structure and Animal Economy," in *The Harvey Lectures under the Auspices of the Harvey Society of New York* (Philadelphia: Lippincott, 1908), 139–69; "Factors of Safety in Animal Structure and Animal Economy," *Journal of the American Medical Association* 48, no. 8 (1907), 655–64. George Crile also used the term *margin of safety* with reference to cancer surgery, but Cannon did not make this link. See Crile and William E. Lower, *Anoci-Association* (Philadelphia: Saunders, 1915), 174, 211, 212. In "Biocracy" (205), Cannon refers to a "factor of safety."

92. L. J. Henderson, "The Theory of Neutrality Regulation in the Animal Organism," *American Journal of Physiology* 21 (1908), 427–48. On Henderson's *buffer* and *factor of safety*, see Parascandola, "Organismic and Holistic Concepts in the Thought of L. J. Henderson," 70. Iris Fry's discussion of *factor of safety* is taken wholesale from Parascandola. Fry, "On the Biological Significance of the Properties of Matter," *Journal of the History of Biology* 29, no. 2 (1996), 160ff.

93. Cannon, "Biocracy: Does the Human Body Contain the Secret of Economic Stabilization?" in *Technology Review* 35, no. 6 (March 1933), 227.

94. On Bernard and freedom, see *CWB* 263, 38, 59.

95. Cannon to Joseph Barcroft, November 30, 1931, 62:809. Barcroft, Cannon Papers.

96. Cannon, "Biocracy," 227.

97. Dale's position on the Haldane-Henderson debate is not clear, but he is cited as being uncomfortable with Haldane's response to Henderson. Hill to L. J. Henderson (March 19, 1929), Henderson Papers 2:2.

98. See Dale, introduction to *Paul Ehrlich*, by Martha Marquardt (New York: Schuman, 1951), xiii–xx; Paul Ehrlich, "On Immunity with Special Reference to Cell Life," in *Collected Papers of Paul Ehrlich*, ed. R. Himmelweit, Martha Marquardt, and Dale (London: Pergamon, 1956–57), 2:178–95.

99. Dale, *Adventures in Physiology, with Excursions into Autopharmacology* (London: Wellcome Trust, 1965), xi.

100. Ibid., x.

101. About the same time as Dale and Patrick Laidlaw, Ackermann ("Über den bakteriellen Abbau des Hisidins" *Zeitschrift für Physiologische Chemie*, 65 [1910], 504–10) obtained histamine by the putrefaction of histidine ("The Physiological Action of a Secal Base and of Imidazolylethylamine." *Zentralblatt für Physiologie*, 24, 163 (1910). Barger and Dale ("o-imin-azolylethylamine a Depressor Constituent of Intestinal Mucosa," *Journal of Physiology* 41 [1910–1911]: 499–503) isolated histamine from intestinal mucosa, but with the advent of the First World War, it was not until 1919 that Dale and Alfred N. Richards ("The vasodilator action of histamine and of some other substances," *Journal of Physiology* 52, 2–3 [1919]: 110–65) published their classic paper on histamine dilatation.

102. E. M. Tansey, "Henry Dale, Histamine, and Anaphylaxis: Reflec-

tions on the Role of Chance in the History of Allergy," *Studies in the History and Philosophy of Science, Part C* 34, no. 3 (2003), 456; Dale, *Adventures in Physiology,* 171.

103. Mark C. Fishman, "Sir Henry Dale and the Acetylcholine Story" *Yale Journal of Biology and Medicine* 45 (1972), 104–18; Herbert S. Gasser, "Henry Dale," *BMJ* (June 4, 1955), 1359–61; Ronald Rubin, "A Brief History of Great Discoveries in Pharmacology," *Pharmacological Reviews* 59, no. 4 (2007), 289–359.

104. Dale, *The Nature and Cause of Wound Shock* (Philadelphia: Lippincott, 1921); Dale and Richards, "Vasodilator Action of Histamine and of Some Other Substances," 110–65.

105. Dale, "The Oliver-Sharpey Lectures on The Activity of the Capillary Blood Vessels," pt. 1, *BMJ* (June 9, 1923), 959; pt. 2 (June 16, 1923), 1006.

106. Dale, *Nature and Cause of Wound Shock,* 27.

107. Dale, "Medical Research as an Aim in Life," (1949), in *Autumn Gleaning,* 132; on Dale's research method and problem construction, see Tansey's commentary on Dale's archival papers, in "Illustrations from the Wellcome Institute Library," *Medical History* 34 (1990), 199–209.

108. Bayliss, "General Discussion on Shock," in *Proceedings of the Royal Society of Medicine* (London: Royal Society of Medicine, 1919), xii; H. A. Hare and Edward Martin, eds., *The Therapeutic Gazette* (Detroit: Swift, 1919), 603–5.

109. Dale declared that "wound shock" was a definition of the "widest kind," and in the attempt to arrive at a consistent definition "we failed, and I believe we were bound to fail." Dale, "Royal Society Croonian Lecture: The Biological Significance of Anaphylaxis," *Proceedings of the Royal Society of London Part B* 91 (1920), 126–46.

110. A. Windaus and W. Vogt, "Synthese des Imidazolyläthylamins. *Berichte Deutscher Chemischen Gesellschaft* 40 (1907), 3691; Dale and Laidlaw, "The *Physiological* Action of /3-Iminazolyl Ethylamine," *Journal of Physiology,* 41 (1910–1911), 318–44; H. O. Schild, "The Multiple Facets of Histamine Research," *Agents and Actions* 11 (1981), 12–19; James F. Riley, "Histamine and Sir Henry Dale," *BMJ* (1965): 1488–90.

111. Dale and Laidlaw, "*Physiological* Action of /3-Iminazolyl Ethylamine."

112. Dale, "On Some Physiological Actions of Ergot," *Journal of Physiology* 34 (1906), 163–206; Dale, "The Occurrence of Ergot and Action of Acetylcholine" *Journal of Physiology* 48 (1914), iii–iv.

113. Dale, the Herter Lectures, Baltimore, MD, November 13–15, 1919, I. "Capillary Poisons and Shock," published August 1920, 31 (354), 257–65; II. "Anaphylaxis," published September 1920, 31 (355), 310–19; III. "Chemical Structure and Physiological Action," published October 1920, 31 (356), 373–80, all in *Bulletin of the Johns Hopkins Hospital.*

114. Dale, "β-*iminazol*-lylethylamin A Depressor Constituent of Intestinal Mucosa," *Journal of Physiology* 41, no. 6 (1911), 499–503; commentary on the article written in 1951 in *Adventures in Physiology,* 123.

115. Charles Richet, "Anaphylaxis (Nobel Lecture, 1913)," *Scandinavian*

Journal of Immunology 31 (1990), 375–88; Richet, *L'Anaphylaxie* (Paris: Alcan, 1912). See also Mark Jackson, *Allergy* (London: Reaktion, 2007), 32.

116. Jackson, "Introduction: Allergy and History," *Studies in the History and Philosophy of Biology and Biomedical Sciences* 34 (2003), 383–98.

117. Dale, "*Capillary* Poisons and Shock," xxxi; Arthur F. Coca and Robert Cooke, "On the Classification of the Phenomena of Hypersensitiveness," *Journal of Immunology* 8 (1923), 163–82; Besredka, *Anaphylaxis and Anti-Anaphylaxis* (London, 1919); Auguste Lumière, *Le problème d'anaphylaxie* (Paris: Doin, 1924); R. Cranston Low, *Anaphylaxis and Sensitisation* (Edinburgh: Green, 1924); Svante Arrhenius, *Immunochemistry* (New York: Macmillan, 1907); Clemens von Pirquet, "Allergie," *Münchener Medizinische Wochenschrift* 30 (1906), 1457–58; and Ilana Löwy, "On Guinea Pigs, Dogs, and Men," *Studies in History and Philosophy of Biological and Biomedical Sciences* 34 (2003), 399–424.

118. Jackson, *Allergy*, 45, 97.

119. Jackson, "Allergy and History," 388.

120. Dale, "The Mechanics of Anaphylaxis," originally presented in French on May 10, 1952 in Paris, reprinted in *Autumn Gleaning*, 210; also anaphylaxis experiments from March to November 1915, PP/HHD/5, Papers of Henry H. Dale (hereafter Dale Archive), Wellcome Library.

121. Dale, "Biological significance of anaphylaxis" (1920), 142. See also Riley, "Histamine and Sir Henry Dale."

122. Dale, "Mechanics of Anaphylaxis," in *Autumn Gleaning*, 204.

123. Ibid., 211.

124. Dale, "Accident and Opportunism in Medical Research" (1948), in *Autumn Gleaning*, 124.

125. Richet, "Anaphylaxis."

126. Dale and Charles H. Kellaway, "On Anaphylaxis and Anaphylatoxins." *Philosophical Transactions of the Royal Society Part B* 211 (1922), 273–315, in Dale, *Adventures in Physiology*, 373.

127. PP/HHD/22, Dale Archive.

128. Rudolf Kobert, *Practical Toxicology for Physicians and Students* (New York: Jenkins, 1897).

129. Bayliss, "On the Physiology of the Depressor Nerve," *Journal of Physiology* 14 (1893), xiv.

130. Dale, "The Action of Extracts of the Pituitary Body." *Biochemical Journal* 4, no. 9 (1909), 427–47, reprinted in *Adventures in Physiology*, 37; Bayliss and Starling, "Mechanism of Pancreatic Secretion," *Journal of Physiology* 28 (1902), 325–52.

131. Fühner and Starling, "Experiments on the Pulmonary Circulation," *Journal of Physiology* 47, nos. 4–5 (1913), 286–304; and Arnold M. Katz, "Ernest Henry Starling, His Predecessors, and the 'Law of the Heart,'" *Circulation* 106 (2002), 2986–92.

132. Dale and Laidlaw, "*Physiological* Action of /3-Iminazolyl Ethylamine."

133. "The Control of Circulation and the Capillary Blood Vessels," p. 1, PP/HHD/21, Dale Archive.

134. Cannon, "A Consideration of the Nature of Wound Shock," in MRC 75.

135. Starling, *Elements of Human Physiology*, 8th ed. (Chicago: Keener, 1907), 10–12, 280–82.

136. "Draft Discourse: Circulation of the Blood in the Capillary Vessels, 1925," p. 2, PP/HHD/20, Dale Archive.

137. "New observations . . . appeared to indicate that the capillaries were not the mechanically passive structures of the accepted teaching." Ibid, 8–9. See also "Control of Circulation and the Capillary Blood Vessels," p. 1, PP/HHD/21, Dale Archive.

138. Dale emphasized that this discovery was made in different labs in several countries during the war. "Circulation of the Blood in Capillary Vessels," p. 1, PP/HHD/20, Dale Archive.

139. "Draft Discourse: Circulation of the Blood in the Capillary Vessels, 1925," pp. 11–12, PP/HHD/20, Dale Archive.

140. Ibid., 26. On circulation as "systemic," see also "Control of Circulation and the Capillary Blood Vessels," p. 3, PP/HHD/21. Krogh's investigations had become particularly well known after the end of the war: they concerned the independent tone of the capillaries, particularly their variable permeability and their capacity to distend abnormally, which is what the British researchers had been concerned with. On Krogh, see J. Kenez, "The Capillaries and Krogh," *Orvosi hetilap* 106 (1965), 177–78; "August Krogh (1874–1949)," *Journal of the American Medical Association* 199, no. 7 (1967), 496–97.

141. In the commentary on his now-famous paper on "histamine shock" with Laidlaw, Dale noted the broadness of the classification of "shock." Dale and Laidlaw, "Histamine Shock," *Journal of Physiology* 52 (1919), 351, reprinted in *Adventures in Physiology*, 290.

142. Dale and Herbert S. Gasser, "The Pharmacology of Denervated Mammalian Muscle, Part I," *Journal of Pharmacology and Experimental Therapeutics* 29, 1 (1926), 53–67.

143. Dale and Richards, "The Vasodilator Action of Histamine and of Some Other Substances," *Journal of Physiology* 52 (1918), 110–65.

144. Dale and Laidlaw, "Histamine Shock," *Adventures in Physiology*, 296. See Haldane's approving adoption of Dale and Laidlaw in *HRE* 289–90.

145. Ibid., 309. Later Dale could not understand why he might have been cautious about forcing a causal connection between histamine and shock (see Dale's "Commentary" on his 1919 paper, "Histamine Shock," *Adventures in Physiology*, 319.)

146. Dale and Laidlaw, "Histamine Shock," *Adventures in Physiology*, 314.

147. Dale, "Conditions Which Are Conducive to the Production of Shock by Histamine," *British Journal of Experimental Pathography* 1 (1920), 103–14, reprinted in *Adventures in Physiology*, 338.

148. "Circulation of the Blood in Capillary Vessels: Abstract of Discourse by H.H. Dale," p. 3, PP/HHD/20, Dale Archive.

149. Ibid.

150. "Draft Paper: The Experimental Study and use of Hormones" (1926), p. 10, PP/HHD/22, Dale Archive.

151. Henry Dale, "Introduction," in Henry Dale, *Adventures in Physiology, with Excursions into Autopharmacology* (London: Wellcome Trust, 1965).

152. Further, as Jackson notes in his survey of allergy, there was considerable distance between the study of hypersensitivity as a concern of pathology and therapy in the early twentieth century and the preoccupation with isolating hormones in the laboratory and describing their mechanisms. The first antihistamine drugs were not introduced until 1937.

153. Warwick Anderson and Ian R. Mackay, *Intolerant Bodies* (Baltimore: Johns Hopkins University Press, 2014), 2; David Napier, *The Age of Immunology* (Chicago: University of Chicago Press, 2002); Warwick Anderson, Myles Jackson, and Barbara Gutmann Rosenkrantz, "Toward an Unnatural History of Immunology," *Journal of the History of Biology* 27 (1994), 575–94; Alfred I. Tauber and Leon Chernyak, *Metchnikoff and the Origins of Immunology* (New York: Oxford University Press, 1991); Anne-Marie Moulin and A. Cambrosio, eds., *Singular Selves* (Paris: Elsevier, 2000); Arthur M. Silverstein, *A History of Immunology* (San Diego: Academic Press, 1989); Ilana Löwy, "The Strength of Loose Concepts," *History of Science* 30 (1992), 371–96; Emily Martin, *Flexible Bodies* (Boston: Beacon, 1994). The now-classic philosophical use of autoimmunity is Jacques Derrida's, in *Rogues* (Stanford, CA: Stanford University Press, 2005); Derrida elaborated on the autoimmunity of democracy in Giovanna Borradori, *Philosophy in a Time of Terror* (Chicago: University of Chicago Press, 2003), 85–136.

154. Jackson, *Allergy*, 19, 48.

155. During the war and shortly after, there were only inklings that histamine was found in almost every tissue, and yet the conceptualization of infection or exposure was not part of these discussions of antigenic stimulation, which in recent years has been taken up through the trope of immunity. What makes Dale's story of histamine so powerful is that it is about both the environment and the organism, with an interior motivation of the body to isolate and react to negative stimuli. See Dale, commentary from 1951 on "Conditions Which Are Conducive to the Production of Shock by Histamine," *British Journal of Experimental Pathography* 1 (1920), 103–14, in *Adventures in Physiology*, 342; Dale and Alfred N. Richards, "The Depressor (Vaso-Dilator) Action of Adrenaline," *Journal of Physiology* 63, no. 3 (1927), 201–10; Dale, "Biological Significance of Anaphylaxis," 126–46, reprinted in *Adventures in Physiology*, 337; Dale and P. Hartley, "Anaphylaxis to the Separated Proteins of Horse-Serum." *Biochemical Journal* 10, no. 3 (1916), 408–33; Henry D. Dakin and Dale, "Chemical Structure and Antigen Specificity," *Biochemical Journal* 13, no. 3 (1919), 248–57.

156. *NHC* 4; see also the critique of mechanism at 8.

157. Ibid., 5.

158. Ibid., 7.

159. Ibid., 9.

160. Ibid., 11.

161. For the reference to Cannon's *Wisdom of the Body*, see *TO*, 55, 397n17. See also disapproving references to Cannon's use of animals on 102, 281, 337. In the late 1930s and early 1940s, Cannon helped Goldstein profes-

sionally (for example, with reviews of *The Organism*), and the two discussed particular papers on nervous disorder. See Goldstein to Cannon, June 5, 1939; Cannon to Goldstein, June 14, 1939, September 8, 1939, January 22, 1940, October 22, 1940; and Goldstein to Cannon, January 25, 1943—all in Cannon Papers 130:1832.

162. Several post hoc appraisals hail Goldstein's dependence on Bernard, despite very little evidence in his actual writing. See Chris Mace, "Psychotherapy and Neuroscience," in *Revolutionary Connections*, ed. Jenny Corrigall and Heward Wilkinson (London: Karnac, 2003), 164; and Walter Riese, "The Structure of Experimental Thought," in *Reach of the Mind*, ed. Marianne L. Simmel (New York: Springer, 1968), 245, 248. Not one reference to Bernard is found in *The Organism* or his later lectures, *Human Nature in the Light of Psycho-Pathology* (Cambridge, MA: Harvard University Press, 1940), and citation here seems to move more radically in the direction of authoritative confirmation than toward genuine recognition of precedent and influence.

163. *TO* 40.
164. Ibid., 295.
165. On patient experience, see *TO* 383–93.
166. *CNP* 182.
167. Goldstein, "Individual and Others," in *HNP*.
168. *TO* 333.
169. Ibid., 326.
170. *CNP* 185.
171. *TO* 49 and 231.
172. Kurt Goldstein, "Zum Problem der Angst," *Allgemeine ärzliche Zeitschrift für Psychotherapie und psychische Hygiene* 2 (1929), 409–37.
173. *TO* 230.
174. Ibid., 240.
175. Ibid., 232.
176. Ibid., 232, 237.
177. Ibid., 240.
178. Ibid., 240.
179. Ibid., 333.
180. See the discussion of these mechanisms in ibid., 281–84.
181. Ibid., 162–63, 167.
182. Ibid., 393.

CHAPTER SEVEN

1. Freud, *Three Essays on the Theory of Sexuality* in *FSE* 7:167, translation amended.
2. Freud, "Triebe Und Triebschicksale" (1915), appearing in English as "Instincts and Their Vicissitudes," in *FSE* 14:122, translation amended. This passage largely repeats the *Three Essays*.
3. Freud, "Instincts and Their Vicissitudes," 120.
4. Freud, "On Narcissism: An Introduction," in *FSE* 14. Nonetheless, Freud maintains the anaclitic model at times, for example in *Introductory Lectures on Psycho-Analysis*, in *FSE* 14:314.

5. To pursue this engagement is not to argue that Freud was replying specifically to these thinkers, although any careful reader of, for example, *Totem and Taboo* knows the weakness of the prevailing image of Freud as a self-involved thinker who showed little interest in the development of neurology and physiology after the ostensible failure of the "Project for a Scientific Psychology" (1895), and who was incapable of following and dealing with developments in related sciences. Historians have also tended to concentrate on the early Freud when considering his relations to the world of neurology and physiology, not least because in his later work Freud appeared aloof. We argue that a lot is hidden in this aloofness.

6. W. H. R. Rivers, "The Repression of War Experience (Dec 4, 1917)," *Proceedings of the Royal Society of Medicine* 11 (1918), 1–17, republished as "The Repression of War Experience," *The Lancet* (February 1918), 513–33. Further citations of "The Repression of War Experience" are to be found as appendix 3 of *RIU*.

7. For example, Mark Jackson relates Freud's "constancy principle" and Gustav Fechner's psychic "principle of stability" to Cannon's "general principles of stabilization" and Norbert Wiener's "control and communication." Jackson, *The Age of Stress* (Oxford: Oxford University Press, 2013), 12, 54.

8. Jean Laplanche, *Life and Death in Psychoanalysis* (1970; Baltimore: Johns Hopkins University Press, 1976), 83.

9. Ibid., 106 and 105.

10. Jacques Lacan, *The Ego in Freud's Theory and in the Technique of Psychoanalysis, Seminar II: 1954–1955* (London: W. W. Norton, 1991), 60.

11. Franz Alexander, "Psychoanalysis Revised" (1940), in *The Scope of Psychoanalysis 1921–1961* (New York: Basic Books, 1961), 128.

12. Alexander, "The Psychosomatic Approach in Medical Therapy" (1954), in *The Scope of Psychoanalysis 1921–1961* (New York: Basic Books, 1961), 346. On his adoption of Cannon's understanding of fear, see 455.

13. Ernest Jones, *Sigmund Freud: Life And Work* (London: Hogarth Press, 1957), 3:288.

14. William A. White, "Psychoanalytic Tendencies," in *Proceedings of the American Medico-Psychological Association and the Seventy-Second Annual Meeting, 1916* (Baltimore: American Medico-Psychological Association, 1916), 282–83.

15. *BPP* 60/73/200.

16. In bringing together the two thinkers, we find inspiration in Patrick Boucheron, *Leonardo e Machiavelli* (Rome: Viella, 2014).

17. William McDougall, *Introduction to Social Psychology* (1908; New York: Methuen, 1960), 32.

18. No copy of McDougall's *Introduction to Social Psychology* survives in any of Freud's libraries (a copy of *The Group Mind* does that postdates the 1916–1917 reference).

19. Freud Museum, London, library item 2156, J. Keith Davies and Gerhard Fichtner, eds., *Freud's Library* (London: Freud Museum, 2006), 440.

20. Rivers to Jones, November 21, 1921, P04/C/H/04–Rivers, CRB/F01/01, Ernest Jones Archive, British Institute of Psychoanalysis. Rivers was famous

for his poor handwriting and gladly complained that people had "lost the art of reading letters."

21. See his "Sociology and Psychology" (1916) in *Psychology and Ethnology* (New York: Harcourt, Brace, 1926), 19; and Rivers, *Psychology and Politics* (London: Kegan Paul, 1923), 42–45.

22. P04/C/H/04–Rivers, CRB/F01/02, Ernest Jones Papers, British Psychoanalytical Society (hereafter Jones Archive). Jones's falling out with Charles Myers in 1926 was even more obnoxious. He requested information about Myers's patients, and in one strange episode, Jones wrote to Myers insisting that he had suppressed a letter sent to *The Lancet* protesting Myers's attack on psychoanalysis—when he himself had typewritten two drafts of that letter, one on official International Psychoanalytical Press letterhead. P04/C/G/07–Myers CMA/F24/06, Jones Archive. Again, only the letters from the breakup survive.

23. Henri Ellenberger, *Discovery of the Unconscious* (New York: Basic Books, 1970), 860.

24. See, e.g., Malcolm Pines's *Circular Reflections* (London: Jessica Kingsley, 1997), 188; and Paul Roazen, *Oedipus in Britain (New York: Other Press, 2000)*, 128. Rivers does not figure in most histories of psychoanalysis; sometimes he is even absent from histories of psychoanalysis in Britain. See also Richard Overy's *The Morbid Age* (London: Allen Lane, 2009), 143.

25. Rivers, "Totemism in Polynesia and Melanesia," *Journal of the Royal Anthropological Institute* 39 (1909), 173, cited in Frazer's *Totemism and Exogamy* (London: Macmillan, 1910). Freud's citation is in *Totem and Taboo* (1913), in *FSE* 13:118.

26. Rivers, "Freud's Psychology of the Unconscious," in *RIU* 160, originally published in *The Lancet*, June 16, 1917.

27. Rivers, "Psychiatry and the War," *Science* 49, no. 1268 (April 18, 1919), 367–68. An example of this "alteration" is offered in G. Eliot Smith and T. Pear, *Shell Shock and Its Lessons* (London: Longmans, Green, 1917), 64 and passim.

28. On the changes of character after the war, see C. S. Myers, "The Influence of W. H. R. Rivers," in Rivers, *Psychology and Politics* (London: Kegan Paul, 1923), 168.

29. Jones, *Sigmund Freud*, 3:11. The same line is repeated by D. W. Winnicott in his necrology for Jones, in Winnicott, "Ernest Jones," *International Journal of Psycho-Analysis* 39 (1958), 300.

30. See the version of the talk published in the *Proceedings of the Royal Society of Medicine* (1918), 18.

31. John Forrester, *Freud at Cambridge* (Cambridge: Cambridge University Press), 97.

32. Jones to Freud, January 25, 1920, and October 17, 1920, in *The Complete Correspondence of Sigmund Freud and Ernest Jones*, ed. R. Paskauskas (Cambridge, MA: Belknap Press of Harvard University Press, 1993), 364–65 and 394–95 (hereafter *Freud-Jones Correspondence*). On the citation of Rivers, see Jones, "War Shock and Freud's Theory of the Neuroses," originally read before the Royal Society of Medicine, Section of Psychiatry, April 9, 1918, and published in the *Proceedings of the Royal Society of Medicine* 9, reprinted

in Jones, *Papers on Psycho-Analysis* (London: Baillière, 1918), and finally in Sándor Ferenczi, Karl Abraham, Ernst Simmel, and Ernest Jones, *Psychoanalysis and the War Neuroses* (London: International Psycho-Analytical Press, 1921), 44–59, quote on 44.

33. Freud, introduction to Ferenczi et al., *Psychoanalysis and the War Neuroses*, 2.

34. Jones reviewed *Instinct and the Unconscious* twice, first in *International Journal of Psycho-Analysis* 1 (1920), 470–76, at length and with extensive criticisms. The second time, referencing the republication of the book with some added essays (in *International Journal of Psycho-Analysis* 3 [1922], 235), he simply wrote, "The criticisms made in this journal [of the first edition] have been practically ignored, so that they apply equally to the present edition." Jones was much more generous toward Rivers's *Dreams and Primitive Culture* in *International Journal of Psycho-Analysis* 1 (1920), 333–34.

35. Rivers dates his preface to July 15, 1920. Ulrike May has pointed to July 1920 as the end date for Freud's revisions.

36. On Freud's understanding of censorship, including a note on Rivers's critique, see Peter Galison, "Blacked-Out Spaces," *British Journal for the History of Science* 45 (June 2012), 235–66.

37. Consider, for example, Freud's use of McDougall's *The Group Mind* in his *Group Psychology and the Analysis of the Ego*, which is interesting because McDougall's book came out in 1920—at about the same time Freud was composing *Group Psychology*. McDougall, *The Group Mind* (New York: Putnam's Sons, 1920); and Freud, *Group Psychology and the Analysis of the Ego*, in *FSE* vol. 18.

38. Freud, "Critical Introduction to Neuropathology"; Katja Guenther, "Recasting Neuropsychiatry," *Psychoanalysis and History* 14, no. 2 (2012), 151–226; and Freud to Fliess, April 27, 1895, in *Complete Letters of Sigmund Freud to Wilhelm Fliess*, ed. Jeffrey M. Masson (Cambridge, MA: Harvard University Press, 1985), 127.

39. Freud to Fliess, May 25, 1895, in *Complete Letters of Sigmund Freud to Wilhelm Fliess*, ed. Jeffrey M. Masson (Cambridge, MA: Harvard University Press, 1985), 129. For an elaboration of this argument, see Stefanos Geroulanos, "The Brain in Abeyance," *History of the Present* 1, no. 2 (Fall 2011), 219–43, esp. 226.

40. Rivers, "Introduction," *RTS* 2:1–2.

41. *RIU* 29, 25.

42. W. H. R. Rivers and Henry Head, "A Human Experiment in Nerve Division," *Brain* 31 (1908), 323–450, republished in Head, *Studies in Neurology* (London: Oxford University Press, 1920), 1:225–328.

43. Head, *Studies in Neurology*, 1:259, 279–80 and passim.

44. Head, *Studies in Neurology*, 2:536.

45. See also "the return of the protopathic" in Young, *Harmony of Illusions*, 63.

46. *RIU* 23.

47. Ibid., 23.

48. For a critique of Rivers's interpretation of the protopathic and the

epicritic (including the distinction between the two), see the discussion in C. S. Breathnach, "W. H. R. Rivers and the Hazards of Interpretation," *Journal of the Royal Society of Medicine* 86 (1993), 413–16; and R. A. Hensen, "Henry Head," *British Medical Bulletin* 33 (1977), 91–96.

49. *RIU* 25, 46.
50. Rivers, "Why Is the Unconscious Unconscious? II," *British Journal of Psychology* 9, no. 2 (October 1, 1918), 236.
51. Ibid., 238.
52. *RIU* 29.
53. Rivers, "Why Is the Unconscious Unconscious? II," 240–41, 244.
54. *RIU* 25.
55. Henrika Kucklick, *The Savage Within* (Cambridge: Cambridge University Press, 1991), 159.
56. Rivers, "Suppression and Inhibition," in *RIU* 25.
57. *RIU* 62, 246, 37, 91.
58. Young, *Harmony of Illusions*, 48.
59. Several contemporary authors objected to this shift of concepts as changing the accepted meaning of *repression* into *suppression*. See D. Bryan, "Functional Nerve Disease," *International Journal of Psycho-Analysis* 1 (1920), 332. See also the comments by W. H. B. Stoddart on Rivers's "Repression of War Experience," *Proceedings of the Royal Society of Medicine* 11 (1918), 19.
60. Rivers, "Repression of War Experience," in *RIU* 185.
61. Ibid., 186.
62. Ibid.
63. Ibid.
64. Ibid., 189 and 197.
65. Ibid., 191.
66. Ibid., 197.
67. Kathleen Frederickson, *The Ploy of Instinct* (New York: Fordham University Press, 2014), 3.
68. Ibid.
69. On animals, see *RIU* 40. On its being biologically innate, see *RIU* 41.
70. *RIU* 52.
71. Rivers, *Psychology and Politics*, 42–45, 49–50.
72. Rivers, "Repression of War Experience," in *RIU* 196.
73. Jones, "War Shock and Freud's Theory of the Neuroses," 44.
74. Ibid., 57.
75. Freud to Jones, February 18, 1919, in *Freud-Jones Correspondence*, 334.
76. Ibid, 333–34.
77. Ulrike May, "Der dritte Schritt in der Trieblehre," *Luzifer-Amor* 51 (2013), 92–169, appearing in English as "The Third Step in Drive Theory," *Psychoanalysis and History* 17, no. 2 (July 2015), 205–72, 209. Citations from English edition.
78. May, "Third Step in Drive Theory," 214; and Ilse Grubrich-Simitis, *Back to Freud's Texts* (New Haven, CT: Yale University Press, 1996), 183.

79. May, "Third Step in Drive Theory," 207.
80. *BPP* 38/46/180.
81. Ibid., 9/5/153. See also the similar passage on ibid., 7/3/151.
82. Ibid., 10/7/154.
83. Ibid., 20/21/163.
84. Ibid., 11/8-9/154-55.
85. More precisely, in *Beyond the Pleasure Principle*, Freud distinguishes between the ego and the repressed (*BPP* 19/20/162). In then-recent essays he had cast repression as occurring in "all three at the border of the system Ucs and Pcs." See, e.g., Freud, "Overview of the Transference Neuroses," in *A Phylogenetic Fantasy*, by Sigmund Freud, ed. Ilse Grubrich-Simitis (Cambridge, MA: Belknap Press of Harvard University Press, 1987, p. 5). Of course, in *The Ego and the Id*, he would famously cast much of the ego itself as unconscious, thereby resolving any ambiguity.
86. As Lydia Marinelli and Andreas Mayer showed in *Dreaming by the Book* (New York: Other Press, 2003), the period 1900–1915 involved the repeated rewriting of the *Interpretation of Dreams* as if by committee, multiplying the symbolic figures and complexes to be identified in dreams. After the removal of Jung from the movement, the excision of many of these symbols and complexes aimed at a control of the theory and the movement. Something analogous can be said about Freud's theory of the drives, in particular about his anxiety that these not accumulate to the point of decentering his basic structure. Rivers and other unorthodox analysts had, like Jung, variegated the instincts (including even hunting instincts) so that the category of instinct would appear to be merely an internal readiness to carry out something. Freud mocked this approach (*BPP* 51/61/191).
87. Freud, *Introductory Lectures on Psycho-analysis*, pt. 3, in *FSE* 16:275. See also his way of moving from the traumatic experience, obsessively repeated by the patient, toward a recognition of the unconscious but not to an internally caused overflow (278).
88. Freud to Jones, February 18, 1919, in *Freud-Jones Correspondence*, 333.
89. Freud, Introduction to *Psychoanalysis and the War Neuroses*, 4, emphasis ours.
90. Laplanche, *Life and Death in Psychoanalysis*, 42–43.
91. Ruth Leys, *Trauma* (Chicago: University of Chicago Press, 2000), 21, and 20, 38.
92. *BPP* 29/33/171.
93. Ibid., 29/32/171.
94. Ibid., 29/33/171.
95. Ibid., 30/33–34/171.
96. Ibid., 29/32/171.
97. Ibid., 29/33/171.
98. Ibid., 35/41/176.
99. Freud never argued for a two-events structure in *Beyond the Pleasure Principle*: what was needed was, strictly speaking, *something more than a single event*. The event could not be traumatic on its own, and either memories

or even internal occasions that had not been properly bound were needed. See also Laplanche, *Life and Death in Psychoanalysis*, 41.

100. BPP 31/36/173.
101. Ibid., 31,12/36,11/173,156.
102. Ibid., 29/33/171.
103. Jean Laplanche and J.-B. Pontalis, *The Language of Psychoanalysis* (London: Hogarth, 1973), 417.
104. Freud, "On Narcissism," in *FSE* 14:67–102.
105. Ibid., 75.
106. See also Freud's admission of the dangers of this redrawing in *Civilization and Its Discontents* (New York: Norton, 1989), 77. In deconstructive studies focusing on tensions in the concept of narcissism, both Laplanche and Mikkel Borch-Jacobsen suggested that the lines dividing self-preservation and ego-libido from object-choice are permeable.
107. See Freud's reference to narcissistic neuroses in his introduction to *Psychoanalysis and the War Neuroses*, 3. See also Freud's reference on *BPP* 33/38/175. The reference to the role of narcissism in injury comes from the second draft of the text.
108. BPP 13, 31/11, 36/156, 173.
109. Ibid., 38/45/179.
110. Ibid., 35/41–42/176–77.
111. Recapitulation is cited on BPP 26, 37/29, 44/168, 178.
112. Ibid., 38/46/180.
113. Ibid., 40/48–49/182.
114. Ibid., 40/48/182.
115. Ibid.; BPP 41/49/182.
116. Ibid., 37/44/179, 39/47/181.
117. Ibid., 41/49/183.
118. Ibid., 39/46–47/180.
119. Ibid., 39/47/181.
120. Lacan, *Ego in Freud's Theory*, 60.
121. Ibid., 61.
122. May, "Third Step in the Drive Theory," 215. Rivers and McDougall, by contrast, insisted on the existence of social drives.
123. May, "Third Step in the Drive Theory," 214.
124. BPP 44/52/185.
125. Ibid., 44/52/185.
126. Ibid., 46/55/187.
127. Ibid., 49/59/189.
128. Ibid., 52/62/192.
129. Ibid., 52/62–63/192; BPP 53/63/193.
130. Patricia Kitcher, *Freud's Dream* (Cambridge, MA: MIT Press, 1992), and for the classic anti-Freudian argument, Frank J. Sulloway, *Freud: Biologist of the Mind* (Cambridge, MA: Harvard University Press, 1979), chs. 7, 10, 11.
131. Sulloway, *Freud*, 259, and chs. 10 and 11. See the critique of Sulloway's reading of Freud's attention to recapitulation and Darwinism in Paul

Robinson, *Freud and His Critics* (Berkeley: University of California Press, 1993), 69–73.

132. This is not to take into account the earlier invocations of cellular biology in the metapsychological essays of the 1910s, notably in "Instincts and their Vicissitudes," 125, and "On Narcissism," 75.

133. BPP 26, 28/29, 32/168, 170.

134. Freud's model of the living vesicle, already in chapter 4 but more emphatically in chapter 6, came alongside similar physiological claims, even commensurability and isomorphism with the physiological form as pursued by Cannon and others. By contrast, his anatomical description of the ego, the unconscious, and so on in *The Ego and the Id* is not of the same order; Freud's reference to a single cell with pseudopods in "On Narcissism" involves the biologistic metaphor that he would pursue further in *Beyond the Pleasure Principle*.

135. Ernest H. Starling, *Elements of Human Physiology* (London: Churchill, 1897), 433.

136. Rivers, "Socialism and Human Nature," in *Psychology and Politics, and Other Essays*, 91.

137. CWB 301.

138. Starling, *Elements of Human Physiology* (London: Churchill, 1892), 29. The passage remains unchanged in the 1897 and 1902 editions (26–27), and the 1907 edition (22–23).

139. Starling, *Elements of Human Physiology* (1892; London: Churchill, 1897), 433, 434.

140. To offer another textbook example, see Friedrich Schenck and August Gürber, *Outlines of Human Physiology* (1899; New York: Henry Holt, 1900), 6, 331.

141. BPP 46/55/187. Freud argued here that the division of protozoa did not amount to eternal life for them (and hence was an exception to the death of every living being), and therefore did not flatly contradict the death drive or the strict division of the life and death drives (BPP 49/59/189) for which he argued. Freud presented Weismann's amoebae as having no death proper, and hence as incapable of sustaining the distinction between life and death drives. He later repeatedly suggested that that was a position that could be overcome. Part of the peculiarity of Freud's formulation of embryology as a basis for the bifurcation of the drives is that eventually he admitted, regarding science and the origin of sexuality, that "we can liken the problem to a darkness into which not so much as a ray of a hypothesis has penetrated." (BPP 57/69/197)

142. For example: "In fact, every cell of the body, like the conscious being, seems to have a power of selection, a power to eschew the evil and chose the good, the good being that which is necessary to its preservation as a unit of the cell community." Starling, *Elements of Human Physiology* (1892), 30.

143. William M. Bayliss, *Principles of General Physiology* (London: Longmans, Green, 1915).

144. BPP 46/55/187. Compare with the strikingly similar passage in Starling, *Principles of Human Physiology* (Philadelphia: Lea & Febiger, 1912), 1347.

145. *BPP* 44/52/185. See also the point on libido "perpetually attempting and achieving a renewal of life" (46/55/187).

146. Ibid., 52/62/192.

147. Ibid., 50/60/190–91.

148. Ibid., 52/62–63/192.

149. Ibid., 53/63–64/193.

150. Because narcissism showed the ego to be libidinally cathected, the self-preservative drives could be distinguished from libido only topographically, not dynamically or in absolute terms. *BPP* 52/63/192. This also meant that the self-preservative drives were now restrained all the more and that the death drive became momentarily tied to a near-empty dualism with the life drives.

151. *BPP* 54/65/194.

152. Ibid., 56/67/196.

153. This understanding of sadism raised a number of minor problems, particularly around masochism, the problem of conjugation, and sexual selection and development (*BPP* 54/66/195), the latter again discussed from a quasi-biological standpoint. Freud concluded a long discussion of the problem of speculation in his "attempt to follow a line of thought and see where it will lead" (*BPP* 59/71/198).

154. *BPP* 39/47/181.

155. Ibid., 55/67/195. See the discussion of the two optima in Laplanche, *Life and Death in Psychoanalysis*, 114.

156. *BPP* 55–56/67/195–96.

157. Ibid., 55/67/195, emphasis ours.

158. Ibid., 59/71/198.

159. Ibid., 63/77/203.

160. See Laplanche's diagrams in *Life and Death in Psychoanalysis*, 114.

161. *BPP* 40/48/182.

162. Ibid., 39/46–47/180.

163. Ibid., 10/7/154, emphasis ours. See also 40/48/182.

164. Ibid., 40/48/181; similarly 46/55/187.

165. Ibid., 55/67/195, emphasis ours.

166. The life drives are linked to "living substance." *BPP* 56, 58/68, 70/196, 198.

167. Ibid., 38/45/179.

168. Ibid., 35/42/177. *They* refers to children but can extend to individuals in general.

169. *TO* 259.

CHAPTER EIGHT

1. John Henderson, *A Life of Ernest Starling* (Oxford: Oxford University Press, 2005), 178–80 and passim.

2. Frank W. Stahnisch and Thomas Hoffmann, "Kurt Goldstein and the Neurology of Movement during the Interwar Years," in *Was bewegt uns?*, ed. Christian Hoffstadt, Franz Peschke, Andreas Schulz-Buchta, and Michael Nagenborg (Freiburg: Projekt, 2010), 285–86. See also Uwe Gerrens, "Psychiater

unter der NS-Diktatur," *Fortscrhitte der Neurologie, Psychiatrie* 67, no. 7 (2001), 330–39.

3. G. Eliot Smith, "Prefatory Note," in *Psychology and Politics*, by W. H. R. Rivers (New York: Harcourt, Brace, 1923), v. See also Rivers's chapters 2–4 in that book.

4. Peter J. Kuznick, *Beyond the Laboratory* (Chicago: University of Chicago Press, 1987), 145, 153–56.

5. John Parascandola, "L. J. Henderson and the Mutual Dependence of Variables," in *Science at Harvard University*, ed. Clark A. Elliott and Margaret W. Rossiter (London: Associated University Presses, 1992), 182.

6. Henderson, *A Life of Ernest Starling*, 105; Charles Lovatt Evans, *Reminiscences of Bayliss and Starling* (Cambridge: Cambridge University Press, 1964), 7.

7. Ernest H. Starling, *Report on Food Conditions in Germany* (Cmd.280), with memoranda by A. P. McDougall and by C. W. Guillebaud (London: His Majesty's Stationery Office, 1919).

8. Starling to Charles Lovatt Evans, October 4, 1918, PP/CLE/D1, Charles Lovatt Evans Papers, Wellcome Library.

9. Starling, *Report on Food Conditions in Germany*, 4. Of course the status of the Central Powers with regard to the regulation of provisions and the "siege" in which they found themselves was a matter of much discussion from the beginning of the war, in both camps. See Alfred E. Zimmern, *The Economic Weapon in the War against Germany* (New York: Doran, 1918).

10. Starling, *Report on Food Conditions in Germany*, 5.

11. Ibid., 8.

12. Ibid., 10–11. On the evaporation of national feeling, see also 13.

13. John M. Keynes, *The Economic Consequences of the Peace* (London: Macmillan, 1920), 218n1.

14. Ibid., 1.

15. Martti Koskenniemi, *The Gentle Civilizer of Nations* (Cambridge: Cambridge University Press: 2004), 4, 112, 224 and passim. On disintegration (and not merely autonomy) as the radical counterconcept to integration, see 243n232.

16. Donna Haraway, *Simians, Cyborgs, and Woman* (New York: Routledge, 1991), 223.

17. Laura Otis, *Membranes* (Baltimore: Johns Hopkins University Press, 1999), 5.

18. Klaus Theweleit, *Male Fantasies*, vol. 2 (Minneapolis: University of Minnesota Press, 1989).

19. Henry Dale, introduction to *Paul Ehrlich*, by Martha Marquardt (New York: Henry Schuman, 1951), xiii.

20. See the classic study by Allan Brandt, *No Magic Bullet* (New York: Oxford University Press, 1987), 97–98.

21. Albert Schäffle, *Bau und Leben des sozialen Körpers* (Tübingen: Lauppsch, 1875–1878); and Paul von Lilienfeld, *Die soziale Psychophysik* (Mitau: Behre, 1877). See I. Bernard Cohen, "An Analysis of Interactions

between the Natural Sciences and the Social Sciences," in *The Natural Sciences and the Social Sciences*, ed. Robert S. Cohen (Dordrecht: Springer, 1994), 53–57. See also Max Weber's discussion of Schäffle in *Economy and Society* (1922; Berkeley: University of California Press, 1978), 15.

22. On psychology and the nation, see Glenda Sluga, *Nation, Psychology, and International Politics, 1870–1919* (London: Palgrave Macmillan, 2007); and Egbert Klautke, *The Mind of the Nation* (New York: Berghahn Books, 2013).

23. Marcel Proust, *Finding Time Again* (London: Penguin, 2003), 79–80.

24. Robert J. Richards, *The Tragic Sense of Life* (Chicago: University of Chicago Press, 2008); Knox Peden, "Alkaline Recapitulation," in *Republics of Letters* 4, no. 1 (2014). Mikkel Borch-Jakobsen, in *The Freudian Subject* (Stanford, CA: Stanford University Press, 1998), emphatically linked Freud to Le Bon. See also the (dated) book by R. A. Nye, *The Origins of Crowd Psychology* (London: Sage, 1975).

25. Bronisław Malinowski offered a reading of Durkheim through the lens of group psychology. Box 22, folder 6, Bronisław Malinowski Papers, London School of Economics and Political Science Archives.

26. Driesch's influence is broad; celebrated today by Jane Bennett as a paragon of progressive thought, Driesch influenced Jan Smuts and right-wing German thought. Smuts, *Holism and Evolution*, 171. See Canguilhem's critique in *Knowledge of Life*, 68–69.

27. Hans Driesch, *Die Überwindung des Materialismus* (Zürich: Rascher, 1935), 59.

28. On Malinowski's opposition to nationalism and his anthropological treatment of international affairs in terms at times appealing to integration, see "Geneva Lectures (at Zimmern's School of International Studies)," 22.2, and "Lecture Notes (War)," 22:4, Malinowski Papers.

29. Reinhart Koselleck, "Krise," in *Geschichtliche Grundbegriffe*, ed. Otto Brunner, Werner Conze and Reinhart Koselleck (Stuttgart: Klett-Cotta, 1982), 617–50, translated as "Crisis," *Journal of the History of Ideas* 67, no. 2 (April 2006), 399. Koselleck argues that the term has moved from medicine to other realms "since the beginning of the nineteenth century" (361).

30. Marcel Mauss, "La nation et l'internationalisme," in *Oeuvres 3* (Paris: Les Éditions de Minuit, 1969), 630–31.

31. Alfred Fried, *Europäische Wiederherstellung* (Zürich: Orell Füssli, 1915), 72, translated as *The Restoration of Europe* (New York: MacMillan, 1916) 83, translation amended.

32. Fried, *The Restoration of Europe*, 11; also 21, 86, and passim.

33. Compare Hobhouse's *Mind and Evolution* (London: Macmillan, 1901) to his *Metaphysical Theory of the State* (London: G. Allen & Unwin, 1918).

34. Hobhouse, *Metaphysical Theory of the State*, 11.

35. Harold Laski, *Studies in the Problem of Sovereignty* (New Haven, CT: Yale University Press, 1917), 6. Laski's claim is in some respects the inverse of Fried's: in his argument unity is only apparent, and it is a consequence of the crisis. Still, our concern is with the crisis-based logic for integration and bodily analogy.

36. Laski, *Authority in the Modern State* (New Haven, CT: Yale University Press, 1919), 262–63.

37. Leonard Woolf, *International Government* (London: George Allen and Unwin, 1916), 150.

38. Vilfredo Pareto, *The Mind and Society* [*Trattato di sociologia generale*] (1916; New York: Harcourt, Brace, 1935), 4:1436, §2068n1.

39. Alfred Zimmern, *Europe in Convalescence* (London: Mills and Boon, 1922), 172–73; for Germany as a "patient," see also 132, 178.

40. Ibid., 242–43.

41. The same, more or less, could be said of J. A. Hobson's use of similar motifs, like "human personality as an organic whole." Hobson, *Problems of a New World* (New York: Macmillan, 1921), 139.

42. Consider also his notes on the disintegration of the Austro-Hungarian Empire and the "hurt" and "struggling" national bodies that emerged from it, in *The Effect of War in South-Eastern Europe* (New Haven, CT: Yale University Press, 1936), 33, 79, 126, and esp. 191. On the Austro-Hungarian Empire's disintegration and figures of bodies and souls, see Sluga, "Bodies, Souls, and Sovereignty," in *Ethnicities* 1 (2001), 207–32.

43. David Mitrany, *The Progress of International Government* (New Haven, CT: Yale University Press, 1933), 102, 137, 89.

44. In the political texts on nationalism and internationalism that he wrote after the Paris Peace Conference, Mauss similarly advocated for his thinly veiled socialism by diagnosing postwar European societies as "dominated" by a "state of absolute economic interdependence." Careful not to write in Durkheim's organicist language, Mauss still returned to bodies and organs, noting at the same time the need for interconnections: "the resupplying of exhausted countries, the reconstruction of devastated ones, have been taken over by international organizations." Marcel Mauss, "Complément à 'La Nation,'" (1920), in *La Nation*, ed. Marcel Fournier and Jean Terrier (Paris: Presses universitaires de France [Quadrige], 2013), 399. Mauss used the term *integration* for domestic and national societies, so his occasional proximity after World War I to an organic or integrationist approach should be taken with a grain of salt, despite his internationalism. G. E. G. Catlin, at the time attached to Fabianism, proposed it as a long-standing basis for political philosophy. See Catlin, *A Study of the Principles of Politics* (New York: Macmillan, 1930), 458. As Karl Popper would note in *The Open Society and Its Enemies* (London: Routledge, 1946), 1:267n68, Catlin referenced Thomas Aquinas, George Santayana, and even J. S. Mill to bolster his claim.

45. Henri Bergson, *The Two Sources of Morality and Religion* (1932; London: MacMillan, 1935), 248.

46. "Disintegrating effect" is in Bergson, *Two Sources of Morality and Religion*, 238. See Philippe Soulez, "Bergson as Philosopher of War and Theorist of the Political"; Alexandre Lefebvre, "Bergson and Human Rights," in *Bergson, Politics and Religion*, ed. Alexandre Lefebvre and Melanie White (Durham, NC: Duke University Press, 2012), 99–125 and 206 (for "faint" faith); and Jimena Canales, "Einstein, Bergson, and the Experiment that Failed," *Modern Language Notes* 120, no. 5 (2005), 1168–91.

47. Jan Smuts, *Holism and Evolution* (New York: Macmillan, 1926). Christopher Lawrence and George Weisz suggest that *holism* was coined by Smuts. See their "Medical Holism," in *Greater Than the Parts*, 2.

48. See especially Smuts, *Holism and Evolution*, 140–41.

49. Mark Mazower, *No Enchanted Palace* (Princeton, NJ: Princeton University Press, 2009), ch. 1.

50. At the domestic level an imaginary of integration was far less prevalent, whether in Germany and France, where right wing politics veered between traditional conservatism and a reactionary revolution that made use of a different register of biological and racial metaphors, or in Britain and the United States, where liberalism and British imperialism guarded against visions of collaborative, networked organization.

51. *TO* 388.

52. *The Organism* could be considered a project tied to the rise and fall of the Weimar Republic, partly because of the upward trajectory of Goldstein's career as a German Jew and because of Goldstein's blend of socialist politics, a liberal concept of the self, a nostalgia for unity, and concurrent commitments to rationalism and holism that might also be called decidedly *weimarsche*. Even more pregnant in this regard are elements that turn *The Organism* into a metaphor for Weimar culture: disease as a disorder that may be stabilized or may suddenly bring about a catastrophic situation; the individual as a whole, not merely a sum of parts, and as capable of autonomy vis-à-vis the milieu; and the therapeutic focus on facilitating, for the injured self, a new if restricted totality. See also Mitchell Ash, "Weimar Psychology" in Peter Eli Gordon and John P. McCormick, *Weimar Thought* (Princeton, NJ: Princeton University Press, 2013), 35–54.

53. On one-world proposals, see Glenda Sluga, *Internationalism in the Age of Nationalism* (Philadelphia: University of Pennsylvania Press, 2013), 84–85.

54. William H. Beveridge, *Unemployment* (London: Longmans, Green, 1909), 38, 44.

55. John M. Keynes, *A Tract of Monetary Reform* (London: Macmillan, 1924), 65.

56. Ernst Wagemann, *Economic Rhythm* (New York: McGraw-Hill, 1930), 58, 217; see also 13. Our thanks to Jamie Martin for pointing us to Wagemann; also see Adam Tooze, *Statistics and the German State* (Cambridge: Cambridge University Press, 2001), chs. 2–4.

57. Wagemann, *Economic Rhythm*, 10.

58. Ibid., 10–11.

59. Ibid., 217.

60. Ibid., 10–11.

61. Wilhelm Röpke, *International Economic Disintegration* (London: William Hodge, 1942), 272–73.

62. Ibid., 7.

63. Ibid., 33.

64. Ibid., 258.

65. Ibid., 261.

66. Ibid., 69.

67. Cynthia Russett, *The Concept of Equilibrium in American Social Thought* (New Haven, CT: Yale University Press, 1968); Barbara S. Heyl, "The Harvard 'Pareto Circle,'" *Journal of the History of the Behavioral Sciences* 4 (1968), 316–334. See also Stephen J. Cross and William R. Albury's useful, albeit politically undifferentiating "Walter B. Cannon, L. J. Henderson, and the Organic Analogy," *Osiris* 3 (1987), 165–92; and Elizabeth Colson's critique of cultural homeostasis, in "Culture and Progress," *American Anthropologist* 78 (1976), 261–71.

68. See the celebration of integration, "a new panoramic unity, new in governmental planning," in "Mr. Hoover at the Turning Point," *New York Times*, March 1, 1931; Hoover's 1931 State of the Union address, *New York Times*, December 9, 1931, and "Congress Mirrors a National Dilemma," *New York Times*, February 22, 1931.

69. See also Joel Isaac, "Making a Case," in *Working Knowledge: Making the Human Sciences from Parsons to Kuhn* (Cambridge, MA: Harvard University Press, 2011), 63–91.

70. L. J. Henderson, *The Order of Nature* (Cambridge, MA: Harvard University Press, 1917).

71. Pareto, *Mind and Society*, 4:1436, §2068n1. Henderson uses this passage in *Blood* (New Haven, CT: Yale University Press, 1928), 362. See the discussion in John Parascandola, "Organismic and Holistic Concepts in the Thought of L. J. Henderson," *Journal of the History of Biology* 4, no. 1 (1971), 104.

72. Henderson exulted interconnections and collaboration with Mayo, the Business School, the McLean Hospital, and the Department of Anthropology in a letter to David L. Edsall, January 21, 1937, Henderson Papers 1:47.

73. Walter B. Cannon, "Biocracy: Does the Human Body Contain the Secret of Economic Stabilization?" *Technology Review* 35, no. 6 (March 1933), 203, 206.

74. Ibid., 227.

75. Ibid., 206.

76. John Dewey, "Authority and Social Change," in *Authority and the Individual* (Cambridge, MA: Harvard University Press, 1937), 181, our emphasis.

77. HNP 232–36.

78. Pitirim Sorokin, *Contemporary Sociological Theories* (New York: Harper, 1928), ch. 4. George Lundberg, *Foundations of Sociology* (New York: Macmillan, 1939), chs. 5–6. See also R. W. Gerard, "Higher Levels of Integration," *Science* 95, no. 2445 (March 27, 1942) 309–13.

79. W. H. Manwaring, "The Organic Theory of the State," *Scientific Monthly* 47, no. 1 (July 1938), 48–50. Manwaring was professor emeritus of bacteriology and pathology at Stanford. See also Franz Neumann, *Behemoth* (1942; Chicago: Ivan R. Dee, 2009), 139–40. Neumann traced the construction of the "blood and soil" narrative out of organicism in part to the writings of the geographer Friedrich Ratzel and political scientist Rudolf Kjellen.

80. On Parsons in relation to Henderson, Cannon, and cybernetics, see Ronald Kline, *The Cybernetics Moment* (Baltimore: Johns Hopkins University Press, 2015), 148; and Lundberg, "Human Social Problems as a Type of

Disequilibrium in a Biological Organism," *American Sociological Review* 13, no. 6 (December 1948), 689–99, citation of Cannon on 690, 691.

81. Gerard's "Higher Levels of Integration" was severely criticized by Alex B. Novikoff, "The Concept of Integrative Levels and Biology," *Science* 101, no. 2608 (March 2, 1945), 209–15.

82. Lewis Mumford, *The Condition of Man* (New York: Harcourt, Brace. 1944), 5–6.

83. *CNP* 254. By the 1950s, Canguilhem had read Cannon more closely and now considered metaphors of social integration very troubling. See *Writings on Medicine* (New York: Fordham University Press, 2012), 67–78. In the 1970s, he dropped some of this opposition; we discuss Canguilhem's engagement with regulation in our "Canguilhem's Critique of Medical Reason," in ibid., 18–21.

84. Georges Bataille, *The Accursed Share*, vol. 1, *Essay on General Economy* (1949; New York: Zone Books, 1991), 23, 177.

85. Ibid., 106–7.

86. Cannon, "Organization for Physiological Homeostasis," *Physiological Reviews*, 9, no. 3 (July 1929), 400.

87. See also Cannon's more recent influence on Francisco Varela's concept of *autopoiesis*; William Bechtel, "Addressing the Vitalist's Challenge to Mechanistic Science," in *Vitalism and the Scientific Image in Post-Enlightenment Life Science*, ed. Sebastian Normandin and Charles T. Wolfe (Heidelberg: Springer, 2013), 347–70, esp. 358.

88. *CBC* 1915: 285, 285–86.

89. Ibid., 288.

90. Ibid., 293.

91. Ibid., 25.

92. Ibid., 309.

93. Ibid., 310.

94. Ibid., 311–12.

95. Canguilhem points out a proximity between Cannon and Bergson in *Writings on Medicine*, 74–76.

96. Despite his left-wing sympathies, a Freedom of Information Act request on Cannon (FOIA No. 1332710-000) did not yield any information.

97. Cannon, *The Body as a Guide to Politics* (London: Watts, 1942), 3, 7–8.

98. Ibid., 5–6.

99. Ibid., 23–24. "Analogue" appears on p. 24.

100. Ibid., 43–44.

101. Ibid., 29.

102. Ibid., 30.

103. Cannon, "Biocracy," 227.

104. On matters of economic and financial disturbance, Cannon writes that "stabilizing agencies do not act on the instant to guarantee security; instead, disruptive factors have full sway." Cannon, *Body as a Guide to Politics*, 27.

105. On dictators, see ibid., 45.

106. *HNP* 221.
107. Cannon, *Body as a Guide to Politics*, 36 and 35.
108. Ibid., 30.
109. On defensive action, see ibid., 30.
110. Ibid., 40.
111. Cannon, "Voodoo Death," *American Anthropologist* 44, no. 2 (April 1942), 169–81.
112. Malinowski, *Argonauts of the Western Pacific* (London: G. Routledge and Sons, 1922), chs. 17–18.
113. Barbara Tepa Lupack, *Literary Adaptations in Black American Cinema* (Rochester, NY: University of Rochester Press, 2002), 226.
114. See his list of recipients and their replies in "Working Papers on Voodoo Death," box 67, folder 902, Cannon Papers.
115. Kingsley Roth to Cannon, June 6, 1935, Cannon Papers 67:899. Cannon acknowledged the problem of believable reporting in "Voodoo Death," 171.
116. This goes counter to the strong cultural studies argument proposed by Otniel E. Dror, "'Voodoo Death': Fantasy, Excitement, and the Untenable Boundaries of Biomedical Science," in *The Politics of Healing*, ed. Robert D. Johnston (London: Routledge, 2004), 71–81. Dror argues that Cannon saw in the correspondence what he wanted to believe and did not appreciate his correspondents' mistrust of the vagaries and prurient myths regarding Voodoo, and its racially oriented cultural baggage. Ignoring the wound shock referent and, more important, the considerable number of letters reporting evidence of unexplained death—most of them from outside the United States and hence less dependent on the Voodoo language—Dror further overemphasizes the "decerebrate cat" as Cannon's epistemic object. But Dror goes too far to suggest that Cannon's point was (even unconsciously) one of a Christian and racial primitivization of others (81), which was then reimported in Hans Selye's language of stress. This was certainly a collateral effect, but Cannon's point was not to show how others were superstitiously attached to these vagaries; it was to demonstrate the capacity of any human body to succumb to emotional and social force. Shock and magic were the two then-standard reference points for such emotional effect. However biased the presuppositions of Cannon and his correspondents were, non-Western medicine and magic provided the medico-anthropological context for recovering the operations of the human body, and it mattered that Cannon's essay was read as an invitation not to look askance at "primitives" but to understand how emotional force caused lesions and functional disorder. This is how Lévi-Strauss (discussed by Dror on 71) and Selye read the work. The absence of wound shock from the historiography has resulted in readings like Dror's.
117. R. F. A. Hoernle to Cannon, March 20, 1935, Cannon Papers 67:900. Hoernle sent Cannon a 1929 report written by a lance corporal in the South African mounted police of an incident of bone-pointing resulting in death. Hoernle to Cannon, June 4, 1935, Cannon Papers 67:900.
118. H. L. Arnold to Cannon, July 11, 1934, Cannon Papers 67:899.
119. Laurie Sharpe to Cannon, November 3, 1934, Cannon Papers 67:901.

120. Roger Choisser to Cannon, December 4, 1934, Cannon Papers 67:898.

121. George W. Harley to Cannon, December 3, 1934, and April 29, 1935, Cannon Papers 67:900. Cannon noted that S. M. Lambert (Suva, Fiji) had answered his letter of October 1, 1934, but the letter is missing from his correspondence, in Cannon Papers 67:902. See also Cannon to Cleland, January 8, 1935, Cannon Papers, 67:901. Cleland confirmed (March 7, 1935) that poison was not an issue in bone-pointing practices in Australia.

122. S. D. Porteus to Cannon, January 5, 1935, Cannon Papers 67:899. In April 1935, Basedow's widow confirmed that her recently deceased husband had believed in the capacity of bone pointing to lead to death; see Nell Basedow to Cannon, April 23, 1935, Cannon Papers 67:901.

123. See the correspondence between Warner and Cannon in Cannon Papers 67:903.

124. Cannon, "Voodoo Death," 175–76.

125. Cannon discussed shock and the draining of blood in detail in ibid., 178–79, 180–81. On systemic failure, see ibid., 178.

126. Ibid., 179.

127. James Siegel diagnosed Lévi-Strauss's attempts to reconcile the physical with the psychical, to tame the unknown (through Cannon and others), as a problem not so much of evidence but of the shortcoming of imagined finality. Siegel, "The Truth of Sorcery," *Cultural Anthropology* 18, no. 2 (2003), 135–55, 153.

128. Norbert Wiener, *Cybernetics*, 2nd ed. (1948; Cambridge, MA: MIT Press, 1961), 2, 3.

129. Ibid., 80.

130. The literature on Wiener's synthesis is huge; among recent works are Andrew Pickering, *The Cybernetic Brain* (Chicago: University of Chicago Press, 2011), which includes a note on the pickup of homeostasis by Ashby, Ostwald, Edgell, and Humphrey on 423n12; Kline, *The Cybernetics Moment*, passim; and David Bates's work on W. Ross Ashby and Kurt Goldstein, including "Unity, Plasticity, Catastrophe: Order and Pathology in the Cybernetic Era," in *Catastrophes*, ed. Andreas Killen and Nitzan Lebovic (Berlin: De Gruyter, 2014), 32–54.

131. Wiener, *Cybernetics*, 8, 1. Cannon and Rosenblueth had published *Autonomic Neuro-Effector Systems* (New York: Macmillan, 1937), and, after Cannon's death, *The Supersensitivity of Denervated Structures* (New York: Macmillan, 1949)

132. Ibid., 114–15. See also the reference to Cannon on 17.

133. Wiener, *The Human Use of Human Beings: Cybernetics and Society* (1950; London: Free Association Books, 1989), 96.

134. Wiener, *Cybernetics*, 95–96, 98.

135. Ibid., 8, 107, 113.

136. Ibid., 156.

137. Ibid., 160.

138. Ibid., 161–62.

139. Ibid., 160.

140. Ibid., 162.
141. Ibid., 160.
142. Ibid., 162.
143. Goldstein lifted Cassirer's concepts of forming and acting wholesale from the *Philosophy of Symbolic Forms* into *The Organism* (*TO* 358).
144. Goldstein Papers, box 1, folder "Ernst Cassirer." See Cassirer to Goldstein, January 5, 1925, in Cassirer, "Two Letters to Kurt Goldstein," *Science in Context* 12, no. 4 (1999), 663, translation amended; Cassirer to Goldstein, March 24, 1925, in "Two Letters," 665. See also Ernst Cassirer, *Philosophy of Symbolic Forms*, vol. 3, *The Phenomenology of Knowledge* (New Haven, CT: Yale University Press, 1957), 239. Cassirer further discusses Schneider on 237–38, 253–54, and 267.
145. As Peter E. Gordon has suggested, the encounter with aphasia would also become a central component of Cassirer's critique of Martin Heidegger after Davos; see his *Continental Divide: Heidegger, Cassirer, Davos* (Cambridge, MA: Harvard University Press, 2010), 254–56.
146. Cassirer, *Philosophy of Symbolic Forms*, 3:xiii–xiv.
147. Ibid., 191. For sensationalism, see 192–93, 217; for associationism, see 235, 268; for Kant, see 194, 195. On his appreciation and critique of Edmund Husserl along similar lines, see 196–97.
148. Ibid., 202.
149. Ibid., 235–36.
150. Ibid., 243.
151. Ibid., 238, quoting from Gelb and Goldstein, "Zur Psychologie des optischen Wahrnehmungs- und Erkennungsvorganges," in *Psychologische Analysen hirnpathologischer Fälle* (Leipzig: Barth, 1920), 128n.
152. Cassirer, *Philosophy of Symbolic Forms*, 3:234.
153. Ibid., 240.
154. Ibid., 227n, 246–47.
155. Ibid., 209.
156. Ibid., 276.
157. Durkheim and Mauss, "De quelques formes primitives de classification," *L'année sociologique* 6 (1903), 22n, reprinted in Mauss, *Oeuvres II* (Paris: Minuit, 1969), 35n73.
158. Mauss detailed Haddon's report of the decay of indigenous institutions and criticized its idea that the purpose of the expedition was as a salvage operation. See Mauss, *Oeuvres II* (Paris: Minuit, 1969), 421–22, and 440.
159. Mauss's long review of Rivers's 1910 *The Todas* appeared in *L'année sociologique* in 1911; see Mauss, *Oeuvres I* (Paris: Minuit, 1969), 491–500. Marcel Fournier dates their meeting to Mauss's extended stay in London in 1912; see Marcel Fournier, *Marcel Mauss* (Princeton, NJ: Princeton University Press, 2008), 164. No correspondence survives.
160. Mauss, "Rapports réels et pratiques de la psychologie et de la sociologie," 294.
161. Mauss's manuscript obituary for Rivers can be found in 2AP 4-2 C/2/e, Marcel Mauss Papers, Muséum national d'histoire naturelle, and in its published version in *Oeuvres III*, 465–72. Mauss's dismissal of Rivers's

posthumously published *Conflict and Dream* (coupled with admiration for *Instinct and the Unconscious*) is in "Le Rêve et la pensée mythique selon Rivers (1925)," in *Oeuvres II*, 287–88. The posthumous critique of Rivers for falling for the "unicentric diffusion of civilization" theory (known as diffusionist and heliolithic theory), proposed by G. Elliot Smith and W. J. Perry (which Rivers controversially supported from his 1914 *History of Melanesian Society* onward) can be found in *Oeuvres II*, 513, 518. But even there, Mauss attempted to save him by suggesting that Rivers never believed Perry's idea that civilization was born specifically in Egypt; elsewhere, he reported on the debate between diffusionism and evolutionism as a debate concerning the use of history in sociology. "Rapports réels et pratiques de la psychologie et de la sociologie" (1924), in *Sociologie et anthropologie* (Paris: Presses universitaires de France, 1950), 288. See also the editor's remarks in *Oeuvres I*, xli, to the effect that disagreements were grounded in Rivers's changing methodology, which distanced him from the Durkheim school.

162. Mauss, *Oeuvres I*, 556.

163. Head and Mauss may have maintained a correspondence as well as communicating through Rivers, but only one letter from Head to Mauss survives. It is dated 1922 and ends "Your ami." See MAS 5.87, Fonds Marcel Mauss, Bibliothèque du Collège de France.

164. Mauss, "The Obligatory Expression of Sentiments (in Australian Oral Funerary Rites)," in *Oeuvres III* (Paris: Minuit, 1969), 269–78, citation in 269. Originally published in *Journal de psychologie* 18 (1921).

165. Ibid., 277.

166. Ibid., 278.

167. Mauss, *Oeuvres II*, 122–23.

168. Ibid., 121.

169. Ibid.

170. Mauss, *Oeuvres III*, 259; C. K. Ogden and I. A. Richards, *The Meaning of Meaning* (New York: Harcourt Brace, 1923); and C. K. Ogden, "Word magic," *Psyche* 18 (1933–1952), 19–126.

171. Mauss, *Oeuvres III*, 260.

172. Ibid., 261.

173. Mauss, "Rapports réels et pratiques de la psychologie et de la sociologie," 289–90.

174. Ibid., 294.

175. Ibid.

176. Ibid., 299–300.

177. Even Mauss's theory of gift giving is fundamentally affected. Without citing Head, Mauss recalled the same reference points as earlier—namely obligation, disinterestedness, and collectivity as essential for the functioning of society and symbolism, to articulate the same conception of the symbolic as permeating all social and moral life. He also cited physiological and psychological holism. Mauss, *The Gift* (London: Routledge, 1942), 42, 80.

178. Mauss, "Rapports réels et pratiques de la psychologie et de la sociologie," 305.

179. Ibid., 290.

180. As Bruno Karsenti argues, this totality undid Durkheim's *Homo duplex*, which separated the individual from the social. Bruno Karsenti, *L'Homme total* (Paris: Presses universitaires de France, 2011), 100–101. Karsenti is the only author to have paid serious attention to Mauss's readings of Head and Cassirer; he does not, however, signal the all but causal relation between the notion of the symbolic as Mauss derived it from Head and the ultimate direction this would take.

181. Mauss, "Les techniques du corps" (1936), in *Sociologie et anthropologie*, 372, translated as "Techniques of the body," *Economy and Society* 2, no. 1 (1973), 70–88, citation on 76 (alternate edition in *Techniques, Technology and Civilization*, ed. Nathan Schlanger [New York: Durkheim Press, 2006], 83), translation amended.

182. Mauss, "Une catégorie de l'esprit humain: la notion de personne, celle de 'moi'," in *Sociologie et anthropologie*, 335. First published in the *Journal of the Royal Anthropological Institute* 68 (1938), 263–81.

183. Warren Breckman correctly argues that Mauss's "most basic innovation" was "his shift from individual symbols to symbolic systems. Mauss is not interested in deciphering isolated elements of this exchange of symbolic goods, but in the dynamic exploration of relations that are generated and mediated by the symbolic system." Breckman, *Adventures of the Symbolic* (New York: Columbia University Press, 2013), 17.

184. French commentators, among them Fournier and Karsenti, routinely distance Mauss from Lévi-Strauss's interpretation in *Introduction to the Work of Marcel Mauss*, emphasizing continuities across his writing and especially continuities of his later work with Durkheim's. But Lévi-Strauss and Breckman are right to emphasize the distance of the later Mauss from Durkheim and his own earlier writing.

185. Claude Lévi-Strauss, *Structural Anthropology* (New York: Basic Books, 1963), 162, translation amended.

186. For early expressions of this criticism, see Patrick Wilcken, *Claude Levi-Strauss* (New York: Penguin, 2012), 185.

187. Lévi-Strauss, *Introduction to the Work of Marcel Mauss* (London: Routledge & Keegan Paul, 1987), 3, originally published in *Sociologie et anthropologie* (Paris: Presses universitaires de France, 1968), 9–44.

188. Lévi-Strauss, *Introduction to the Work of Marcel Mauss*, 12.

189. "The Idea of Death," "Techniques of the Body," and "A Category of the Human Mind: The Notion of the Person."

190. Lévi-Strauss, "La notion de structure en ethnologie (i)" (1952), in *Anthropologie structurale* (Paris: Plon, 1958), 316–17, translated as "Social Structure," in *Structural Anthropology* (New York: Basic Books, 1963), 288, translation amended.

191. Lévi-Strauss, "La notion de structure en ethnologie (i)," 307; "Social Structure," 280, translation amended.

192. Lévi-Strauss, "The Sorcerer and His Magic" (1949), in *Structural Anthropology*, 167–68.

193. On language and the difference between Lévi-Strauss and Mauss, see in particular Siegel, *Naming the Witch* (Stanford, CA: Stanford University

Press, 2006), 44. Siegel convincingly argues that Lévi-Strauss bent the accounts he ostensibly reproduced in "The Sorcerer and His Magic" to fit his structural model.

194. Lévi-Strauss, "Sorcerer and His Magic," 181.
195. Ibid., 181–82.
196. On Lévi-Strauss and the site of language, see Stefanos Geroulanos, *Transparency in Postwar France* (Stanford, CA: Stanford University Press, 2017), chs. 13–14.
197. Lévi-Strauss, *Introduction to the Work of Marcel Mauss*, 12.
198. Lévi-Strauss, "The Sorcerer and his Magic," 184.

CHAPTER NINE

1. *SWB*, citation from Job on 685.
2. Ibid., 685.
3. Ibid., 688.
4. Ibid., 685.
5. Ibid., 687, 688.
6. Ibid., 688.
7. Ibid., 690.
8. Ibid., 686.
9. Ibid., 690.
10. Ibid.
11. On the motivations and development of the welfare states in the interwar period, see the standard work by Susan Pedersen, *Family, Dependence, and the Origins of the Welfare State* (Cambridge: Cambridge University Press, 1995), 24 and ch. 2. See Karen Offen, "Body Politics," in *Maternity and Gender Policies*, ed. Gisela Bock and Pat Thane (New York: Routledge, 1991). Regarding modern industrial demands on health, see Henry Sigerist, *Medicine and Human Welfare* (New Haven, CT: Yale University Press, 1941), 132, 134. Sigerist also published a then-contemporary history of medical welfare, "From Bismarck to Beveridge," *Bulletin of the History of Medicine* 13 (1943), 365–88.
12. Theodore M. Porter, *The Rise of Statistical Thinking* (Princeton, NJ: Princeton University Press, 1986).
13. Adam Tooze, *Statistics and the German State* (Cambridge: Cambridge University Press, 2001), chs. 2–3.
14. Sigerist, *Medicine and Human Welfare*, 139.
15. Ibid., 145.
16. Sigerist, *Socialized Medicine in the Soviet Union* (New York: W.W. Norton, 1937). See also Sigerist's translation and introduction of Johann Peter Frank's address, "The People's Misery: Mother of Diseases" (1790), *Bulletin of the History of Medicine* 9, no. 1 (1941): 81–100.
17. Sigerist, *Socialized Medicine in the Soviet Union*, 310, 307.
18. George Weisz, *Divide and Conquer* (New York: Oxford University Press, 2005), 191.
19. Fielding H. Garrison, *Introduction to the History of Medicine*, 3rd ed. (Philadelphia: Saunders, 1921), 589–91. "Labyrinth of specialties" is François

Dagognet's term in *Philosophie biologique* (Paris: Presses universitaires de France, 1955), 3.

20. George Weisz, *Chronic Disease in the Twentieth Century* (Baltimore: Johns Hopkins University Press, 2014), ch. 2.

21. Ibid., 31–32; and Jeremy Greene, *Prescribing by Numbers* (Baltimore: Johns Hopkins University Press, 2007), 1–17.

22. Harry M. Marks, *The Progress of Experiment* (New York: Cambridge University Press, 1997), 17–19.

23. Short critical accounts of the recurrence of Hippocratism are offered by Erwin H. Ackerknecht, *Therapeutics* (New York: Hafner, 1973), 157; and Georges Canguilhem, "The Idea of Nature in Medical Theory and Practice," *Writings on Medicine* (New York: Fordham University Press, 2012), 27–33 and passim.

24. Ivan Illich, *Medical Nemesis* (New York: Pantheon, 1976), 11. Our study adds to the rich histories of this fragmentation. See Anne Harrington, *The Cure Within* (New York: Norton, 2008); and the contributions to Christopher Lawrence and George Weisz, eds., *Greater Than the Parts* (New York: Oxford University Press, 1998).

25. Arturo Castiglioni, "Neo-Hippocratic Tendency of Contemporary Medical Thought," *Medical Life* 41 (1934), 115.

26. René Leriche, "Conclusion générale?" in *Encyclopédie française*, ed. René Leriche, vol. 6, *L'Être humain* (Paris: L'Encyclopédie française, 1936), 6.76-6, 6.76-7.

27. Bernhard Aschner, *Die Krise der Medizin*, 6th ed. (Stuttgart: Hippokrates, 1934), 16, discussed in Carsten Timmermann, "Weimar Medical Culture" (PhD diss., University of Manchester, 1999), 10. Timmermann considers Aschner's critique of mechanistic medicine (*Die Krise der Medizin*, 5–7) on 90–93. Aschner would become well known in the United States through his *The Art of the Healer* (New York: Dial, 1942). See also Timmerman, "Constitutional Medicine, Neoromanticism, and the Politics of Antimechanism in Interwar Germany," *Bulletin of the History of Medicine* 75, no. 4 (2001), 717–39.

28. Lawrence and Weisz, "Medical Holism," in *Greater than the Parts: Holism in Biomedicine, 1920–1950*, ed. Lawrence and Weisz (*New York: Oxford University Press, 1998)*, 1.

29. Allan Young, *The Harmony of Illusions* (Princeton, NJ: Princeton University Press, 1995), ch. 2.

30. Cannon, "Organization for Physiological Homeostasis," *Physiological Reviews* 9, no. 3 (July 1929): 399, reprinted in *CWB* 20–21.

31. *CWB* 240, 243.

32. Canguilhem, "Idea of Nature," 29, citing Cannon, *CWB* 240, 242.

33. Stephen J. Cross and William R. Albury, "Walter B. Cannon, L. J. Henderson, and the Organic Analogy," *Osiris* 3 (1987), 178.

34. Cannon, "Reasons for Optimism in the Care of the Sick," *New England Journal of Medicine* 199, no. 13 (September 27, 1928), 597, reprinted with minor modifications in *CWB* 240–41.

35. Mark Jackson, *The Age of Stress* (Oxford: Oxford University Press,

2013), 11; Hans Selye, *The Story of the Adaptation Syndrome* (Montreal: Acta Medica, 1952); Selye, *The Physiology and Pathology of Exposure to Stress* (Montreal: Acta Medica, 1950); Jackson, "Evaluating the Role of Hans Selye and the Modern History of Stress," in *Stress, Shock, and Adaptation in the Twentieth Century*, ed. David Cantor and Edmund Ramsden (Rochester: University of Rochester Press, 2014), 21–47.

36. Selye, *Stress of Life* (New York: McGraw-Hill, 1956), 46. Only paragraphs later, he returned to his claims to originality, seemingly caught up with convincing himself: "Besides, Cannon never proposed the term *stress* as a scientific name for anything in particular; it does not even appear in the subject index of his book, and, as far as I know, he used it only figuratively in one semipopular lecture." Selye, *Stress of Life*, 47.

37. Selye, *The Stress of My Life* (Toronto: McClelland & Stewart, 1977), 221. Jackson notes the overlap between Selye's early studies of alarm reaction, George Crile's studies of shock, and Cannon's studies of shock as well as fight or flight. Jackson, *Age of Stress*, 84; Sándor Szabo, Yvette Tache, and Arpad Somogyi, "The Legacy of Hans Selye and the Origins of Stress Research," *Stress*, 15, no. 5 (2012), 472–78.

38. Selye, *Stress of My Life*, 222; letters between Cannon and Selye (Box 136: 1927, Cannon Papers), also cited in Jackson, *Age of Stress*, 87.

39. Jackson, *Age of Stress*, 83; Selye, "The General Adaptation Syndrome and the Diseases of Adaptation," *Journal of Clinical Endocrinology* 6 (1946), 117–230.

40. Selye, *Textbook of Endocrinology* (Montreal: Acta Medica, 1947), 839–41.

41. He even noted in a footnote that while Henry Dale and Otto Loewi had won the Nobel Prize for their work on the neurohumoral transmission of nerve impulses, Cannon deserved the same honor but died before his contribution was appreciated (appreciation that Selye thought had finally come to Cannon thanks to Selye's own celebrity). Selye, *Stress of Life*, 149.

42. Ibid., 288. As Jackson notes, this list might include Crile's conception of self-preservation. Jackson, *Age of Stress*, 57. See also Peter C. English, *Shock, Physiological Surgery, and George Washington Crile* (Westport, CT: Greenwood, 1980). But Crile, like Cannon, "preferred to conceptualize emotions in more complex terms involving not merely visceral reflexes but also central control." Jackson, *Age of Stress*, 71.

43. Selye, *Stress in Health and Disease* (Boston: Butterworth, 1976), 15, 13.

44. Ibid., 13.

45. Selye, *Stress*, 11-minute film (Montreal: National Film Board of Canada, 1956); Szabo et al. "Legacy of Hans Selye"; Selye, "Letter: A Syndrome Produced by Diverse Nocuous Agents," *Nature* 138, no. 3479 (May 18, 1936), 32.

46. C. H. Waddington, *The Strategy of the Genes* (London: George Allen & Unwin, 1957). See the notes and commentary in John Bowlby-drafted extensive notes and commentary. PP/BOW/M.220, Wellcome Library.

47. Roy Porter, "The Patient's View," *Theory and Society* 14, no. 2 (1985), 175–98.

48. Pierre Abrami, "Cours de pathologie médicale," *La Presse médicale* 103 (December 23, 1936), 2081.

49. Walter Langdon-Brown, "Return of Aesculapius," *The Lancet* (October 7, 1933), 821.

50. Abrami, "Les troubles fonctionnels en pathologie (Leçon d'ouverture du Cours de Pathologie Médicale)," *La presse médicale* (December 23, 1936), 2082.

51. Sigmund Freud, "The Dynamics of Transference" (1912), in *FSE* 12:97–108.

52. It may be unfair to insist on the difference between the 1917 and 1911 definitions because Freud had already indicated that the origin of the transferential attachment belonged not to the psychoanalyst but to dynamics already half-formed in the patient; nevertheless the difference is significant. FSE 16:442, 447–48.

53. Freud, *Introductory Lectures on Psycho-Analysis*, in *FSE* 16:445.

54. Franz Alexander, *The Scope of Psychoanalysis* (New York: Basic Books, 1961). In an essay titled "The Psychosomatic Approach to Medical Theory" (1954), Alexander, a psychoanalyst and physician, discussed the uneasy connection between Freud's methods for studying psychophysiological phenomena and the work of Groddeck and others, noting in particular the significance of Cannon's scientific experimental methods on the emotions and digestion as a key point of contrast. Alexander wrote that for Groddeck and his contemporaries, "attempts consisted in arbitrary, sometimes fantastic, interpretations of organic symptoms attributing even to visceral functions specific symbolic meaning. They could neither physiologically nor psychologically be verified by observation" (347).

55. Anaïs Nin, *The Diary of Anaïs Nin* (1966; Bronx, NY: Ishi, 2011), 1:105, on the "laboratory of the soul" reference.

56. René Allendy, *L'orientation des idées médicales* (Paris: Au Sans Pareil, 1929); and Weisz, "A Moment of Synthesis," in *Greater Than the Parts*, 74–75.

57. Canguilhem, "A la gloire d'Hippocrate, père du tempérament," *Libres propos* (August 20, 1929), republished in Canguilhem, *Oeuvres complètes* (Paris: Vrin, 2011), 1:248–51, 248. See Giuseppe Bianco "The Origins of Canguilhem's Vitalism," in *Vitalism and the Scientific Image in Post-Enlightenment Life Science*, ed. Sebastian Normandin and Charles Wolfe (Heidelberg: Springer, 2013), 247–48.

58. Bernard Aschner, *The Art of the Healer* (New York: Dial, 1942), 297.

59. Georg Groddeck, *Natura sanat, medicus curat: Der gesunde und kranke Mensch* (Leipzig: Hirzel, 1913). Canguilhem, *Writings on Medicine*, 33; see also Freud, "'Wild' Psycho-Analysis" (1910), in *FSE* 11:219–28.

60. Groddeck, *Natura sanat, medicus curat*, 191, translation amended.

61. Groddeck, *The Book of the It* (New York: New American Library, 1961), 107.

62. Ibid, 108–9.

63. Ibid., 226–27.

64. Ibid., 237

65. Ibid., 229.

66. Ibid., 227. Groddeck repeated the expression "disease is a vital expression of the organism" for emphasis.
67. Ibid., 236–37.
68. Ibid., 109.
69. Freud, *Introductory Lectures on Psycho-Analysis*, in FSE 431–32.
70. Adhémar Gelb to Sigmund Foulkes (Fuchs), October 22, 1928, PP/SHF/B/1, Sigmund Heinrich Foulkes, Archives and Manuscript Collection, Wellcome Library.
71. Sigmund Fuchs, "The Position of Contemporary Biology," *Imago* 26 (1939), 556–57.
72. Daniel Lagache, "Note sur le langage et la personne," in *Les hallucinations verbales* (Paris: Presses universitaires de France, 1977), 205–14, and preface at viii.
73. Canguilhem's notebook from Lagache's course on Goldstein survives in "Psychologie pathologique, cours de Daniel Lagache, Clermont-Ferrand, 1941–42," CAPHES/ENS GC.11.1, Canguilhem Archive. See also Lagache's review of Canguilhem, "Le normal et le pathologique d'après Canguilhem," *Revue de métaphysique et de morale* 51, no. 4 (October 1946), 355–70. On Goldstein and Canguilhem, see Claude Debru, *Georges Canguilhem, science et non-science* (Paris: Éditions Rue d'Ulm, 2004), 49–63; Charles T. Wolfe, "Was Canguilhem a Biochauvinist?" in *Medicine and Society*, ed. Darian Meacham (Dordrecht: Springer, 2015), 197–212; and Geroulanos and Meyers, *Experimente im Individuum* (Berlin: August, 2014), ch. 3.
74. On Binswanger and Goldstein, see Simon Taylor, "The Modern Condition: The Invention of Anxiety, 1840–1970" (PhD diss., Columbia University, 2014), chs. 3–4.
75. *Journal of Individual Psychology* 15, no. 1 (May 1959).
76. William Rappleye to R. A. Lambert, December 21, 1934, RF/1.1/200A/78/739, Rockefeller Archive Center.
77. F. Golla to Dr. O'Brien, December 27, 1934, and Lambert to Golla, December 29, 1934, RF/1.1/200A/78/739, Rockefeller Archive Center.
78. Karl Lashley, foreword to *The Organism*, by Kurt Goldstein (New York: American Book Co., 1939), v–vi.
79. Montefiore and the Rockefeller Foundation wrangled for years over his job, whose precariousness could not be institutionally resolved with a stroke of a pen. See RF/1.1/200A/78/940, Rockefeller Archive Center.
80. Marianne L. Simmel, ed., *The Reach of the Mind: Essays in Memory of Kurt Goldstein* (New York: Springer, 1968), 5.
81. For contrast, see Yvon Belaval, *Les conduites d'echec* (Paris: Gallimard, 1953), in which Belaval routinely reads Freud and Goldstein against one another. Goldstein's archive includes a complete translation of the book into German, in Goldstein's handwriting. See box 15, Goldstein Papers.
82. *TO* 259.
83. Ibid., 163.
84. *HNP* 120.
85. Goldstein, *After-Effects of Brain Injuries in War* (New York: Grune & Stratton, 1942), 25; *TO* 333, 327.

86. *TO* 333, 328 (italicized in 1939 edition, at p. 431).
87. Ibid., 325.
88. Ibid., 340.
89. Ibid., 380. He repeated this parallel between physician and educator in *HNP* 234.
90. *TO* 281–84.
91. Ibid., 162–63, 167.
92. "Social Insurance and Allied Services," report by Sir William Beveridge presented to Parliament by Command of His Majesty, November 1942 (Cmd.6404) (London: His Majesty's Stationery Office, 1942).
93. Michel Foucault, "The Crisis of Medicine or the Crisis of Antimedicine?" *Foucault Studies* 1 (2004), 6.
94. René Dubos, *Mirage of Health* (1959; New York: Harper & Row, 1971), 117.
95. Ibid., 266.
96. Michael Balint, "The Doctor, His Patient, and the Illness," (1955), in *Problems of Human Pleasure and Behaviour* (New York: Liveright, 1957), 204–5, emphasis ours; Balint, *Treatment or Diagnosis* (Philadelphia: Lippincott, 1970).
97. Illich, *Medical Nemesis*, 221.
98. Ibid., 275.
99. Ibid., 70.
100. Ibid., 42, citing L. Edelstein, *The Hippocratic Oath* (Baltimore: Johns Hopkins University Press, 1943). See also ibid., 96.
101. Ibid., 83–84.
102. Ibid., 146–47. Illich cites Achille Souques, "La douleur dans les livres hippocratiques," *Bulletin de la Société française de l'histoire de médecine* 31 (1937), 209–14, 279–309; 32 (1938), 178–86; 33 (1939), 37–38, 131–44; and 34 (1940), 53–59, 79–93.
103. Ibid., 151. Illich cites Charles Richet, "Douleur," in *Dictionnaire de physiologie* (Paris: Alcan, 1902), 173–93. See LSP 30.
104. Pierre Klossowski argues this point in an essay titled "La plus belle invention du malade." Klossowski, *Nietzsche et le cercle vicieux* (Paris: Mercure de France, 1969), 290.
105. Rollo May, "The Delphic Oracle as Therapist," in *The Reach of Mind*, ed. Marianne L. Simmel, 211–18. Goldstein cited May's *Concept of Anxiety* in the 1951 edition of *HNP* 246.
106. Jeffrey Mishlove, "Rollo May," In *Thinking Allowed* (Tulsa, OK: Council Oak, 1995), 117–23. Compare to Illich, "Body History," *The Lancet* 328, no. 8519 (December 6, 1986), 1325–27.
107. May, *The Courage to Create* (New York: W.W. Norton, 1972), 100.
108. Such difficulties may account for the effacement of "self-actualization" from his technical works—for example, in Goldstein and Sheerer's *Abstract and Concrete Behavior* (1941) and Goldstein's *After-Effects of Brain Injuries in War* (1942—although it is crucial in his 1951 *Human Nature in the Light of Psychopathology* and in his other post-1945 works.
109. See our *Experimente im Individuum*, conclusion.

110. Canguilhem, "Is a Pedagogy of Healing Possible?" *Writings on Medicine*, 53–66.
111. Ibid., 63.

CHAPTER TEN

1. It is not that after World War I this logic vanishes. Rivers, for one, kept it in the background. W. H. R. Rivers, "Socialism and Human Nature" (1921), in *Psychology and Politics* (London: K. Paul, Trench, Trubner, 1923), 91). Cannon also did, as when he noted that homeostatic regulation confirmed "the idea that the history of the individual summarizes the history of the race, or that ontogeny recapitulates phylogeny." *CWB* 301. But the analytical and interpretive value of this logic gave way, and phylogenetic recapitulation participated here almost only as an explanation of historical derivation.

2. On physiology emerging like Minerva, see Charles Singer, *A History of Biology to about the Year 1900* (1950). Minerva was a trope: Arturo Castiglioni used it as well for "the new medicine" of the late nineteenth century in "Neo-Hippocratic Tendency of Contemporary Medical Thought," *Medical Life* 41 (1934), 115.

3. See chapter 3 of our *Experimente im Individuum*; Uta Noppeney, "Kurt Goldstein," *Journal of the History of the Neurosciences* 10, no. 1 (2001), 67–78, links the same thinkers but on different grounds.

4. Maurice Merleau-Ponty, *Humanism and Terror* (Boston: Beacon, 1969), 130, quoted in Isabel Gabel, "Biology and the Philosophy of History in Mid-Twentieth-Century France" (PhD diss., Columbia University, 2015), 118.

5. *TO* 388.

6. Merleau-Ponty, "The War Has Taken Place" in *Sense and Non-Sense* (Evanston, IL: Northwestern University Press, 1964), 144, 147.

Index

Abrahams, Adolphe, 43
Abrami, Pierre, 303
acidosis, 57, 59, 68, 156. *See also* shock, wound
Ackerknecht, Erwin, 18, 32
Ackermann, Dankwart, 187
Aesculapius, 303
Agamben, Giorgio, 167
aggression, viii, 9, 69, 156, 209, 210, 222, 241, 258, 273
Alexander, Franz, 209
Allendy, René, 305, 307, 310, 315
allergy, 186, 188, 189, 191. *See also* anaphylaxis; autopharmacology
Althusser, Louis, 283
Ambulance Américaine (American Ambulance Hospital), 32, 35, 50, 54
American Association for the Advancement of Science, 268, 270
American Committee for Democracy and Intellectual Freedom, 269
American Expeditionary Forces, 51, 55
American Medico-Psychological Association, 210
anaclisis, 227
anaphylaxis, 186–91. *See also* autopharmacology; histamine
animal experimentation, 4, 12, 15, 16, 19, 32, 37, 43, 56, 139, 140–45, 148, 153, 157, 188
anoci-association, 32, 48, 65
anoxemia, 156, 177
anthropology, 16, 17, 26, 31, 213, 248, 249, 255, 267; structural, 249, 275, 286, 289. *See also* ethnography, ethnology
antihumanism, 283, 291, 321
antimedicine, 310, 313
anti-Pasteurianism, 305
antiseptic (Carrel-Dakin solution), 40
anxiety, 28, 37, 82, 111, 128, 129, 143, 144, 148, 151, 195, 198, 199–202, 205, 208, 211, 219, 225–28, 297, 303, 319

aphasia, 7, 67, 101, 102, 112, 113, 115, 116–21, 123–26, 208, 279, 280, 282, 317; Goldstein on, 126–29; symptoms of, 120, 121, 126, 130, 132
aphasia, Head's categories: nominal, 122–24, 126; semantic, 122, 126; syntactical, 122, 126; verbal, 122, 123, 126. *See also* Goldstein, Kurt; Head, Henry
aphasiology, 118, 119, 279, 280
Armistice (World War I), 35, 63, 213
Artaud, Antonin, 305
Aschner, Bernard, 305
atomism, atomistic symptomatology, 23, 101, 102, 126, 196, 198
autoimmune reaction, 190, 194
autoimmunity, 190, 191, 194
autopharmacology, 69, 165, 186, 191, 194–95. *See also* autoimmunity

Babinski, Joseph, 89, 91
bacteriology, 253
balance, viii, 122, 164, 169, 182, 183, 190, 266, 277, 320; chemical, 156; fluid, 163; homeostatic, 302. *See also* equilibrium, and disequilibrium; homeostasis
Balint, Michael, 311, 312
Banks Islands, 212
Barcroft, Joseph, 164, 180, 185
Barger, George, 187, 188
Barré, Jean Alexandre, 91
Barth, Karl, 255
Bartlett, Frederic, 94, 95
Basedow, Herbert, 274
Bastian, Adolf, 26
Bastian, Charlton, 121
Battle of Mons, 40
Battle of the Somme, 51, 100
Bayliss, William, 3, 5, 13–15, 18, 20, 34, 35, 44, 47–49, 55, 56, 58–60, 63, 74, 138, 141, 181, 186, 188, 189, 191, 210, 235, 293

Beaumont, William, 141
Bergson, Henri, 259
Bernard, Claude, 4, 6, 13, 15, 18, 19, 24, 97, 105, 147, 155, 161–66, 171–76, 183, 185, 186, 198, 202, 225, 265, 292, 300, 301, 305
Bertalanffy, Ludwig von, 167, 172
Bethe, Albrecht, 7
Bethe, August, 112
Béthune, 49, 51
Beveridge, William, 261, 312; Beveridge Report, 311
Binet, Alfred, 85
Binswanger, Ludwig, 308
Binswanger, Otto, 91
biochemistry, 165–68, 171, 173–77, 206
biocracy, 184, 205, 264, 269, 271, 278, 320
blood, 4, 34, 39, 43–49, 54–59, 63, 69, 71, 72, 74, 149–51, 156, 162, 164, 168, 170–72, 175–78, 181–83, 189–93, 263, 271, 289, 293–94, 321; pressure, 39, 45–49, 54–59, 63, 94, 146, 168, 173, 189, 191–93, 289; transfusion, 34, 63. *See also* capillaries; Henderson, Lawrence J.; shock, wound
Blumenberg, Hans, 337n88
Boas, Franz, 17, 22, 162, 255, 269
body, as analogue to the body politic, xi, 30, 32, 255–63, 264, 266, 270–73, 281, 295, 321
Boer War, 65
Boulogne, 34, 35, 63, 64, 67
Bowlby, Anthony, 34, 70, 71
British Expeditionary Force (World War I), 34, 40, 50, 52, 60, 61
British Physiological Society, 14, 138, 174, 215
British Psychoanalytic Society, 213
brittleness, 28, 31, 54, 64, 155, 158, 209, 317, 318
Britton, Sydney, 153
Broca, Paul, 26

INDEX

Brown Dog controversy, 15; Cannon, 138–60. *See also* experimentation: animal
buffer, 170–72, 178, 183, 184. *See also* margin of safety
Burdon-Sanderson, John, 12
Burroughs Wellcome & Company, 186, 191

Cabot, Richard, 92
Caius College, Cambridge, 3
Canguilhem, Georges, 19, 30, 107, 135, 165, 166, 172, 198, 266, 299, 308, 315, 320, 337n88
Cannon, Walter Bradford, vii, viii, ix, x, 5, 10, 13–16, 21, 26, 28, 30–35, 44, 48–51, 54–59, 63, 64, 68, 69, 74, 84, 85, 87, 138, 139–59, 165, 166, 181–86, 192, 195, 197, 203–6, 208–11, 218, 226, 233, 247–49, 254, 263–78, 286–92, 295, 297–303, 306, 310, 313, 314, 317, 319–21
capillaries, 46, 48, 57, 186, 191–95
Carrel, Alexis, 37, 40, 50, 53, 54, 64
case studies, case histories, xi, 31, 56, 78–108, 118, 126, 141, 143, 145–46, 149, 306, 315, 317
Casper, Stephen T., 86
Cassirer, Ernst, 30, 102, 132, 249, 278–80, 283, 314, 320
Castiglioni, Arturo, 298
Casualty Clearing Station, 33, 50
catastrophic situation (Goldstein), 10, 105, 127, 129, 198, 200, 204, 260
catharsis, 119
Catlin, G. E. G., 259
cell theory, in relation to integration, 253
Charcot, Jean-Martin, 38, 89
chemo-physiological phenomena, 168–76, 187. *See also* Henderson, Lawrence J.; Turck, Fenton
Cheyne, Watson, 66
Cleland, J. B., 274
Clermont-Ferrand, 308

Cobb, Humphrey, 37
collapse. *See* disintegration
Columbia University Clinic for Psychoanalytic Training and Research, 308
communism, 30, 261, 319
Compiègne, 50, 54
consciousness, 3, 74, 96, 115, 116, 125, 126, 133–36, 165, 185, 203, 215–20, 222–26, 243, 279, 280, 283, 285, 314, 321; in psychoanalysis, 215–18, 220, 222, 224, 226, 243
constancy, 25, 155, 157, 163, 175, 177, 182, 185, 202, 209, 221, 232, 241, 264, 269, 310
constitutional alarm, 65, 66
Cowell, E. M., 38, 39, 44–48, 52, 55, 58, 60–63
Craige, John Houston, 273
Craiglockhart Hospital (Edinburgh), 80, 94
Crile, George Washington, 31–35, 37, 39–45, 47–50, 52–57, 64–68, 72, 76, 149, 233
Cushing, Harvey, 32, 45
cybernetics, xii, 30, 31, 146, 158, 209, 210, 248, 249, 267, 276, 278, 318, 322

Dakin, Henry, 40
Dale, Henry, ix–x, 34, 44, 47–51, 55, 56, 59, 60, 63, 68, 76, 154, 165, 186–95, 203, 204, 206, 208, 317, 319
Dardanelles Campaign, 50
Darwin, Charles, 147–49, 152
Darwinism, 24, 117, 140, 210
death drive, 207–13, 217, 220–23, 227–34, 236–43
Déjerine, Joseph Jules, 89
Delbet, Pierre, 53
Dewey, John, 85, 264
diaschisis, 67, 119
digestion, 138–55, 256
Dijon, 49, 51

disintegration, x, xii, 5, 8, 9, 23, 28–30, 66, 113, 114, 126, 129, 134–36, 156, 163, 165, 186, 196, 197, 200, 201, 203, 207, 209, 227, 242, 248, 259, 260, 262, 268, 276, 317–20. *See also* equilibrium, and disequilibrium; integration
dissociation, 74, 82, 96, 215
Dix, Otto, 107, 108
Donati, Mario, 298
dreams, 41, 42, 50, 89, 228, 264
Driesch, Hans, 167, 255
drives: drives contrasted to instincts, 217, 220, 221, 222; drive to self-actualization (Goldstein), 200–202; in Freud, 3–4, 31, 207–13, 217–20, 222, 223, 227–29, 230, 231–32, 236–43. *See also* instincts, Rivers on
dualism, mind-body, 11, 31, 71, 133, 136, 165, 302; monism, 133, 136
Dubos, René, 311, 313
Durkheim, Émile, 253, 280–83

Ebbecke, Ulrich, 193
Edsall, David L., 171
ego, 18, 208, 209, 212, 217, 219–24, 226, 227, 231–33, 237, 240, 242, 248
Ehrlich, Paul, 186, 253
Ellenberger, Henri, 212
embryology, 167, 196, 232, 236
Emergency Committee in Aid of Displaced Foreign Scholars (Rockefeller), 308
emotions, viii, 5–6, 8, 10, 15, 16, 20, 22, 26, 28, 39, 40, 52, 58, 68, 74–75, 82, 92, 121, 138–40, 142, 144–59, 164, 184, 197, 203, 210, 275, 278, 291, 303, 310; Cannon on emotions and bodily changes, 138–40, 142, 144–59, 164, 203, 267, 268; Darwin on expression of emotions, 140, 147–49, 152; James-Lange theory of, 139, 140, 197

Empire Hospital, 115
endocrinology, xi, 5, 8, 44, 187, 301
environment, milieu, viii, x, xi, 10, 14, 17, 21, 27, 28, 40, 69, 94, 102, 118, 127, 130–32, 145–50, 155–59, 161–203, 208, 216, 226, 229–30, 238, 250, 272, 277, 288, 299, 311; internal environment (*milieu intérieur*), 4, 19, 20, 24, 25, 29, 162–64, 183, 226, 265–66, 270, 299, 302
epicritic. *See* protopathic and epicritic (Head and Rivers)
epigenetics, xi, 302
Eppendorf Hospital, 89
Epstein, Julia, 87
equilibrium, and disequilibrium, 4, 5, 6, 10, 23, 58, 76, 98, 105, 129, 130–33, 152, 154, 159, 162, 164, 165, 170, 182, 183, 196, 198, 231, 255, 263, 265–66, 290, 299, 318
ergot, 187
Erichsen, John Eric, 66
Esser, Johannes, 100
ethnography, ethnology, 7, 16, 17, 25, 26, 93, 146, 149, 153, 154, 157, 212, 214, 273, 274, 280, 284–86, 288. *See also* anthropology
ethnopsychology (*Völkerpsychologie*), 16
Evans, Andrew, 16
Evans, Charles Lovatt, 13, 249
evolutionary biology, 12, 15, 19, 24, 25, 28, 94, 146–49, 154, 163, 167, 214–16, 229, 233–34, 242, 253–55, 256–57, 259–60, 266, 317; integrationism's evasion and revision of, 14, 28, 113, 117, 146–49, 150, 155, 167, 266, 267, 290. *See also* emotions
exaemia, 48, 76. *See also* shock, wound
exhaustion, 6, 32, 33, 35, 37, 40–43, 47, 50, 54, 57, 64, 68, 72–76, 89, 301
experimentation: animal, 4, 12, 15, 19, 24, 37, 56, 114, 141, 142,

INDEX 413

144, 157; self, 13, 116, 180–81, 214. *See also* radial nerve; X-rays (Röntgen rays)

facial injury, 46, 80, 100, 112, 259
fatigue, 39, 43, 53, 58, 64, 67, 68, 76, 89, 94, 144, 149, 150, 168, 171, 263
Fechner, Gustav, 209, 230, 232, 241
feedback, 20, 30, 276; failure of, 291; negative, 158, 277, 312
fight-or-flight response, 30, 147, 149, 150, 151, 152, 159, 184, 204, 211, 216, 220, 226, 302; Cannon interested in its preconditions, 150–52; Crile on, 149–50; Freud on (flight or defense), 211, 219, 220; Jones on, 216, 219; MacDougall on, 150–52, 210, 211, 216. *See also* Cannon, Walter Bradford
Fletcher, Walter, 248
Fliess, Wilhelm, 214
fluid matrix, 69, 139, 146, 155, 157, 158, 181, 183–85, 190, 195, 264
Foucault, Michel, 88, 283, 311
Frankfurt Neurological Institute, 80, 100, 115, 132
Fraser, John, 57
Frazer, James, 210
Frédéricq, Léon, 164
Freud, Sigmund, 3–5, 17, 18, 31, 38, 40, 50, 96, 200, 206–43, 254, 259, 266, 303–10, 314
Fromm-Reichmann, Frieda, 308
Fuchs (Foulkes), Sigmund, 307
Fulton, John, 181

Galdi, Francesco, 298
Galen, 18, 28
Gall, Franz Joseph, 116
gangrene, 32, 53, 59
Garrison, Fielding H., 296
Gelb, Adhémar, 7, 80, 88, 100–102, 115, 116, 119, 126, 132, 279, 307
Gesell, Robert, 58
Gillies, Harold, 100

Ginzburg, Carlo, 86, 87
Goldstein, Kurt, ix, x, 7, 10, 20, 22, 23, 27–31, 80, 87, 88, 93, 100–108, 111–16, 119, 124, 126–37, 165, 166, 195–206, 208, 243, 249, 254, 260, 265, 272, 278–80, 283, 285, 287, 288, 297, 303, 306–10, 313–15, 317–21
Government Hospital for the Insane (St. Elizabeth's), 210
Graves, Robert, 111, 112
Gray, H. M. W., 40, 59, 60
Great Retreat, 40, 43
Groddeck, Georg, 305–7, 310, 315
Grubrich-Simitis, Ilse, 220
gum acacia solution, 34, 35, 63, 64
Guthrie, George, 66

Haddon, Alfred Cort, 16, 17
Haeckel, Ernst, 180, 228, 229–34
Haldane, John S., 13, 20–22, 26, 27, 164, 165, 166, 172–83, 186, 189, 203, 247, 260, 265, 292
Hallett, Christine, 62
Hampstead, 49, 50, 63
Haraway, Donna, 253
Harrington, Anne, 101
Harrison, Ross, 292
Harvard Fatigue Laboratory, 168, 171, 263
Harvard Tercentenary, 264
Harvey, William, 12, 292–95
Harvey Society, 184
Head, Henry, ix, x, xii, 7, 13, 14, 18, 27, 31, 37, 111–40, 145, 150, 161, 165, 195, 196, 214, 249, 278–87, 295, 298, 317, 321
Hegel, G. W. F., 256
Heidegger, Martin, 200, 201, 283
Helmholtz, Hermann von, 12, 13, 210
hémoclasiques crises, 190
hemorrhage, 40, 44, 46, 52, 56–59, 65, 67; postpartum, 187; punctiform, 72. *See also* blood
Henderson, Lawrence J., 65, 66, 167, 168–76, 180–84, 186, 189, 196, 202, 248, 263, 264, 278, 299, 300

Henderson, Yandell, 48, 172, 173, 182
Hennen, John, 66
Hill, A. V., 167, 172, 181, 292
histamine, 8, 55, 57, 59, 68, 69, 76, 165, 187–204, 208, 317. *See also* autopharmacology; Dale, Henry
Hobhouse, Leonard, 256, 257
holism: defined by Smuts, 260; Goldstein's and holistic approaches to patienthood, 265, 305, 310; Haldane's, 176, 177, 179; integrationism's difference from, 20, 22, 23, 67, 113, 126, 146, 166, 176, 184, 201, 253; and Nazism, 255
homeostasis, v, vii, viii, x, 28, 69, 138–39, 145, 155–56, 158, 165, 181–86, 197, 209–10, 221, 231, 241, 249, 263, 265, 267–69, 271–73, 276–78, 291, 298–300, 302, 317, 318; debate over homeostatic theories before Cannon, 167–81; Freud, Lacan, and psychoanalytic homeostasis, 231, 241, 243, 249 (*see also* constancy); social "homeostasis," 264, 269, 271, 277, 278, 289; Wiener on, 276–78
Hooper, A. N., 57
Hoover, Herbert, 263
hormonal self, 8, 163, 297, 298
hormones, 4, 5, 14, 18, 20, 22, 39, 138, 139, 142, 144–46, 151, 164, 179, 183, 187, 210, 257, 262, 277, 293, 295, 298, 317
Hörstadius, Sven, 167
humanism, medical, xi, 29, 248, 292, 297, 311–15. *See also* individuality
Huneman, Philippe, 83, 84
hunger, 5, 142, 146, 147, 207, 215, 222, 227, 240, 278; air, 49, 57, 63; sexual, 223
Hurst, Arthur, 80, 88–91, 95, 96, 100, 102, 106, 108
Huxley, Thomas Henry, 20, 25

hypothalamic-pituitary-adrenal axis, 301
hysteria, 82, 90

Illich, Ivan, 297, 311–15, 319
individuality, x, xi, 4, 9, 10, 22, 23, 30, 31, 55, 66, 79–80, 84, 85, 105, 120, 127, 162–67, 171–72, 177–81, 184–86, 195–99, 201–6, 208, 211, 236, 239, 241–43, 256, 264, 282, 284–85, 291, 297, 309, 312, 316–21; nonliberal individualism, x–xi, 10–11, 273, 319, 320, 321
instincts, Rivers on, 94, 208, 213, 215, 216, 217–18, 233, 236; compared to drive, 217, 220, 221, 222. *See also* drives
Institute for Business-Cycle Research, 261
Institute of Experimental Psychology (Moscow), 196
integration: in Goldstein, 133–36; in Head, 133–34, 136; integrationism, 5, 7, 22, 23, 24, 67, 165, 168, 196, 210, 251, 252, 255, 288, 290, 295; integrative medicine, xi; in Sherrington (integrative action), 7, 19, 24–25, 113–14, 126, 128, 134, 210. *See also* disintegration

Jackson, John Hughlings, 25, 113, 117, 118
James, William, 85, 151, 153, 265, 268, 272
Janet, Pierre, 89, 266
Jonas, Hans, 30
Jones, Ernest, 209, 211–15, 218, 219, 223, 228, 232, 237, 242
Jung, Carl G., 215, 227
Jünger, Ernst, 111, 112

Kant, Immanuel, 279, 282
Kardiner, Abram, 308
Keith, N. M., 55

Kellogg-Briand Pact, 259
Keynes, John Maynard, 251, 252, 261
Kobert, Rudolf, 191
Koffka, Kurt, 130
Königsberg, 115
Koselleck, Reinhart, 255
Krogh, August, 193, 292

Labour Party, 212, 247
Lacan, Jacques, 209, 231, 239
La fixité du milieu intérieur est la condition de la vie libre et indépendante (Bernard), 162–64, 183
Lagache, Daniel, 307, 308
Lakeside Hospital (Cleveland), 31, 32
Langdon-Brown, Walter, 28, 303, 304, 310
language, 30, 31, 112, 119–21, 124, 126, 133; Head's theory of, 116–25, 134, 135–37; Saussure and integrationists, 30–31, 286, 287
La Penne, 50, 54, 167, 184
Laplanche, Jean, 209, 224, 239
Lashley, Karl, 7, 20, 112, 308, 310
Laski, Harold, 256
law of the heart (Starling), 15, 293, 295
Le Bon, Gustave, 17, 254
Leriche, René, 51, 59, 60, 80, 88, 91, 93, 97–100, 106, 108, 203, 298, 310, 313
Lévi-Strauss, Claude, 30, 249, 275, 278, 279, 286–91, 320, 321
Lewis, Thomas, 193, 292
Lilienfeld, Paul von, 253
liquid matrix, 183
localizationism. *See* atomism, atomistic symptomatology
Loewi, Otto, 186, 187, 269
London Hospital, 115
London Shock Committee, 34, 39, 44, 48, 50–53, 55–60, 62–66, 68, 69, 74, 154
Loughran, Tracey, 72

Low, Barbara, 209
Ludwig, Carl, 12
Lundberg, George, 265
Luria, Alexander, 7, 20, 23, 112, 165, 196, 197, 317
Lycée Pasteur, 32

MacCurdy, J. T., 218, 219
Malinowski, Bronisław, 17, 255, 273, 283
Manning, Frederic, 100
manometer, 142
margin of safety, 158, 159, 181, 183, 184, 272, 273, 275, 277
Maritain, Jacques, 255
Mauss, Marcel, 17, 30, 249, 259, 278–90, 320
May, Rollo, 313, 319
May, Ulrike, 220
Mayo, Elton, 263
McDougall, William, 16, 18, 29, 93, 149, 150, 152, 210, 214, 216, 218, 254
McGuire, Hunter Holmes, 65
mechanism, 18–25, 98, 141, 171, 175–76, 180, 197, 318. *See also* holism; integration: integrationism; vitalism
Medical Congress of Rome, 162
Medical Research Committee, 43, 175
Meltzer, S. J., 184
Merleau-Ponty, Maurice, 102, 320, 321
metapsychology, 4, 214, 231, 232
method, epistemological and historical, 337n88
Meyer, Adolf, 84, 85
milieu. *See* environment, milieu
Mill, John Stuart, 10, 273
Miners' Union, 177, 247
Mitrany, David, 258, 261
Monakow, Constantin von, 119
monism, 133, 136, 308
Montefiore Hospital, 308
Morris, Edwin, 65–67

motor-kinetic disorder, 102
Mott, Frederick W., 71–74, 80, 82, 91, 95
Müller, Johannes, 6, 12
Mumford, Lewis, 265
Münsterberg, Hugo, 16
Murphy Island, 75
Murray Island, 93
Myers, Charles S., 16, 40, 68, 71, 72, 75, 80–83, 87, 90, 91, 93–95, 100, 106, 138, 213, 214

Nancy, Jean-Luc, 85
narcissism, 209, 218, 220, 227, 228, 232, 237, 241
neogrammarian theory, 135
neo-Hippocratism, 297, 298, 301, 307
Neuilly-sur-Seine, 32, 50
Neumann, Franz, 265
neurasthenia, 37, 42, 72, 80, 82, 89
neurology. *See* Goldstein, Kurt; Head, Henry; radial nerve; Sherrington, Charles
Nicoll, Maurice, 215
nitroid crises, 190
nonconscious, bodily, x, 159
Nonne, Max, 89, 90
normal, concepts of the, x, 20, 21, 27, 29, 37, 84, 105, 121, 123, 127, 129, 133, 187, 198, 199, 201, 203, 255, 276, 280, 287, 299, 308, 310, 320. *See also* norms; pathological-to-normal reasoning
norms, x, 9, 37, 38, 39, 106, 127, 133–36, 145, 183, 194, 198–99, 201–2, 249, 295, 309–11; bodily norms at war, 38–39; Goldstein and Canguilhem on, 9, 106, 127, 133, 134–36, 198–99, 201–2, 205, 320; social, 37, 165, 309–11, 317, 320

occupational therapy, xi
Ogden, C. K., 282
oligaemia, 48
ordoliberalism, 262

pain, 5, 11, 22, 29, 45, 52, 81, 93, 95, 96, 106, 146, 147, 150, 152–56, 214, 217, 221, 228, 270, 294, 312, 318; surgery of, 80, 97, 98–100
Pareto, Vilfredo, 170, 257, 263, 265, 299
Paris Peace Conference, 258
Parkinson, James, 91
Parsons, Talcott, 263, 265
pathogeny, 83
pathological-to-normal reasoning, 9, 12, 39, 58, 129, 132, 136, 137, 147, 188, 198, 203, 204, 266, 276, 280, 291, 298, 308
Pavlov (Pawlow), Ivan, 3, 14, 16, 138, 247
performance: Goldstein on, 24, 101–2, 105, 106, 115, 125–27, 129–31, 134, 144, 198, 200, 202, 206; Hurst on, 90; Mauss on, 290; Wiener on, 276
personalized medicine, xi, 319
phantasm, 100
phenomenology, 279, 320
phylogeny, phylogenesis, 37, 149, 158, 180, 209, 214, 215, 228, 233, 234, 241, 254, 265, 317
physicochemical, 19, 21, 22, 157, 167–72, 182, 267
physiology, viii–ix, 3–33, 49, 58, 64, 71–72, 94, 99, 116, 127, 136, 138–59, 162–67, 169–86, 192, 196–97, 233–36, 255, 261, 263, 268–69, 282–83, 291–93, 296–98, 317; "new physiology," 6, 18, 21, 26, 27, 30, 31, 64, 175, 234, 243, 298
Pike, F. H., 68
Pinel, Philippe, 84
pleasure principle, 207–9, 211, 218, 220–31, 235, 237–43
Poncet, Antonin, 97
Porter, Roy, 85, 303
Porter, Theodore, 296
Porter, William Townsend, 39, 48, 49, 50, 54, 55, 64–66

Portier, Paul, 189
positivism, 26, 293
protopathic and epicritic (Head and Rivers), 117, 125, 130, 133, 214–16, 218
protoplasm, 169, 171, 174–76
psychoanalysis, viii, 17, 31, 93, 96, 207–43, 304–6, 307–8, 311
psychotechnics, 16

Quénu, Édouard, 51

race: degeneration, 10; psychology, 254
radial nerve, 13, 116, 214. *See also* experimentation: self
rage, 146, 147, 150–58, 289; sham, 153, 275
railway spine, 51, 66, 92
Rank, Otto, 218
Raymond, Fulgence, 89
repression, 94–97, 216, 219, 222–24; vs. suppression (in Rivers's work), 96, 216
Ribot, Théodule, 285
Richards, Alfred N., 55, 59, 193
Richards, I. A., 282
Richet, Charles, 164, 189, 191
Rivers, W. H. R., 16, 17, 26, 27, 29, 72, 75, 88, 91–97, 116, 117, 125, 138, 207–19, 221, 222, 226, 230, 231, 233, 236, 242, 243, 247, 270, 273, 281, 283–86, 304, 314, 319
Rockefeller Institute for Medical Research, 37, 40, 50, 154
Röpke, Wilhelm, 262
Rosenblueth, Arturo, 139, 276
Royal Anthropological Institute, 212
Royal Army Medical Corps (RAMC), ix, 27, 38, 80, 188, 212
Royce, Josiah, 168
Russett, Cynthia, 263
Russo-Japanese War, 65, 66

sadism, 222, 237, 241
Sassoon, Siegfried, 94
Saussure, Ferdinand de, 30, 31, 286, 287
Schneider, Johann, 101, 102, 126, 279, 280
Schweninger, Ernst, 305
secretin, 3, 189, 293. *See also* hormones
self-actualization, 30, 134–36, 195, 198, 201–5, 243, 260, 297, 303, 306–11, 313, 314, 319
Seligman, Charles, 93, 214
Selye, Hans, 301, 302, 311, 313
sensory-perception disorder, 102
sepsis, 35, 40, 53, 59, 60, 65, 67; battlefield, 32. *See also* antiseptic (Carrel-Dakin solution)
sex, 212, 229, 250
sexual drives, 207–8, 218–20, 222, 223, 227–29, 230, 231–32, 236–43; Freud criticized on, 200, 215; union, 238
sexuality, 3, 217, 236, 253; infantile, 216. *See also* Freud, Sigmund; sexual drives
Shapin, Steven, 84, 251
Sharpey-Schafer, Edward, 12, 14, 292
Sherrington, Charles, 7, 14, 16, 19, 20, 24, 25, 27, 31, 50, 111, 113, 114, 124, 133, 134, 163, 210, 233, 317
shock, wound, 31, 34–77, 87, 91, 139, 154, 188, 191–92, 208, 226, 253, 274–75; nervous/traumatic, 48, 65, 66; primary, 46, 52, 58; secondary, 52, 56, 58, 61; shell, 17, 27, 35, 36, 48, 67–68, 72–75, 76–77, 80–83, 89–92, 106, 208, 219, 221, 226; spinal, 66–68, 82; surgical, 33, 37, 39–40, 47, 48, 50, 52, 54, 55, 56, 65, 76. *See also* trauma (psychic)
Sigerist, Henry E., 296
Singer, Charles, 14, 16, 18, 19, 161
Smith, Adam, 269
Smith, J. Augustus, 273
social hygiene, 16, 253
Société psychanalytique de Paris, 305

sociology, 16, 168, 171, 253, 263, 267, 274, 280–83, 312
soldier's heart, 35, 40, 43, 53, 67, 177
sorcery, 274, 275, 288, 289
Sorokin, Pitirim, 265
Southard, E. E., ix, 27, 80, 88, 90–92, 106, 108, 210
Soviet Union, 196, 247, 248, 269, 296
Spanish influenza epidemic, 296
Spemann, Hans, 167
Spencer, Herbert, 117, 149
stabilization, 10, 24, 139, 145, 156, 163, 172, 198, 261; battlefield, 45, 62. See also shock, wound; Thomas splint
standardization, 106, 295–97
Starling, Ernest H., 3–5, 11–16, 18, 22, 28, 30, 36, 44, 55, 138, 141, 144, 163, 166, 183, 186, 189, 192, 210, 233–36, 247, 249–52, 254, 292–95, 297, 315, 317, 320
statistics, 158, 249, 279, 296, 309; aggregation, 86, 171, 199; paradigm, 84, 87; vital, 47
stress, 301, 302, 311, 313
suffering, 7–10, 22, 27, 33, 35, 44, 47, 51, 53, 69, 79–81, 89, 93, 98, 99, 111, 112, 121, 153, 154, 198, 200, 204, 208, 215, 240, 258, 275, 277, 289, 294, 302, 303, 312, 313, 318
suppression, Rivers's concept of, 96, 216–17, 222. See also repression
supra-individual, 236, 253, 271, 309
surgery, 8, 33, 36, 40, 47–48, 51, 53, 54, 55, 57, 63, 65, 66, 80, 92, 97–100, 106, 116, 141, 157, 298
Sydenham, Thomas, 83
symbolic, viii, 7, 8; Cassirer on symbolic forms, 132, 279–80; Head on symbolic thought, 119–22, 124, 125, 136; Lévi-Strauss on symbolic function, 289–91; Mauss on the symbolic, 239, 280–89
syphilis, 67, 90–92, 253, 259, 306

Temkin, Owsei, 83
Thomas splint, 44, 61, 66
Titchener, E. B., 84
Tonks, Henry, 100, 112
tonus, 129–36, 138, 197; *Tonus* (film), 102, 105, 129–31
Torres Strait expedition, 16, 17, 81, 93, 94, 281
toxemia, 51, 52, 55, 61, 68, 69, 74, 76
trauma (psychic), ix, 36, 40, 42, 71, 72, 75, 82, 90, 94, 97, 107–8, 113, 208, 209, 215–17, 218–26, 228, 233, 236, 237, 239, 242, 243
Turck, Fenton, 167, 172
Turner, Grey, 45
Tycos instrument, 56

unconscious: "bodily," x, 129, 133, 159, 204, 285; Freudian, 4, 220, 222, 226, 242, 243; Rivers on (protopathic), 94–95, 214–16, 218
Union of Socialist Physicians, 247
US Army Neuropsychiatric Training School, 90
US Civil War, 65, 66
US National Research Council, 50, 51

vasodilation, 189, 191–94; vasoconstriction, 98, 189, 191. See also histamine
vasomotor center, 54, 59, 64, 74, 76, 97–100; dysfunction, 100
Veyne, Paul, ix, 337n88
vigilance, Sherrington and Head on, 124–25, 130, 133–36
Virchow, Rudolf, 162, 253, 254
vis medicatrix, 257, 261, 292, 297–300, 302
vitalism, 18–21, 25, 167, 175–80, 196, 197. See also mechanism
vivisection, 15, 37, 141, 144; anti-vivisection, 141. See also experimentation: animal
von Uexküll, Jakob, 166, 167, 196

Wagemann, Ernst, 261, 262
Wallace, Cuthbert, 34, 62, 63
Warner, W. Lloyd, 274
War Neuroses (film), 89–90
Washburn, A. L., 142
Washington, DC, 210
Weimar Statistical Office, 261
Weismann, August, 232–36
Weisz, George, 296, 298, 305
welfare state, xi, 29, 247–48, 295, 311, 320, 401
Wellcome, Henry, 187
Wellcome Physiological Research Laboratories, 186
Wernicke, Carl, 115
White, William A., 210
Whitehead, Alfred North, 168, 263
Whitla, William, 43, 70
Wiener, Norbert, 30, 249, 276–78, 291, 320
Wieviorka, Michel, 86
Windaus, Adolf, 188
Wise, M. Norton, 86
Woolf, Leonard, 257
word-blindness, word-deafness. *See* aphasia: symptoms of
Wundt, Wilhelm, 16

X-rays (Röntgen rays), 5, 13, 14, 56, 89, 138, 141, 145, 146, 157, 214

Zimmern, Alfred, 258, 320